High Entropy Alloys

Saurabh S. Nene
Editor

High Entropy Alloys

A Machine-Generated Literature Overview

Springer

Editor
Saurabh S. Nene
Department of Metallurgical and Materials
Engineering
Indian Institute of Technology Jodhpur
Jodhpur, India

ISBN 978-981-99-7172-5 ISBN 978-981-99-7173-2 (eBook)
https://doi.org/10.1007/978-981-99-7173-2

This Springer imprint is published by the registered company Springer Nature Singapore Pte Ltd.
The registered company address is: 152 Beach Road, #21-01/04 Gateway East, Singapore 189721, Singapore

Foreword

In recent years, high-entropy alloys (HEAs) have emerged as an exciting area of research. HEAs have unique microstructures and impressive properties due to their unique design approach involving multiple principal elements. Over the years, many HEA systems have been reported to combine excellent strength-ductility, synergy, high hardness, thermal stability, resistance to fatigue, fracture, corrosion, oxidation, wear, and irradiation. These properties portray HEAs as promising candidates for various applications, such as structural, functional, magnetic, thermoelectric, and biomedical. The field has been extended to other non-metallic materials expanding its scope as high-entropy materials (HEMs).

In the early stages of development, HEMs were initially designed with a more exploratory approach. However, in recent years, HEMs have been developed with clear goals, carefully selected components, and tailored microstructures. Recent years have seen the HEM field bloom into one of the most promising fields of research, with more than 6000 scientific publications, including excellent review articles and textbooks that cover many aspects of current research on HEMs.

A significant increase has been observed in efforts to predict phase formation in HEAs through thermodynamic calculations and attempts to calculate multicomponent phase diagrams through Calphad approaches. Several new HEA thermodynamic databases have been released by commercial Calphad tools to assist researchers in multicomponent calculations. Researchers have also explored artificial intelligence/machine learning (AI/ML) as a tool to understand the phase formation and properties of HEMs, in addition to first principle calculations. Further, in addition to the existing routes for the synthesis of HEMs, new routes, such as carbothermal shock synthesis, plasma-related synthesis, microwave-assisted synthesis, combustion-assisted synthesis and laser-related synthesis, are used in order to prepare HEMs with unique properties and 3D printing through additive manufacturing routes developed recently have widened the scope of the field.

AI/ML is emerging as an important tool to design newer HEM compositions, with desired microstructure and properties. I am very happy to see that a high-quality compilation of recent HEM research pertaining to key aspects, such as artificial intelligence and machine learning (AI/ML)-based design, additive manufacturing, functional properties, structural performance, mechanical, fatigue, creep, welding processes, and corrosion behaviour has been included in this book. It has the distinct edge over the other books published in this field, since it uses for the first time an AI-based algorithm for obtaining summaries of the papers included in each section. A next generation book writing experience is thus obtained by combining extreme human mind power and high computing power. I am delighted to see this approach, and I am sure that the book will be an excellent resource for anyone working in this field and for those entering this exciting field, regardless of their academic background.

Prof. B. S. Murty
Director, IIT Hyderabad
Hyderabad, India

Preface I

Auto-summaries can be generated by either an abstractive or extractive auto-summarization:

- An extraction-based summarizer identifies the most important sentences of a text and uses the original sentences to create the summary.
- An extraction-based summarizer creates new text based on deep learning. New phrases are created to summarize the content.

The auto-summaries you will find in this book have been generated via an extractive summarization approach.

Each chapter was carefully edited by [Saurabh S. Nene]. The editors selected the papers which were then auto-summarized. The editors have not edited the auto-summaries due to the extraction-based approach, and have not changed the original sentences. You will find the editors'; reviews and guidance on the auto-summaries in their chapter introductions.

In machine-generated books, editors are defined as those who curate the content for the book by selecting the papers to be auto-summarized and by organizing the output into a meaningful order. Next to the thoughtful curation of the papers, editors should guide the readers through the auto-summaries and make transparent why they selected the papers.

The ultimate goal is to provide a current literature review of Springer Nature publications on a given topic in order to support readers in overcoming information overload and to help them dive into a topic faster; to identify interdisciplinary overlaps; and to present papers which might not have been on the readers' radar.

Please note, that the selected papers are not used to train a LLM while the auto-summaries are created.

Jodhpur, India Saurabh S. Nene

Preface II

Metallurgy was known mainly for the work on steels, Cu, and Al-based alloys till a new class of materials like Mg and Ti-based lightweight alloys were invented. The development of new materials was a result of increased demands of human life to make it better and safer. Not only other classes of alloys but also the field of steels displayed dramatic evolutions with respect to their design as well as performance to satisfy the human needs. Such examples include advanced high-strength steels, PH-stainless steels, and deformed and partitioned steels.

It is clear from the above discussion that any invention in materials occurred based on application till the concept of High Entropy Alloys (HEAs) was put on the floor by Prof. J. W. Yeh and Prof. B. S. Murty more than a decade ago. The concept of HEAs is purely based on the scientific fundamentals of materials thermodynamics, such as stabilization of high entropy phases by maximization of configurational randomness in alloys. The concept of HEAs has also evolved with time as many researchers have entered into it and thus further refining from the conventional definition of HEAs having at least five elements in equal proportions to having all elements present in concentrated amounts. The contribution of Dr. Dan Miracle and Prof. Dierk Raabe in founding the physical metallurgy-based theories related to HEAs along with the founding researchers played a crucial role in bringing HEAs to their present state.

The first book named *High Entropy Alloys* written by Prof. Murty, Prof. Yeh, and Prof. Ranganathan is the first attempt to read HEAs not as a research topic but more as a fundamental topic, thus, it became very famous as a textbook on HEAs all over the globe for all the young minds attempting to enter this field. As mentioned in that book, the field of HEAs is still among the few that have seen not only a maximum number of publications but also publications in most of the materials-related domains, such as structural materials, functional materials, and computational materials.

This book is a unique attempt to bring these all-scientific contributions published in reputed journals published by Springer Nature on a single platform and put together in the form of a reference book for all enthusiastic minds who are planning to start their research career in the exciting field of HEAs. Moreover, this book uses AI-based algorithms to generate summaries of the published papers thereby putting forward a new way of book writing as well as reading experience to the materials community.

I believe this book will be a useful addition in the domain of HEAs and will benefit all researchers in this field to elevate the quality of their research output.

Jodhpur, India Saurabh S. Nene

Contents

About the Editor

Dr. Saurabh S. Nene is currently working as an Associate Professor in the Department of Metallurgical and Materials Engineering (MME), IIT Jodhpur, from June 2023. His academic journey started at College of Engineering, Pune (COEP), where he finished his B.Tech. in Metallurgical Engineering (2010) followed by completing M.Tech. in Materials Science from IIT Bombay in 2012. He started his Ph.D. immediately in 2012 under a joint venture of IIT Bombay, India, and Monash University, Australia, and finished it in 3.5 years (2016). He subsequently moved to Centre for Friction Stir Processing (CFSP), University of North Texas (UNT) in March 2017 as a post-doctoral research associate and subsequently joined Department of MME, IIT Jodhpur, in August 2020 as an Assistant Professor.

Chapter 1
Introduction

1.1 Introduction by Editor

The field of metallurgy and materials science is strongly founded on four major pillars namely, structural materials, functional materials, computational materials engineering and process metallurgy. Each of these pillars are stand-alone serve as a field of research. However certain materials not only interrelate these pillars but also complement their needs due to exciting properties in them. Some of these materials include carbon nano-tubes, graphene, steels and recently discovered high entropy alloys (HEAs).

The major philosophical difference between conventional and HEA design is the absence of matrix element and increase in the number of constituent elements in the alloy system in later compared to former. The design of HEAs is being done using various computational methods such as conventional CALPHAD approaches as well as recently trended artificial intelligence (AI) and machine learning (ML) approaches. The unconventional nature of HEAs makes it tougher to computationally predict the microstructural evolution in comparison with the conventional alloys [1].

This new class of alloys was named as "***High Entropy Alloys (HEAs)***" with a notion that, these materials exhibit very high configurational entropy (> 1.5R) due to presence of, more than five constituent elements in equal or nearly equal amounts [2–4]. This attempt was made to enrich the materials with the benefits of conventional alloys and metallic glasses in synergy (i.e., obtain a crystalline material with minimized solute partitioning during solidification) such that single phase solid solution can form in the microstructure upon solidification. However, as this field evolved with time, researchers realized that, obtaining multiple high entropy phases in HEAs is much beneficial than just maximizing entropy for obtaining single solid solution phase for improving their fabrication ability as well mechanical performance. As a result, HEAs containing multiple high entropy phases and inherent core effects were developed and gave rise to so called as ***non-equiatomic HEAs*** [2–6].

S. S. Nene (ed.), *High Entropy Alloys*, https://doi.org/10.1007/978-981-99-7173-2_1

Non-equiatomic HEAs (popularly considered AlxCoCrNiFeMn system) showed advantage of having multiple phases formed in microstructure giving rise to activation of multiple deformation modes instead of just solid solution strengthening effect. With evolution in the understanding of HEA design deeper, Lu et al. [7] merged this strategy with the classical eutectic alloy theory and developed novel eutectic HEAs (e.g. $AlCoCrFeNi_{2.1}$ HEA) which showed exceptional yield strength in spite of having typical colony type eutectic structure composed of B.C.C. and L_{12} phases. At the similar time, Li et al. [8] proposed a concept of metastable HEAs which for the first-time incorporated transformation/twinning induced deformation at room temperature in HEAs. This was considered as a major breakthrough in the journey of HEAs which was further elevated by the design of microstructurally flexible HEAs by Nene et al. [9] wherein sensitive phase stability was obtained in HEA under synergistic action of alloy chemistry, strain and temperature. ***Flexible HEA*** design provided a HEA with lowest stacking fault energy (g = 6 mJ/m^2) in the metallic systems so far [9, 10].

Recent work by Miracle and co-workers [11, 12] on high temperature alloys include mainly design of ***refractory HEAs*** which have shown extremely good strength with increase in temperatures. However, they have shown limitations such as very high cost and density along with poor tensile behavior in most of the cases. Development of ***high entropy superalloys*** have been recently done in order to improve the fatigue and creep resistance of turbine blades during service [13, 14]. Recent work by Nene and co-workers [15] also demonstrated total elongation more than 550% at 700 °C and conventional strain rate of 10^{-3} s^{-1} for dual phase HEA after friction stir processing thereby confirming superplasticity in the alloy. Another important domain wherein HEAs shown extreme potential is corrosion resistance [16] tuning with compositional as well as processing variations. The main efforts are put towards minimizing the corrosion rate since elimination of corrosion is thermodynamically impossible [16].

HEAs have also found to be potential candidates for advanced manufacturing techniques such as fusion based as well as solid state additive manufacturing [17] wherein near net shape can be obtained using CAD model. Use of HEAs in attempting dissimilar welding with light weight alloys such as Al alloys is found to eliminate the issue of intermetallic formation thereby improving the weld strength and thus the efficiency [18]. Another field where HEAs displayed interesting behavior is conventional as well as rapid solidification processes in terms of obtaining wrought like cast structures thereby minimizing the detrimental columnar dendrite morphology [19].

At last, recent research work on developing HEAs for catalytic applications displayed huge potential by showing effective catalytic control during the water purification or hydrogen generation reactions [20]. Moreover, HEAs found to be promising candidates as hydrogen storage materials in the form of metal hydrides compared with other complex non-formable materials. HEAs were also found to be superior electrolytes in lithium batteries and can also show enhanced superconductivity and magnetic performance compared to other conventional metallic systems [20–22] (Fig. 1.1).

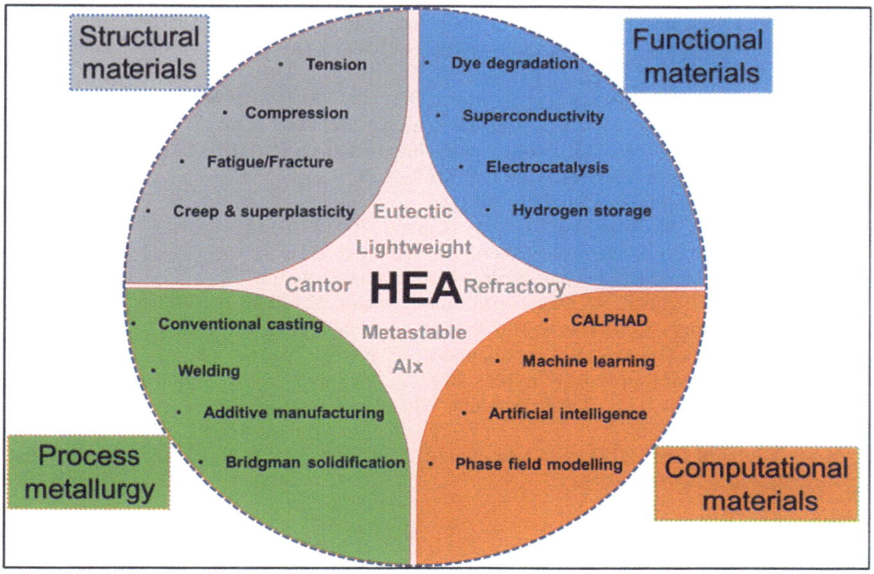

Fig. 1.1 Schematic showing overall structure of human–machine collaborated book on HEA

This all discussion therefore elaborates that HEAs are prepatent candidates in all four pillars on which Metallurgy and Materials Science stream is founded and hence is a point of attraction for doing research at present in materials science community. This book is the first attempt to summarize the high-quality work done on HEAs on artificial intelligence (AI) and machine learning (ML) based HEA design, additive manufacturing, functional properties and structural performance of HEAs including mechanical, fatigue, creep, welding process and corrosion behavior [21]. The uniqueness of this book in comparison with the other conventional reference books is that it uses for the first time AI based algorithm for obtaining summary of the papers included in the respective sections. Thus, it is an attempt to use extreme human mind power with the high computing ability of machine to obtain a next generation book writing experience. At last, additional comments added by the author in italics at the end of specific summaries explain the information to the readers.

References

1. P. Bajaj, A. Hariharan, A. Kini et al., Mater. Sci. Engg. A, 772 (2020) 138633.
2. B.S. Murty, J.W. Yeh, S. Ranganathan, High Entropy Alloys, Butterworth-Heinemann (Elsevier), 2014 1st Edition.
3. J.W. Yeh, S.K. Chen, S.J. Lin et al., Adv. Engg. Mater., 6 (2004) 299–303.
4. M-H. Tsai & J-W. Yeh, Mater. Res. Lett. 2 (2014), 107–123.
5. A.S. Sharma, S. Yadav, K. Biswas, B. Basu, Mater. Sci. Eng. R, 131 (2018) 1–42.

6. R.S. Mishra, R.S. Haridas, P. Agrawal, Mater. Sci. Eng., A, 812 (2021) 141085.
7. Y. Lu, Y. Dong, S. Guo, et al., Scientific Repts, 4 (2014) 6200.
8. Z. Li, K.G. Pradeep, Y. Deng, et al., Nature 534 (2016) 227–230.
9. S.S. Nene, M. Frank, P. Agrawal et al., Mater. Des., 194 (2020) 108968.
10. R.S. Mishra, S.S. Nene, Metall. Mater. Trans. A, 52 (2021) 889–896.
11. O.N. Senkov, G.B. Wilks, D.B. Miracle et al., Intermetallics, 18 (2010) 1758–1765.
12. O.N. Senkov, Miracle Mater. Res, 33(19), 3092–3128.
13. C. T. Sims, Superalloys, (1984) 399–419.
14. J. Chen, X. Zhou, W. Wang et al., J. Alloys Compd. 760 (2018) 15–30.
15. S.S. Nene, K. Liu, S. Sinha et al., Materialia 9 (2020) 100521.
16. S.S. Nene, M. Frank, K. Liu et al., Scripta Mater. 166 (2019) 168172.
17. P. Agrawal, S. Thapliyal, S.S. Nene et al., Addit. Manuf. 32 (2020) 101098.
18. S.S. Nene, S. Gupta, C. Morphew, R.S. Mishra, Materialia 11 (2020) 100740.
19. A.R. Balpande, S. Agrawal, X. Li et al., under preparation (2023).
20. L-H. Liu, N. Li, M. Han et al., Rare Met. 40 (2022) 125–131.
21. S.S. Nene, S. Sinha, D.K. Yadav, A. Dutta, Metallurgical Aspects of High Entropy Alloys, 1005 (2024) 175849.
22. X Ma X Ding R Chen 2022 RSC Adv. 12 11272 11281 X. Ma, X. Ding, R. Chen et al., RSC Adv., 12 (2022) 11272–11281.

Chapter 2
Fundamentals of High Entropy Alloys

2.1 Introduction by Editor

High-entropy alloys (HEAs) are a relatively new class of materials that push the boundaries of traditional alloy design strategies and provide vast compositional space. This multifaceted field combines thermodynamics, metallurgy, phase stability, crystallography, and mechanical behavior to fulfill the need to achieve unique synergies caused by high compositional complexity. Yet, even in their nascent stages, the definition, and characteristics of HEAs have not been without conflict and debate, adding an additional layer of complexity to this intriguing field. This chapter discusses the intricate aspects of HEAs such as thermodynamics, the four core effects, alloy design strategies, and rebutting misconceptions related to HEAs.

High-entropy alloys (HEAs) represent a paradigm shift in materials science, pushing the boundaries of traditional alloy design strategies and unlocking vast compositional space. This multifaceted field combines principles from thermodynamics, metallurgy, phase stability, crystallography, and mechanical behavior to exploit the unique synergies arising from high compositional complexity. Yet, despite their relative novelty, the definition and characteristics of HEAs have been subject to debate and conflict, adding layer of complexity to this intriguing field. This chapter aims to unravel the intricate aspects of HEAs, including thermodynamics, the four core effects influencing their behavior, alloy design strategies, and addressing common misconceptions surrounding these fascinating materials.

HEAs have emerged as a promising class of materials characterized by their unconventional composition comprising multiple principal elements in near-equimolar proportions. The sheer diversity of elements in HEAs offers unprecedented opportunities for tailoring properties to meet specific application requirements. Traditional alloy design strategies, which typically focus on one or two primary elements with minor additions for alloying purposes, are transcended by HEAs, which embrace a broader compositional space, challenging conventional metallurgical principles.

S. S. Nene (ed.), *High Entropy Alloys*, https://doi.org/10.1007/978-981-99-7173-2_2

In the context of HEAs, the phrase "high entropy" refers to the significant amount of configurational entropy that results from the wide range of atomic configurations present in the alloy matrix. This phenomenon, driven by entropy, not only increases the stability of the alloy but also encourages the development of distinct microstructures, such as solid solutions with a single phase or compositions with multiple phases. At the heart of HEA research lies the interplay between thermodynamics and kinetics. Thermodynamic principles dictate phase stability and microstructural evolution in these complex systems, while kinetic factors influence the formation of metastable phases and the kinetics of phase transformations. The pursuit of thermodynamically stable phases amidst the vast compositional landscape of HEAs presents both challenges and opportunities, driving researchers to explore novel alloying strategies and predictive models to understand phase behavior and stability.

Four core effects fundamentally influence the behavior of HEAs: the entropy effect, configurational entropy, the cocktail effect, and the sluggish diffusion effect. Each of these effects plays a critical role in determining HEAs' unique properties and performance. The "entropy effect" is primarily driven by the high configurational entropy inherent in HEAs. This high entropy results from the nearly equimolar proportions of multiple principal elements, which significantly increases the number of possible atomic configurations. Consequently, this elevated configurational entropy stabilizes solid solution phases over intermetallic compounds, suppressing phase transformations that typically occur in conventional alloys. This stabilization is crucial because it allows HEAs to maintain their structural integrity and performance under a wide range of conditions, including high temperatures and mechanical stresses. "Configurational entropy" refers to the measure of disorder introduced by the presence of multiple elements in equimolar or near-equimolar proportions. This disorder leads to the formation of configurationally complex microstructures that are not present in traditional alloys. The high configurational entropy contributes to the stability of the solid solution phases and enhances the mechanical properties of HEAs. The presence of multiple elements in a disordered arrangement can impede dislocation motion, resulting in improved strength and hardness. Additionally, the complex microstructures can enhance the alloy's resistance to wear and deformation, making HEAs suitable for demanding applications. The "cocktail effect" describes the synergistic interactions among the various elements in HEAs. Unlike traditional alloys, where the addition of elements often serves to modify the properties of a base metal, HEAs benefit from the collective behavior of multiple principal elements. These interactions can lead to properties that surpass those of conventional alloys. For instance, the combination of different atomic sizes, bonding characteristics, and electron configurations can result in enhanced mechanical properties, such as increased strength and ductility, as well as superior corrosion resistance and thermal stability. The cocktail effect underscores the potential of HEAs to achieve a balanced set of properties tailored to specific application needs. The "sluggish diffusion effect" arises from the slower diffusion kinetics observed in HEAs. The presence of multiple principal elements with different atomic sizes and bonding energies creates a complex energy landscape, which inhibits the movement of atoms. This sluggish diffusion prevents the rapid formation of intermetallic compounds and phase separation, which are common in traditional alloys. As a result, HEAs exhibit exceptional thermal stability and resistance to high-temperature degradation. The sluggish diffusion effect also contributes to the alloy's ability to retain its microstructural integrity over prolonged periods, making HEAs ideal for applications involving high-temperature environments and long service life.

Apart from these four core effects, metastability (sensative phase stabilization) is another feature of HEAs which drive not only formation of configurationally random phases but also with unique characteristics, e.g., γ phase in metastable HEA demonstrated activation of bidirectional TRIP effect leading to excellent deformability. Thus, phase metastability can be used as a powerful tool in the design of next generation HEAs along with other core effects.

Alloy design strategies for HEAs encompass a wide range of approaches aimed at tailoring composition, microstructure, and properties to meet specific application requirements. Examining binary phase diagrams is an important initial step in investigating the extensive range of compositions in high-entropy materials. This chapter explores the methods and knowledge acquired by utilizing binary phase diagrams to discover probable compositions with high entropy. These strategies include the selection of alloying elements based on their size, electronegativity, and mixing enthalpy to promote solid solution strengthening and phase stability. Additionally, hierarchical microstructural design, such as introducing secondary phases or controlling grain size and morphology, can further enhance mechanical and functional properties. Researchers can expedite the identification of new HEAs with customized features by using thermodynamic databases, phase diagram calculations, and computational modelling. These tools allow for systematic screening and prioritization of potential alloy systems, which can then be further validated through experimental testing. Computational modelling and simulation techniques play a crucial role in guiding alloy design by predicting phase stability, microstructural evolution, and mechanical behavior in HEAs.

Despite the promising attributes of HEAs, misconceptions and controversies surround these materials. One common misconception is the belief that HEAs are inherently brittle due to their high compositional complexity. However, research has shown that properly designed HEAs can exhibit exceptional mechanical properties, including high strength, ductility, and fracture toughness. Another misconception relates to the scalability and cost-effectiveness of HEA production. While initial synthesis and processing of HEAs may present challenges, advancements in manufacturing techniques and process optimization have made large-scale production feasible and economically viable.

In conclusion, high-entropy alloys represent a transformative material class that challenges conventional alloy design strategies and offers unprecedented opportunities for tailoring properties to meet diverse application requirements. By unravelling the intricate aspects of HEAs, including thermodynamics, core effects, alloy design strategies, and addressing misconceptions, this chapter aims to provide insights into the potential of these fascinating materials and shape future directions in materials research. As research in HEAs continues to evolve, they are poised to revolutionize various industries, offering enhanced performance, durability, and sustainability in a wide range of applications [1–58].

2.2 Machine Generated Summaries

Disclaimer: The summaries in this chapter were generated from Springer Nature publications using extractive AI auto-summarization: An extraction-based summarizer aims to identify the most important sentences of a text using an algorithm and uses those original sentences to create the auto-summary (unlike generative AI). As the constituted sentences are machine selected, they may not fully reflect the body of the work, so we strongly advise that the original content is read and cited. The auto generated summaries were curated by the editor to meet Springer Nature publication standards. To cite this content, please refer to the original papers.

Machine generated keywords: compositional, phase diagram, diagram, compositional complexity, compositional space, vast, space, article, alloy design, number, remelted, review, mix, relatively, sluggish

Alloy Design Strategies and Future Trends in High-Entropy Alloys [51]

This is a machine-generated summary of:

Yeh, Jien-Wei: Alloy Design Strategies and Future Trends in High-Entropy Alloys [51]

Published in: JOM (2013)

Link to original: https://doi.org/10.1007/s11837-013-0761-6

If you want to cite the papers, please refer to the original.

For technical reasons we could not place the page where the original quote is coming from.

Abstract-Summary

"High-entropy alloys (HEAs) are newly emerging advanced materials."

"To conventional alloys, HEAs contain multiple principal elements, often five or more in equimolar or near-equimolar ratios."

"The basic principle behind HEAs is that solid-solution phases are relatively stabilized by their significantly high entropy of mixing compared to intermetallic compounds, especially at high temperatures."

"Wise alloy design strategies for suitable compositions and processes to fit the requirements for either academic studies or industrial applications thus become especially important."

Introduction

"The alloy design, alloy production, and alloy selection are almost confined by one-element or one-compound concept."

"This alloy concept obviously limits the degree of freedom in alloy's composition and thus limits the development of special microstructure and properties."

"In recognition of this tradition, a brand-new alloy concept of high-entropy alloys (HEAs) to break the confinement have been proposed and explored since 1995 [1, 2]. Up until now, it has been demonstrated that the explored alloys in this virgin field are feasible to be synthesized, processed, and analyzed contrary to the misconceptions based on traditional experiences."

"It is still very important to present basic concepts relating with HEAs in advance, including the origin of high entropy, definition, four core effects of HEAs, and misconceptions on HEAs, which are helpful in alloy design and research for HEAs."

Definition and the Reason to Use High Entropy

"Because ΔH_f can be regarded as the energy required to destroy about one twelfth of all bonds in the solid, a mixing entropy of R per mole due to mixing in an alloy state would be quite large to decrease the mixing free energy by an amount of RT at high temperatures, e.g., RT = 8.314 kJ/mol at 1000 K, and to compete with the mixing enthalpy of intermetallic compounds arising from the interaction between unlike metallic atoms."

"From the configurational entropy equation, we can obtain that an element with a concentration of 5 at.% would contribute a mixing entropy of 0.05Rln0.05 = 0.15R, which is just 10% of minimum requirement of 1.5R for HEAs."

"Under the above definition and practical upper bound of 13 principal elements, at an arbitrary choice of a group of 13 metallic elements, we can obtain a total of 7099 possibilities for designing equal-mole HEAs systems from 5 to 13 elements2: We may design an equal-mole AlCoCrCuFeNi alloy."

Four Core Effects of HEAs

"As the name implied, high entropy is the unique and most important of HEAs because its effect could enhance the formation of solution phases and makes the microstructure much simpler than previously expected."

"To elucidate a high entropy effect in enhancing the formation of solid-solution phases and inhibiting the formation of intermetallic compounds, HEAs composed of constituent elements with stronger bonding energies between each other are considered."

"To include the effect of atomic size difference, Zhang and others [3] first proposed the forming trend of disordered solid solutions, ordered solid solution, intermediate phases, and BMG by comparing ΔS_{mix}, ΔH_{mix}, and atomic size difference (δ)."

"This alloy design relied on the selection of equimolar ferromagnetic elements (Fe, Co, and Ni) for forming ductile fcc phase with higher atomic packing density than bcc, and suitable addition of nonmagnetic elements (Al and Si having slightly anti-parallel magnetic coupling with Fe, Co, and Ni) to increase lattice distortion."

Several Misconceptions on HEAs

"Several misconceptions about HEAs are as follows: (I) HEAs have low entropy and complex microstructure, (II) HEAs are hard and brittle, and (III) HEAs are expensive and difficult to fabricate."

"HEAs have a high mixing entropy at such states compared with those of conventional alloys."

"If an alloy has a high mixing entropy, then simple solid-solution phases will form at high temperatures due to the large $T\Delta S_{mix}$."

"[4] Conversely, if multielement alloys do not have a high mixing entropy at high temperatures, then intermetallic phases would form at high temperatures."

"HEAs are like conventional alloys: Some are hard and brittle, while some are softer and ductile."

"Fcc phases in HEAs in general possess lower hardness but excellent ductility."

"A high entropy effect tends to make all these phases become multielement solid solutions and to improve their mechanical properties."

Alloy Design Strategies for HEAs

"There are several routes to develop HEAs: (I) use combinatorial materials synthesis technique; (II) use computational materials science: ab initio simulation, molecular dynamics (MD), finite element, New PHACOMP, CALPHAD method, etc.; (III) use Taguchi method to optimize the properties; (IV) start from promising alloy systems; and (V) use alloy design principles of materials science."

"As for MD simulation, a many-body tight-binding potential model has been applied to study the effect of the number of elements and size difference on the amorphous structure of HEAs."

"The Taguchi method could be applied to an HEA system to observe those compositions that display better properties and performance."

"This route needs better understanding on the features of elements such as melting point, atomic size, crystal structure, valence electron, electronegativity, density, elastic constants, interactions in unlike atom pairs, thermodynamics and kinetics, microstructure and properties relationship, and four core effects of HEAs."

Future Trends of HEAs

"Because materials science and solid-state physics are mainly based on the research on conventional materials with one or two principal elements, what happens in HEAs would be interesting in many aspects and become valuable academic issues."

"More research on promising alloy systems for better mechanical, chemical, and physical properties is required."

"More research on the performance challenges unattainable by traditional alloys or materials is required, such as room-temperature superconductors, alloys exceeding the performance and temperature capability of superalloys, and tools with very high hot hardness and long lifespan."

"More research on high-entropy nitrides, carbides, oxides, and their combinations is required."

"A certain amount of such research [5, 6] has indicated that four core effects of HEAs also hold true in such high-entropy ceramics (HECs), and many promising properties are obtained for diffusion barriers, hard coatings, and functional coatings."

"By this effort, one might get closer to use or modify HEAs and HECs to fit the application requirements."

Conclusions

"High-entropy materials have four core effects: high entropy, sluggish diffusion, severe lattice distortion, and cocktail effects, and could provide a wide spectrum of properties."

"Like traditional alloys, HEAs have potential applications in different fields and might replace traditional materials."

"Basic science, new HEAs, new HECs, and new applications are awaiting further research."

High-Entropy Alloys [52]

This is a machine-generated summary of:

George, Easo P.; Raabe, Dierk; Ritchie, Robert O.: High-entropy alloys [52]

Published in: Nature Reviews Materials (2019)

Link to original: https://doi.org/10.1038/s41578-019-0121-4

If you want to cite the papers, please refer to the original.

For technical reasons we could not place the page where the original quote is coming from.

Abstract-Summary

"For the past decade and a half, however, a new alloying strategy that involves the combination of multiple principal elements in high concentrations to create new materials called high-entropy alloys has been in vogue."

"A few high-entropy alloys have already been shown to possess exceptional properties, exceeding those of conventional alloys, and other outstanding high-entropy alloys are likely to be discovered in the future."

"We review recent progress in understanding the salient features of high-entropy alloys."

"We also address the vast compositional space that remains to be explored and outline fruitful ways to identify regions within this space where high-entropy alloys with potentially interesting properties may be lurking."

Introduction

"This was a counter-intuitive notion because the conventional view—likely based on binary phase diagrams in which solid solutions are typically found at the ends and compounds near the centres—was that the greater the number of elements in concentrated alloys, the higher the probability that some of the elements would react to form compounds."

"An alloy created by combining two pure elements A and B can form a solid solution or one or more intermetallic phases, depending on the relative free energies of the following reactions: where ΔG_{mix}, ΔH_{mix} and ΔS_{mix} are the Gibbs free energy, enthalpy and entropy of mixing, respectively; ΔG_f, ΔH_f and ΔS_f are the corresponding values for the formation of an intermetallic compound with AB stoichiometry; and T is the absolute temperature."

Single-phase, high-entropy alloys

"The results show that the alloy is a single-phase, FCC, solid solution with no indication of clustering or short-range ordering, although if such local ordering were experimentally proven to exist, density functional theory (DFT) calculations suggest that it would have a profound effect on critical properties such as the stacking-fault energy and dislocation mobility, and, thus, it would be a major factor in controlling mechanical properties [7]."

"Several important features of the Cantor alloy were revealed [8]: this alloy exhibits a strong temperature dependence of strength below 473 K, a weaker temperature dependence at elevated temperatures up to 1273 K and a modest strain-rate dependence at low homologous temperatures; its strength and ductility both increase with decreasing temperature down to the cryogenic range, with ultimate strengths and elongations in excess of 1 GPa and 60%, respectively, at 77 K; its work-hardening rate increases strongly with decreasing temperature, which postpones the onset of necking and, thereby, increases the tensile ductility; and it fractures transgranularly by ductile microvoid coalescence, which is consistent with the high ductilities observed."

Uncharted high-entropy alloys space

"Since at least the mid-1960s, combinatorial approaches, which involve high-throughput synthesis and characterization of multiple-alloy compositions, have been used in materials science to rapidly determine phase diagrams in multi-component spaces and screen physical and mechanical properties [9–14]."

"Disadvantages of small-scale combinatorial methods include limited product size, sometimes too small for a comprehensive screening of functional and/or mechanical properties (such as strain hardening, ductility or fracture toughness), and resulting materials that are too artificial in terms of microstructure, morphology, size, distribution and orientation of the phases present."

"Allowing for such compositional deviations from the original HEA rules is justified: first, the mixing entropy curves have a shallow shape so that deviations from equimolar compositions do not alter the configurational entropy substantially; second, compositional tuning of the stacking-fault energy, or other key thermodynamic parameters, is far more important than entropy maximization; third, when striving for mechanical properties, avoiding brittle, intermetallic phases is more of an imperative than obeying static, alloy design rules."

Prospects and future opportunities

"As we discussed, the high-entropy phenomenon has been principally driven by three main factors: the exceptional mechanical properties of a few specific alloys, most notably the FCC CrCoNi-based alloys, especially for cryogenic temperature applications; the search for new refractory HEAs that can operate with sustained strength at ever-increasing temperatures, a difficult pursuit because of the limited ductility and oxidation resistance of these materials; and the prospect of discovering new materials with unprecedented properties, or combinations of properties, from the seemingly unbounded list of possible multiple-principal-element materials, although this is a non-trivial exercise in light of the methods currently available to venture into uncharted materials, composition and microstructural space."

"Special emphasis should be placed on the roles of interstitial elements such as B known for their beneficial effects on interface cohesion in conventional alloys, on reaching higher precision in the prediction of phase transformations and on low-temperature equilibrium and non-equilibrium phases and microstructure states, because some high-entropy alloys (HEAs) target cryogenic applications."

Pt-Induced Atomic-Level Tailoring Towards Paracrystalline High-Entropy Alloy [53]

This is a machine-generated summary of:

He, Xingjia; Zhang, Yu; Gu, Xinlei; Wang, Jiangwei; Qi, Jinlei; Hao, Jun; Wang, Longpeng; Huang, Hao; Wen, Mao; Zhang, Kan; Zheng, Weitao: Pt-induced atomic-level tailoring towards paracrystalline high-entropy alloy [53]

Published in: Nature Communications (2023)

Link to original: https://doi.org/10.1038/s41467-023-36423-1

If you want to cite the papers, please refer to the original.

For technical reasons we could not place the page where the original quote is coming from.

Abstract-Summary

"Based on the vast composition space and the complex atomic interactions in the high-entropy alloys (HEAs), we present an "atomic-level tailoring" strategy to create the paracrystalline HEA."

"The addition of atomic-level Pt with the large and negative mixing enthalpy induces the local atomic reshuffling around Pt atoms for the well-targeted local amorphization, which separates severe-distorted crystalline Zr-Nb-Hf-Ta-Mo HEA into the high-density crystalline MRO motifs on atomic-level."

"Such an enthalpy-guided strategy in HEAs can provide the atomic-level tailoring ability to purposefully regulate structural characteristics and desirable properties."

Introduction

"Under the thermodynamic control, it is difficult to achieve the uniform and size-controllable precipitation of crystalline phase in MGs because of the complexity of crystallization processes [15], particularly for a paracrystalline state that is closely associated with the accurately uniform precipitation of the crystalline MRO but devoid of long-range order (LRO)."

"In recent research, Wu and others [16] have demonstrated that in the Cr-Fe-Co-Ni HEA the uniform introduction of the glass-forming elements B and Si, tending to diffuse and gather at grain boundaries for about 1nm-thickness amorphous tissue, has yielded a nanocrystal-glass dual-phase high-entropy composite during the co-sputtering process."

"Based on enthalpy-guided strategy, Pt–Zr-enriched amorphous groups surrounding foreign Pt atoms successfully separate severe-lattice-distorted Zr–Nb–Hf–Ta–Mo grains into crystalline MRO, forming the paracrystalline HEA."

Results

"It seems that the Pt-induced local composition variation dominates the local amorphization, and thus the targeted introduction of local amorphization achieved by the precise control of Pt atoms can accurately and uniformly separate the high-crystalline HEA into paracrystalline HEA."

"Different from the large-scale SBs in $Zr_{11}Nb_{10}Hf_{15}Ta_{16}Mo_{17}Pt_{31}$ MG suffering from severe strain localization, the existence of high-population crystalline MRO motifs enables paracrystalline HEA to deform more homogeneously in the form of much-smaller nanoscale SBs."

"Incorporating a trace amount of Pt element, possessing the large and negative ΔH_{mix} between other constituent elements, into the severe-distorted crystalline Zr–Nb–Hf–Ta–Mo HEA can trigger the local composition reconfiguration around Pt atoms to satisfy the local amorphization conditions: the large and negative ΔH_{mix} and the high distortion."

Methods

"The composition of the films was controlled by tuning the sputtering power applied to the Pt target, and a ~150 nm-thickness Ti interlayer was deposited on the Si wafers to improve the adhesion between the alloys and the Si substrates."

"The deposition was carried out for 120 min with an applying current of 450 mA for ZrNbHfTaMo compound target supplied by direct current power and variable sputtering power for the Pt target (0 W, 10 W, 100 W) supplied by radio-frequency power, and the thicknesses for all obtained films are ranging from 2.0 μm to 2.8 μm."

"The composition gradient was controlled by tuning the RF sputtering power applied to the Pt target and each HEA layer was sustained at ~250 nm."

"The cross-sectional samples for indented samples were prepared with a FIB instrument (FEI Strata 400S) and observed via JEM 2100 F HRTEM."

Influence of Compositional Complexity on Species Diffusion Behavior in High-Entropy Solid-Solution Alloys [54]

This is a machine-generated summary of:

Seoane, Axel; Farkas, Diana; Bai, Xian-Ming: Influence of compositional complexity on species diffusion behavior in high-entropy solid-solution alloys [54]

Published in: Journal of Materials Research (2022)

Link to original: https://doi.org/10.1557/s43578-022-00545-x

If you want to cite the papers, please refer to the original.

For technical reasons we could not place the page where the original quote is coming from.

Abstract-Summary

"Detailed comparative molecular dynamics simulations of the diffusion process in a model quinary equiatomic FeNiCrCoCu FCC alloy are presented."

"To assess the role of compositional complexity, the results are compared to corresponding simulations in each of the pure individual components of the alloy as well as the corresponding "average atom" potential, with similar properties to the alloy but no compositional randomness."

"The results are compared with experimental data for self-diffusion in similar high-entropy alloys."

Introduction

"Interdiffusion coefficients were not necessarily lower in FCC Al–Co–Cr–Fe–Ni–Mn alloys; therefore, no sluggish diffusion was observed in this FCC HEA."

"Osetsky and co-workers [17] argued that coupled percolation and composition-dependent barriers for vacancy jumps within different subsystems in medium- and high-entropy alloys can lead to sluggish diffusion."

"Chen and others [18] reported interdiffusion results in Al–Co–Cr–Fe–Ni–Ti HEA indicating that the sluggish diffusion effect exists for only some of the components present in the complex alloy."

"The existence of sluggish diffusion in HEAs is still controversial."

"The measured diffusivities in HEAs are outcomes of synergetic effects of compositional complexity, lattice mismatch, and interactions of different alloying elements (the so-called "cocktail" effect [19])."

"We propose a different method to clearly isolate the effects of compositional complexity in the diffusion behavior of HEAs."

Methods

"Varvenne and others [20] showed that the average atom potential can be quantitatively accurate for a wide range of random alloy properties."

"Nöring and Curtin [21] studied the finite-temperature thermodynamic properties of the average atom potential to determine if the average atom potential can also represent the important finite-temperature properties of the true random alloy."

"Using a thermodynamic integration approach, they found that the complex alloy's lattice constant as well as elastic constants are well-predicted by the average atom potential over a wide temperature range."

"The data show that the overall properties predicted by the average atom potential are indeed in quantitative agreement with the average results obtained for the HEA alloy, within 10%."

"The mean migration energy in the HEA was calculated through the NEB method for each atom type over 5000 different surrounding neighbor jumps to the vacancy, also 25,000 cases total."

Results

"The results show that the vacancy formation energy obtained for the AA (1.43 eV) material is very close to the average of the vacancy formation energies for the five pure components, as given by the rule of mixtures (1.44 eV)."

"The results show that the vacancy migration energy obtained for the AA material is very close to the average of the vacancy migration energies for the pure components, given by the rule of mixtures."

"The values of the migration energies from the dynamic calculations show the same result: the migration energy obtained for the complex alloy is very similar to that of the AA material, and also similar to the prediction of the rules of mixtures based on the values for each of the pure components."

Discussion

"The important difference in vacancy formation and migration energies is that, while the averages are the same, the values obtained for the complex alloy present a statistical distribution."

"It is possible that some other configurations could still lead to non-Gaussian distributions of vacancy migration and formation energies, and thus result in sluggish diffusion."

"Their pure FCC elemental state results are Fe 1.58, Cr 1.61, Co 1.70 and Ni 1.89 eV. Our values are Fe 1.61, Cr 1.41, Co 1.36 and Ni 1.61 eV. For their HEA, the values are Fe 1.89, Cr 1.62, Co 1.85 and Ni 1.41 eV. For our HEA they are Fe 1.48, Cr 1.42, Co 1.34 and Ni 1.53 eV. This is a reasonable agreement, given that our HEA composition includes Cu which has a much lower vacancy formation energy than the other elements."

Conclusions

"We proposed an approach for evaluating the existence of sluggish diffusion in the complex equiatomic FeNiCrCoCu model HEA through molecular dynamics simulations."

"The new approach compares vacancy diffusion for the HEA with that in the corresponding AA (Average Atom) potential, which averages the equilibrium properties of the atoms."

"We encountered no sluggish diffusion, on average."

"Comparison with the average diffusion values of the alloy components as inspired by the rule of mixtures also showed no sluggish diffusion overall, with some components diffusing faster and other slower than in the corresponding pure FCC structures."

Base Alloy Concept in the Development of High-Entropy Materials [55]

This is a machine-generated summary of:

Krapivka, M. O.; Myslyvchenko, O. M.; Karpets, M. V.: Base Alloy Concept in the Development of High-Entropy Materials [55]

Published in: Powder Metallurgy and Metal Ceramics (2018)

Link to original: https://doi.org/10.1007/s11106-018-9932-x

If you want to cite the papers, please refer to the original.

For technical reasons we could not place the page where the original quote is coming from.

Abstract-Summary

"The requirements to be met by a high-entropy alloy to be used as a base one for the development of new high-entropy materials have been formulated."

"The influence of 12 elements (Al, Ti, Zr, Hf, V, Nb, Ta, Mo, W, Re, Cu, Y) on the phase composition of the base alloy at the equiatomic content of each additive was studied."

"The maximum solubility of each element in the base alloy was determined."

"The feasibility of using the base alloy to develop new high-entropy materials with the approaches applied to develop conventional alloys is shown."

Introduction. Problem statement

"High-entropy alloys (HEAs) [23], which do not have a base element, can be considered an alternative."

"The classical example of high-entropy alloys with the maximum entropy includes those consisting of 5 to 13 elements; however, as they have no predominant element, it is rather difficult to predict their phase composition."

"The structurization and phase formation processes in high-entropy alloys substantially differ from those in conventional alloys, thus generating interest in the integral characteristics of alloys to be found to classify them according to the phase composition."

"According to statistics, the percentage of single-phase high-entropy solid solutions in the total amount of high-entropy alloys is small [24] and two-phase alloys are more frequent; they contain B2, μ, σ, or Laves phases in a disordered solid solution, which significantly influences the mechanical properties of such alloys, often improving their characteristics."

Experimental Procedure

"The ingots were remelted six times to homogenize the composition, the cooling rate for an ingot with the above weight was 100 K/sec."

"For X-ray diffraction, we used a DRON-UM1 diffractometer with a monochromatic Cu-K_α radiation source."

"The X-ray diffraction patterns were plotted by step-by-step scanning."

"Of diffraction patterns, texture in the samples was considered within the March–Dollase model."

Experimental Results and Discussion

"Taking into account the difference in the atomic sizes of Ti and elements of the base alloy, as well as its high solubility in the fcc lattice, titanium can be considered an effective hardening element for the solid solution lattice."

"Each of the elements analyzed dissolves in the lattice, leading to hardening of the solid solution of the base alloy and to precipitation of the second phases (bcc, μ, σ, C14, C15, etc.), whose amounts can be controlled."

"We should only note that if several doping additives are simultaneously introduced into the base alloy in amounts not exceeding the solubility limits for each of them, while the total electron concentration of the melt does not decrease below 7.55 e/a, a single-phase solid solution with extremely strong lattice distortion with different-sized atoms can be obtained in most cases."

Conclusions

"We have formulated requirements to be met by base alloys in the development of new high-entropy materials."

"The CrMnFeCoNi alloy with an fcc lattice that we propose for the development of new high-entropy alloys has been studied in detail."

"It is shown that the use of these doping elements allows significant hardening of the solid solution of the base alloy and precipitation of other phases (bcc, μ, σ, C14, C15, etc.) in controlled amounts."

"The concept that we propose for developing the base alloy allows us to use the same principles as for the production of conventional alloys from pure components

(Fe, Ni, Cr, etc.), but at the same time to obtain high entropy alloys that differ from conventional ones in a number of characteristics (in particular, HEAs are characterized by strong hardening of the solid solution, high thermal stability, and lower energy of interphase boundaries)."

High Entropy Alloys Mined from Binary Phase Diagrams [56]

This is a machine-generated summary of:

Qi, Jie; Cheung, Andrew M.; Poon, S. Joseph: High Entropy Alloys Mined From Binary Phase Diagrams [56]

Published in: Scientific Reports (2019)

Link to original: https://doi.org/10.1038/s41598-019-50015-4

If you want to cite the papers, please refer to the original.

For technical reasons we could not place the page where the original quote is coming from.

Abstract-Summary

"A phenomenological method analysing binary phase diagrams to predict HEA phases is presented."

"The hypothesis is that the HEA structural stability is encoded within the phase diagrams."

"We introduce several phase-diagram inspired parameters and employ machine learning (ML) to classify 600+ reported HEAs based on these parameters."

"Both the overall HEA prediction and specifically single-phase HEA prediction rate are above 80%."

"To validate our method, we demonstrated its capability in predicting HEA solid solution phases with or without intermetallics in 42 randomly selected complex compositions, with a success rate of 81%."

"The presented search approach with high predictive capability can be exploited to interact with and complement other computation-intense methods such as CALPHAD in providing an accelerated and precise HEA design."

Introduction

"Empirical approaches that utilised atomistic and thermodynamic parameters [3, 25–27] were first introduced to investigate the compositional regions of HEA phases, but with only limited success."

"Another model using a database of over 2000 multicomponent alloy compositions from a high-throughput sputter deposition experiment, applied a regression method to interrogate the HEA phase formation tendency [28]."

"Models using neural networks were trained based on atomistic and thermodynamic parameters to predict the HEA phase categories without specific phase formation information [29, 30]."

"High entropy provides the driving force for a HEA system to form a single solid solution phase."

"Rather than forming single solid solution phase HEAs, different solid solution phases can coexist with occasionally IM formation occurring."

"Departing from current approaches, we present herein a phenomenological method using binary phase diagrams to predict the compositional space of HEA phases."

Database Partitioning

"The HEAs included in our model have phases classified as: disordered FCC (A1), disordered BCC (A2), disordered HCP (A3), mixed disordered FCC + BCC (A1 + A2), ordered BCC (B2), B2 mixed with disordered solid solution phases specifically A1, A2, and A3 (B2 + SS), and either Sigma or Laves IM mixed with the other phases (IM+)."

"The set of HEAs included in the IM+ phase have at least Sigma or Laves phase."

"Level 2 is Level 1 with the addition of the B2 + SS HEAs."

"Level 3 is Level 2 with the addition of IM+ HEAs."

HEA Phase Formation Parameters

"The parameters, introduced below, and elaborated on in detail in the method section, provide the basis for quantifying HEA phase formation tendencies."

"Phase evolution may still exist below this temperature because of the high kinetic energies of the atoms."

"A phase formation temperature (T_{pf}) is defined where rapid phase evolution ceases."

"This temperature, the kinetic energy of atoms is not high enough to transform the phase within the brief time of cooling."

"Following the above discussion, information from individual binary phase diagrams is combinatorially used within the model."

"That the tendency for a pair of elements to form a specific phase is directly determined by its binary phase field percentage."

"This effect is included in the model with the phase separation parameter (PSP)."

Visualisation of Phase Regions in Parameter Space

"For the Level 1 phases, there are correlations between the calculated parameters PFP_{A1}, PFP_{A2}, PFP_{A3}, and PSP with the A1, A2, A3, and A1 + A2 phase formation."

"Some A2 HEAs form even with small PFP_{A2} values because their B2 phase field can transfer into the A2 phase to prompt A2 formation."

"For the Level 2 phases, the five parameters are PFP_{A1}, PFP_{A2}, PFP_{A3}, PFP_{B2}, and PSP."

"PFP_{A3} and A3 HEAs are not plotted here because PFP_{A3} does not affect the formation of B2 + SS phase and A3 HEAs are trivial to predict with PFP_{A3} as shown in Level 1."

"Seven parameters PFP_{A1}, PFP_{A2}, PFP_{A3}, PFP_{B2}, PFP_{Sigma}, PFP_{Laves}, and PSP are used to separate the phase regions of A1, A2, A3, A1 + A2, B2 + SS, and IM + HEAs."

"Level 1 shows separation between all single phase HEAs in the PFP_{A1}, PFP_{A2}, PFP_{A3}, and PSP parameter space."

HEA Phases Prediction Using Machine Learning

"ML is employed to analyse the complex parameter space of HEA phase formation."

"Through ML composition-phase correlations are determined and new HEA compositions are predicted."

"The overall success rates with 50–90% training sets are 89–90% for the as-cast HEAs or 87–89% including the annealed HEAs."

"The high prediction success rates prove that these parameters are sufficient for describing the disordered solid solution phase formation behaviour."

"PFP_{B2} is added as a fifth ML feature to predict the B2 + SS HEAs in Level 2."

"The overall and the B2 + SS phase prediction success rates are near 85% for both the as-cast HEA set and the set including the annealed HEAs."

"The IM+ phase prediction success rates are 73–78% for the as-cast HEAs or 71–77% including the annealed HEAs."

Model Validation

"To show that the model avoids overfitting with ML and can expand the current phase regions, 42 new HEAs were synthesised."

"The phases of these elemental combinations, which do not exist in the current collected database, are then predicted by the model."

"The numbers of new HEAs in different predicted phases are approximately proportional to the numbers of different HEA phases in the database."

"Many synthesised HEAs are outside the current known phase regions in order to show the ability to expand the phase region."

Discussion

"A method predicting the phase formation of HEAs based solely on the binary phase diagrams is demonstrated and validated."

"The initial success of using PFP_X and PSP, defined using binary phase diagrams, in predicting the corresponding single phase and mixed phase HEAs, prompted us to apply this method to include more phases."

"Compared with the other large database statistical approaches, Tancret and others combined Gaussian Process using nine thermodynamic and atomistic parameters with CALPHAD to predict the formation of over 60 single solid solution phase HEAs [31]."

"Our method is based solely on binary phase diagrams for which there exist plentiful easily accessible data.2.Phase region expansion ability: New HEAs are predicted with a high success rate outside the regions where 614 HEA phases are currently known.3.Parameter-phase relevance: Unlike the traditional thermodynamic parameters, our parameters directly determine the formation of the corresponding phases."

Method

"The T_m is calculated from the liquidus temperatures in binary phase diagrams."

"The method of calculating binary phase field percentage, X_{i-j}, uses line segments at T_{pf}."

"The binary phase separation percentage for atomic pair i and j, $Separation_{i-j}$, is calculated using the line segment method at T_{pf}."

"For an as-cast HEA, the phase transformation evolves at various temperatures above T_{pf} as it cools from the molten state."

"For the high temperature annealed HEAs, the phases formed during annealing at these high temperatures are locked in during rapid quenching."

"T_{pf} is the annealing temperature and the phase formation tendency is determined from the line segment percentages of the binary phase fields present."

"These trees are used to predict the phases of HEAs in the test set based on their features."

"The study is limited to 614 HEAs formed in the as-cast state or those annealed at temperatures higher than 0.7 T_m."

Some Unique Aspects of Mechanical Behavior of Metastable Transformative High Entropy Alloys [57]

This is a machine-generated summary of:

Mishra, R. S.; Nene, S. S.: Some Unique Aspects of Mechanical Behavior of Metastable Transformative High Entropy Alloys [57]

Published in: Metallurgical and Materials Transactions A (2021)

Link to original: https://doi.org/10.1007/s11661-021-06138-3

If you want to cite the papers, please refer to the original.

For technical reasons we could not place the page where the original quote is coming from.

Abstract-Summary

"This short review paper examines the uniqueness of metastable high entropy alloys (HEAs) that transform during plastic deformation."

"One key aspect is the adaptive nature of deformation mechanisms because of activation of different micromechanisms during plasticity."

"This activation of deformation micromechanisms is attributed to transformation-induced plasticity in γ-f.c.c."

Introduction

"The work was further extended by design of interstitial HEA ($Fe_{49.5}Mn_{30}Co_{10}Cr_{10}C_{0.5}$) by minor addition of carbon in DP-HEA to incorporate effect of precipitation strengthening along with classic transformation-induced plasticity (TRIP) [32]. Formation of deformation twins was observed in equiatomic FeMnCoCrNi and $Al_xCoCrMnNi$ HEAs [33–35]. Addition of TRIP mechanism to HEAs extended the extent of deformation micromechanisms in HEAs which not only overcome the conventional strength-ductility tradeoff but expand the exceptional mechanical properties to fatigue and dynamic deformation [33–35]. Our work on design of metastable HEAs by addition of Si in FeMnCoCr matrix further lowered the stacking fault energy [36–39]. We further modified these alloys with minor additions of Cu and Al [40, 41]. These alloy design efforts impacted not only the high strength-ductility combination but also increased yield strength, fatigue limit and corrosion resistance [36–42]. We define the alloy design approach as chemistry driven metastability."

"HEAs [36–39]. Coupling the alloy design with friction stir processing, an innovative high temperature severe plastic deformation technique, opened up microstructural engineering pathways."

"We define this as processing induced metastability [39]. The change in grain size also impacts the metastability of the alloy during subsequent deformation."

"Paper, we propose that the combined effect of alloy chemistry and microstructural engineering can be captured better by the term, high enthalpy states."

"The metastability embedded in these high enthalpy states of HEAs influence the kinetics and extent of deformation induced transformation and subsequent twinning and/or c/a ratio evolution in the ε-h.c.p."

Framework of High Enthalpy States in Metastable High Entropy Alloys

"Pure iron has four allotropic phases; α-Fe with b.c.c."

"crystal structure at low temperatures and pressures, γ-Fe with f.c.c."

"crystal structure at higher temperatures, δ-Fe with b.c.c."

"crystal structure at even higher temperatures, and at very high pressures (> 10 GPa), ε-Fe with h.c.p."

"Alloying approaches in iron have produced a vast array of steels with α-b.c.c."

"phase [43]. The physical metallurgy of steels mostly revolves around the phase transformation from γ-f.c.c."

"The richness of steel metallurgy has led to twinning-induced plasticity (TWIP) and TRIP steels that exhibit attractive strength-ductility combinations [43]. But in all these years, investigation of plasticity in ε- h.c.p."

"Metastability engineered HEAs opened exploration of a new deformation-induced phase transformation domain that was otherwise unexplored in the conventional steels/alloys."

"Introduction of high level of dislocations or grain boundaries in the γ-f.c.c."

Compositional Tailoring of Metastable High Entropy Alloys

"Our initial work on metastable HEAs was built on the work of Li and others [44] which pointed out that variation of Mn content at the expense of Fe content in the Fe-Mn-Co-Cr system can engineer γ-f.c.c."

"We extended this effort with a focus on making γ phase more metastable in Fe-Mn-Co-Cr matrix by (a) modification in alloy chemistry—as highlighted in this section, and (b) application of a high temperature severe plastic deformation technique (friction stir processing)—discussed in the next section."

"Work on the TRIP assisted HEAs [45] and shape memory alloys showed that, addition of Co and Si can increase the driving force for the γ→ε transformation and thus we conducted thermodynamic simulations using ThermoClac TCHEA2 database predicting the $\Delta G^{\gamma \rightarrow \varepsilon}$ at room temperature for $Fe_{45-x}Mn_{20}Cr_{15}Co_{20}Si_x$ system to investigate the effect of Si addition in the presence of Co. Thermodynamic predictions showed that, Si addition in synergy with Co increases the driving force γ→ε transformation at room temperature in Fe-Mn-Co-Cr system [45]. This increased driving force is an indication of increased γ phase matrix metastability and thereby making extensive transformation to ε phase easier."

"Minor dissolution of γ phase stabilizers, such as Al or Cu, can increase its dislocation storage ability before it transforms to ε phase under stress."

Microstructural Flexibility: Integration of Alloy Design with Severe Plastic Deformation

"Change in the process parameters during friction stir processing, does not intrinsically change the phase stability of the α and β phases in this titanium alloy [46]. In contrast, when the TRIP HEA is friction stir processed, merely changing the tool rotation rate changes the phase fraction during cooling to ambient temperature."

"Each processing condition leads to a different final microstructure encompassing the phase fractions and the grain size."

"This broader range of microstructural space for the same alloy chemistry is microstructural flexibility [38, 39]. The responsive phase stabilization observed in the Si containing HEAs can be linked to increased metastability of the γ matrix."

"pct) regained the dominance of γ phase in the microstructure under similar processing condition."

"Higher RPM processing condition led to dominance of ε phase whereas low RPM processing led to formation of γ rich microstructure."

Mechanical Behavior of Metastable High Entropy Alloys

"In spite of having different ε phase fraction (f_ε) and average grain size (d_{avg}) prior to deformation, all HEAs displayed ductility values greater than 25 pct and strength more than 1.1 GPa."

"What are the other factors governing the critical stress or strain required for the onset of γ phase transformation to ε phase?"

"The synergistic action of phase stability and processing conditions decides the activation of preferred deformation mechanisms and results in excellent strain hardening during deformation [36–42, 47, 48]. It is tempting to think about the effect of high enthalpy states on the onset of transformation."

"In a conventional alloy, at any given fatigue stress, some fraction of grains in the microstructure will start deformation."

"The higher strain hardening rate of TRIP HEAs will lead to faster percolation of deformation to other grains."

[Section 6]

"The TRIP HEAs discussed in this paper broadly show a progression of deformation mechanisms depending on the starting microstructure."

"A remarkable observation has been the evolution of c/a ratio in the ε phase during processing and subsequent deformation."

"Metastable HEA alloy design approach provides an opportunity to fine tune the SFE of an alloy in such a manner that all these critical deformation mechanisms can become active hierarchically during the overall stress-strain response."

"This requires a deeper understanding of each micromechanism, including its dependence on thermodynamic and microstructural factors [36–42]. Our earlier finding that the c/a ratio of ε phase in TRIP HEAs evolves with alloy chemistry, processing strain and subsequent deformation is scientifically quite intriguing and at this stage barely investigated."

"alloys, the c/a ratio is considered critical in activation of twinning and non-basal ⟨c+a⟩ slip activities."

"The TRIP HEAs do show very significant non-basal slip activities which increase with strain."

Concluding Remarks

"The metastable high entropy alloys that exhibit deformation-induced phase transformation show some very unique mechanical behavior."

"Different micromechanisms get triggered during the stress-strain response of these alloys."

"Combination of alloy design to manipulate stacking fault energy and innovative processing steps to activate different high enthalpy states allows for exciting scientific opportunities."

"Activation of different set of micromechanisms under tension and compression enables unique fatigue response with crack-tip transformation."

Alloy Design and Properties Optimization of High-Entropy Alloys [58]

This is a machine-generated summary of:

Zhang, Y.; Yang, X.; Liaw, P. K.: Alloy Design and Properties Optimization of High-Entropy Alloys [58]

Published in: JOM (2012)

Link to original: https://doi.org/10.1007/s11837-012-0366-5

If you want to cite the papers, please refer to the original.

For technical reasons we could not place the page where the original quote is coming from.

Abstract-Summary

"This article reviews the recent work on the high-entropy alloys (HEAs) in our group and others."

"HEAs usually contain five or more elements, and thus, the phase diagram of HEAs is often not available to be used to design the alloys."

"We have proposed that the parameters of δ and Ω can be used to predict the phase formation of HEAs, namely $\Omega \geq 1.1$ and $\delta \leq 6.6\%$, which are required to form solid-solution phases."

"Bridgman solidification was also used to control the microstructure of the CoCrFeNiAl alloy, and its plasticity was improved to be about 30%."

Introduction

"With the fast development of the new technologies and theories for developing advanced materials, the number of constituent principal elements for metallic alloys is increased from one to three or more."

"The intermetallic-based alloys, e.g., Fe_3Al- or TiAl-based alloys, usually contain two dominant elements, and the contents of other elements are very low."

"Compared with the conventional metallic alloys based on one or two major elements, HEAs generally have five or more major metallic elements, and each has an atomic percentage between 5% and 35% [49, 50]. Since HEAs possess a very high configurational entropy of mixing, solid-solution phases can be more stable than intermetallic compounds or other complex-ordered phases during solidification."

"Establishing a reasonable phase formation rule for HEAs is essential to guide alloy design."

"The property optimization for the existing HEAs can widen their potential applications."

Experimental Procedures

"Alloy ingots were prepared by arc melting the mixture of high-purity metals with the purity better than 99 wt% under a Ti-gettered, high-purity argon atmosphere on a water-cooled Cu hearth."

"The prepared alloy ingot was then remelted under high vacuum (5×10^{-3} Pa) and suction cast into the water-cooled copper mold to obtain cylindrical rods with diameters of 3 mm and 5 mm."

"The alloys were heated to the semisolid state by adjusting the heating power and holding for 15 min, and then, Bridgman solidification was carried out with withdrawal velocities varying from 200 μm/s to 1,800 μm/s through a temperature gradient of ~70 K/mm into a water-cooled Ga-In-Sn liquid metal."

"Cylindrical samples of Φ3 mm × 6 mm and Φ5 mm × 10 mm were prepared for compressive tests using a MTS 809 materials testing machine at room temperature with a strain rate of 2×10^{-4} s^{-1}."

Results and Discussion

"Compared with conventional metallic alloys with one or two principal components, HEAs possess significantly higher entropy of mixing (ΔS_{mix}) in the liquid state, which could effectively increase the extent of confusion in alloy systems and facilitate simple solid-solution phase formation."

"It is seen that all alloys mainly contain a bcc solid-solution phase."

"All the alloys exhibit high yield strength, high fracture strength, and obvious plastic deformation, which may stem from the solid-solution strengthening effect and second-phase strengthening effect."

"Both Nb0 and Nb0.1 alloys exhibit the reflections of a single bcc solid-solution phase."

"It implies that the stability of the solid-solution phase decreases with the Nb addition; thus, ordered Laves phase appears in Nb-contained alloys."

Conclusions

"By statistically analyzing the parameters δ and Ω for reported multicomponent HEAs, a criterion $\Omega \geq 1.1$ and $\delta \leq 6.6\%$, for forming high-entropy stabilized solid-solution phases, has been proposed, which can be used to assist in developing advanced metallic alloys."

"$TiZrNbMoV_x$ and $CoCrFeNiAlNb_x$ HEAs mainly contain a simple bcc solid-solution phase."

"The mechanical properties of single-phase bcc AlCoCrFeNi HEA were measured at different temperatures."

"It is concluded that the yielding strengths and fracture strengths increase and the fracture strains change very gently as the temperatures decrease from 298 K to 77 K."

References

1. K.H. Huang and J.W. Yeh (Master's thesis, National Tsing Hua University, 1996).
2. J.W. Yeh, Ann. Chim. Sci. Mater. 31, 633 (2006).
3. Y. Zhang, Y.J. Zhou, J.P. Lin, G.L. Chen, and P.K. Liaw, Adv. Eng. Mater. 10, 534 (2008).

4. C.J. Tong, Y.L. Chen, S.K. Chen, J.W. Yeh, T.T. Shun, C.H. Tsau, S.J. Lin, and S.Y. Chang, Metall. Mater. Trans. A 36A, 881 (2005).
5. K.H. Cheng, C.W. Tsai, S.J. Lin, and J.W. Yeh, J. Phys. D Appl. Phys. 44, 205405 (2011).
6. S.Y. Lin, S.Y. Chang, Y.C. Huang, F.S. Shieu, and J.W. Yeh, Surf. Coat. Technol. 206, 5096 (2012).
7. Ding, J., Yu, Q., Asta, M. & Ritchie, R. O. Tunable stacking fault energies by tailoring local chemical order in CrCoNi medium-entropy alloys. Proc. Natl. Acad. Sci. USA 115, 8819–8924 (2018).
8. Gali, A. & George, E. P. Tensile properties of high- and medium-entropy alloys. Intermetallics 39, 74–78 (2013).
9. Kennedy, K., Stefansky, T., Dvy, G., Zackay, V. F. & Parker, E. R. Rapid method for determining ternary-alloy phase diagrams. J. Appl. Phys. 36, 3808–3810 (1965).
10. Hanak, J. J. The "multiple-sample concept" in materials research: synthesis, compositional analysis, and testing of entire multicomponent systems. J. Mater. Sci. 5, 964–971 (1970).
11. Xiang, X.-D. et al. A combinatorial approach to materials discovery. Science 268, 1738–1740 (1995).
12. Specht, E. D. et al. Rapid structural and chemical characterization of ternary phase diagrams using synchrotron radiation. J. Mater. Res. 18, 2522–2527 (2003).
13. Collins, P. C., Banerjee, R., Banerjee, S. & Fraser, H. L. Laser deposition of compositionally graded titanium-vanadium and titanium-molybdenum alloys. Mater. Sci. Eng. A 352, 118–128 (2003).
14. Rar, A. et al. PVD synthesis and high-throughput property characterization of Ni-Fe-Cr alloy libraries. Meas. Sci. Technol. A 16, 46–53 (2005).
15. Xie, Y. et al. Supercluster-coupled crystal growth in metallic glass forming liquids. Nat. Commun. 10, 915 (2019).
16. Wu, G. et al. Crystal-glass high-entropy nanocomposites with near theoretical compressive strength and large deformability. Adv. Mater. 32, 2002619 (2020).
17. Y.N. Osetsky, L.K. Beland, A.V. Barashev, Y.W. Zhang, On the existence and origin of sluggish diffusion in chemically disordered concentrated alloys. Curr. Opin. Solid State Mater. Sci. 22(3), 65–74 (2018)
18. S.Y. Chen, Q. Li, J. Zhong, F.Z. Xing, L.J. Zhang, On diffusion behaviors in face centered cubic phase of Al–Co–Cr–Fe–Ni–Ti high-entropy superalloys. J. Alloy. Compd. 791, 255–264 (2019)
19. J.W. Yeh, S.K. Chen, S.J. Lin, J.Y. Gan, T.S. Chin, T.T. Shun, C.H. Tsau, and S.Y. Chang, Adv. Eng. Mater. 6, 299 (2004).
20. C. Varvenne, A. Luque, W.A. Curtin, Average-atom interatomic potential for random alloys. Phys. Rev. B 93, 104201 (2016)
21. W.G. Nöhring, W.A. Curtin, Thermodynamic properties of average-atom interatomic potentials for alloys. Modell. Simul. Mater. Sci. Eng. 24(4), 045017 (2016)
22. He, J. Y. et al. Effects of Al addition on structural evolution and tensile properties of the FeCoNiCrMn high-entropy alloy system. Acta Materialia 62, 105–113 (2014).
23. J.W. Yeh, Y.L. Chen, S.J. Lin, and S.K. Chen, Mater. Sci. Forum 560, 1 (2007).
24. O. N. Senkov, J. D. Miller, D. B. Miracle, et al., "Accelerated exploration of multi-principal element alloys with solid solution phases," Nat. Commun., 6, 1–6 (2015).
25. Gao, M. C., Yeh, J.-W., Liaw, P. K. & Zhang, Y. High-Entropy Alloys Fundamentals and Applications. (Springer International Publishing, 2018).
26. Gao, M. C. et al. Computational modeling of high-entropy alloys: Structures, thermodynamics and elasticity. J. Mater. Res. 32, 3627–3641 (2017).
27. Gao, M. C. et al. Thermodynamics of concentrated solid solution alloys. Curr. Opin. Solid State Mater. Sci. 21, 238–251 (2017).
28. Kube, S. A. et al. Phase selection motifs in High Entropy Alloys revealed through combinatorial methods: Large atomic size difference favors BCC over FCC. Acta Mater. 166, 677–686 (2019).
29. Islam, N., Huang, W. & Zhuang, H. L. Machine learning for phase selection in multi-principal element alloys. Comput. Mater. Sci. 150, 230–235 (2018).

30. Huang, W., Martin, P. & Zhuang, H. L. Machine-learning phase prediction of high-entropy alloys. Acta Mater. 169, 225–236 (2019).
31. Tancret, F., Toda-Caraballo, I., Menou, E. & Rivera Díaz-Del-Castillo, P. E. J. Designing high entropy alloys employing thermodynamics and Gaussian process statistical analysis. Mater. Des. 115, 486–497 (2017).
32. Z. Li, C.C. Tasan, H. Springer, B. Gault, D. Raabe, Sci. Rep., 2017, vol. 7, pp. 40704.
33. R.S. Mishra, N. Kumar, M. Komarasamy, Mater. Sci. Tech. 2015, vol. 31, pp. 1259–1263.
34. Gludovatz, B. et al. A fracture-resistant high-entropy alloy for cryogenic applications. Science 345, 1153–1158 (2014).
35. M.A. Hemphill, T. Yuan, G.Y. Wang, J.W. Yeh, C.W. Tsai, A. Chuang, P.K. Liaw, Acta Mater., 2012, vol. 60, pp. 5723–5734.
36. M. Frank, Y. Chen, S.S. Nene, S. Sinha, K. Liu, K. An, R.S. Mishra, Mater. Today Comm., 2020,vol. 23, pp. 100858.
37. S. S. Nene, K. Liu, M. Frank, R.S. Mishra, R.E. Brennan, K. Cho, Z. Li, D. Raabe, Sci. Rep., 2017, vol. 7, pp. 16167.
38. S.S. Nene, M. Frank, K. Liu, R.S. Mishra, B. McWilliams, K.C. Cho, Sci. Rep., 2018, vol. 8, pp. 9920.
39. S.S. Nene, M. Frank, K. Liu, S. Sinha, R.S. Mishra, B. McWilliams, K.C. Cho (2018) Scripta Mater. 154: 163–167.
40. S. Nene, M. Frank, P. Agrawal, S. Sinha,, K. Liu, S. Shukla, R.S. Mishra, B. McWilliams, K.C. Cho, Mater. Des., 2020, vol. 194, pp. 108968.
41. S.S. Nene, M. Frank, K. Liu, S. Sinha, R.S. Mishra, B. McWilliams, K.C. Cho, Scripta Mater., 2019, vol. 166, pp. 168 - 172.
42. K. Liu, S.S. Nene, M. Frank, S. Sinha, R.S. Mishra, App. Mater. Today, 2019, vol. 15, pp. 525–530.
43. H.K.D.H. Bhadeshia: Physical Metallurgy of Steels in Physical Metallurgy, D.E. Laughlin and K. Hono, eds., Elsevier Publishing, New York, 2014, 5th ed., pp. 2157–214.
44. Li, Z., Pradeep, K. G., Deng, Y., Raabe, D. & Tasan, C. C. Metastable high-entropy dual-phase alloys overcome the strength–ductility trade-off. Nature 534, 227–230 (2016).
45. S.F. Liu, Y. Wu, H.T. Wang, W.T. Lin, Y.Y. Shang, J.B. Liu, K. An, X.J. Liu, H. Wang, Z.P. Lu, J. Alloys. Compd., 2019, vol. 792, pp. 444–455.
46. J. Su, J. Wang, R.S. Mishra, R. Xu, J.A. Baumann (2013) Mater. Sci. Eng. A 573: 67–74.
47. R.S. Mishra, P.S. De, N. Kumar, Friction stir processing in: R.S. Mishra, P.S. De, N. Kumar (eds.), Friction stir welding and processing, Springer Publishing, New York, 2014, pp. 259–296.
48. S. Sinha, S.S. Nene, M. Frank,, K. Liu, R.A. Lebensohn, R.S. Mishra, Acta Mater., 2020, vol. 184, pp. 164–178.
49. J.W. Qiao, E.W. Huang, F. Jiang, T. Ungar, G. Csiszar, L. Li, Y. Ren, P.K. Liaw, and Y. Zhang, Appl. Phys. Lett. 97, 171910 (2010).
50. Chuang, M. H., Tsai, M. H., Wang, W. R., Lin, S. J. & Yeh, J. W. Microstructure and wear behavior of AlxCo1.5CrFeNi1.5Tiy high-entropy alloys. Acta Materialia 59, 6308–6317 (2011).
51. Yeh, Jien-Wei Alloy Design Strategies and Future Trends in High-Entropy Alloys. JOM (2013). https://doi.org/10.1007/s11837-013-0761-6
52. George, Easo P.; Raabe, Dierk; Ritchie, Robert O. High-entropy alloys. Nature Reviews Materials (2019). https://doi.org/10.1038/s41578-019-0121-4
53. He, Xingjia; Zhang, Yu; Gu, Xinlei; Wang, Jiangwei; Qi, Jinlei; Hao, Jun; Wang, Longpeng; Huang, Hao; Wen, Mao; Zhang, Kan; Zheng, Weitao Pt-induced atomic-level tailoring towards paracrystalline high-entropy alloy. Nature Communications (2023). https://doi.org/10.1038/s41467-023-36423-1
54. Seoane, Axel; Farkas, Diana; Bai, Xian-Ming Influence of compositional complexity on species diffusion behavior in high-entropy solid-solution alloys. Journal of Materials Research (2022). https://doi.org/10.1557/s43578-022-00545-x
55. Krapivka, M. O.; Myslyvchenko, O. M.; Karpets, M. V. Base Alloy Concept in the Development of High-Entropy Materials. Powder Metallurgy and Metal Ceramics (2018). https://doi.org/10.1007/s11106-018-9932-x

56. Qi, Jie; Cheung, Andrew M.; Poon, S. Joseph High Entropy Alloys Mined From Binary Phase Diagrams. Scientific Reports (2019). https://doi.org/10.1038/s41598-019-50015-4
57. Mishra, R. S.; Nene, S. S. Some Unique Aspects of Mechanical Behavior of Metastable Transformative High Entropy Alloys. Metallurgical and Materials Transactions A (2021). https://doi.org/10.1007/s11661-021-06138-3
58. Zhang, Y.; Yang, X.; Liaw, P. K. Alloy Design and Properties Optimization of High-Entropy Alloys. JOM (2012). https://doi.org/10.1007/s11837-012-0366-5

Chapter 3
Design and Synthesis of High Entropy Alloys

3.1 Introduction by Editor

The selection of elements and their optimized concentration in the composition of HEAs is critical, as the microstructure and mechanical properties depend on it. Computational methods, such as CALPHAD, which use databases based on thermodynamics, kinetics, and phase stability, make the HEA design more focused. In this chapter, various computational methods that provide precise information on the thermal conductivity, electrical resistivity, and phase compositions along with the solidification behavior of HEAs are discussed. Moreover, the implementation of artificial intelligence and machine learning in HEA design has proven to be an effective way to predict material characteristics such as mechanical properties, stacking fault energies, and microstructures. Numerous fabrication routes for HEAs, such as casting (arc melting, Bridgman solidification) and additive manufacturing (selective laser melting, mechanical alloying, and wire arc additive manufacturing) have been explained.

Designing or developing an alloy with specified properties targeting desired application requires an appropriate system (elements) selection and subsequent composition formulation. The composition plays a crucial role in defining the properties of an alloy, leading to the possibility of a phase or phases that can induce properties to satisfy the targeted requirement. In cases where the composition alone is not sufficient to attain the targeted phase or morphology, the fabrication technique implemented becomes critical.

A traditional designing approach via first principal calculations has always been a prominent way in understanding the behavior of an alloy on an atomic or molecular level from a physicist's scale. This becomes a limitation when an entity is actually in existence at a production scale. To improve this understanding role of thermodynamic factors in determining the nature of metals and alloys, computational methods, such as CALPHAD (CALculation of PHAse Diagrams), and modeling techniques like

S. S. Nene (ed.), *High Entropy Alloys*, https://doi.org/10.1007/978-981-99-7173-2_3

Artificial intelligence/Machine learning (AI/ML) have been implemented in the HEA design.

CALPHAD approach can predict the probable thermodynamics and kinetics that an alloy may undergo to form the desired condition or product. This field of metallurgy uses cumulative research-generated databases that can project a near-accurate possibility of the existence of a phase or properties in a defined set of conditions for an alloy. For a large amount of data generated almost every day for HEAs, it becomes difficult to trace the properties of an alloy that can ultimately lead to a targeted product. In such instances, tools like AI/ML becomes effective methodologies for data sorting and alloy designing perspectives. Designing an algorithm based on pre-existing data helps researchers implement it to achieve targeted or near-target properties in the alloys for the application.

The data used in these predictions are very limited to the alloy available with limited properties tested or analyzed. This becomes a major drawback where the attained targeted properties are significantly different from the predicted ones; for example, the predicted pseudo-binary phase diagrams for a targeted HEA can be significantly different from the one predicted from the attained phases in equilibrium. This is primarily due to unavailable data for the possible existing composition in conventional alloys or HEAs or can be due to the system's novelty itself. Ultimately the computational aspect for developing an alloy delivering for targeted properties becomes far more difficult than expected. In such instances experimental analysis that includes an actual alloy fabrication and characterization for the same conditions becomes significant.

As we see, the data selection based on the properties is still limited and is again reliable on actual fabrication and testing of an entity, the alloy composition and the fabrication technique implemented can become important in defining the properties post-alloy designing. The properties in such instances may vary based on the approach implemented for fabrication. For instance, a property requirement in one direction (anisotropy) may demand a different casting or processing technique over a conventional casting which will give you isotropic properties for an alloy. In the same way, advanced fabrication methods like additive manufacturing also known as 3D printing can be of significant importance in fabricating near net shapes of complex nature which are otherwise difficult to manufacture using conventional methods.

A conventional fabrication technique like vacuum arc melting or induction melting with conventional cooling can be effective in understanding the general properties of an alloy in as-cast condition or after homogenization. The study for solidification in combination with computational approaches showed a closed replication with the actual solidification behavior as seen in the alloy. This relationship which almost resembles the conventional alloy solidification behavior has made the understanding of the directional path relatively easy.

Direct energy deposition (DED) which includes a series of direct energy deposition techniques based on the form of material induced on the substrate. The possibility of material in specific form can be its mechanical limitation like formability, processability or physical limitation that includes melting point of an alloy. This gives an edge for development of various fabrication techniques based on the availability of raw material.

To understand the designing and fabrication correlation, the following chapter summarizes various computational methods that provide precise information on the thermal conductivity, electrical resistivity, and phase compositions along with the solidification behavior of HEAs. Moreover, the implementation of artificial intelligence and machine learning in HEA design has proven to be an effective way to predict material characteristics such as mechanical properties, stacking fault energies, and microstructures. Numerous fabrication routes for HEAs, such as casting (arc melting, Bridgman solidification) and additive manufacturing (selective laser melting, mechanical alloying, and wire arc additive manufacturing) have been summarized subsequently [1–231].

3.2 Machine Generated Summaries

Disclaimer: The summaries in this chapter were generated from Springer Nature publications using extractive AI auto-summarization: An extraction-based summarizer aims to identify the most important sentences of a text using an algorithm and uses those original sentences to create the auto-summary (unlike generative AI). As the constituted sentences are machine selected, they may not fully reflect the body of the work, so we strongly advise that the original content is read and cited. The auto generated summaries were curated by the editor to meet Springer Nature publication standards. To cite this content, please refer to the original papers.

Machine generated keywords: solidification, calphad, predict, phase diagram, calculate, substrate, manufacture, phase selection, calphad method, binary, manufacturing, diagram, laser, dendrite, selection

3.3 Computational Based Approaches of HEA Design

Machine generated keywords: calphad, phase diagram, calculate, predict, calphad method, binary, diagram, ternary, lave, thermodynamic, use calphad, gibb, quaternary, database, lave phase

Development of Phase-Based Databases via CALPHAD Method for the Design of High-Entropy Alloys [207]

This is a machine-generated summary of:

Zhang, C.; Zhu, J.; Cao, W. S.; Chen, S. L.; Liang, S. M.; Zhang, F.: Development of Phase-Based Databases via CALPHAD Method for the Design of High-Entropy Alloys [207]

Published in: Journal of Phase Equilibria and Diffusion (2022)

Link to original: https://doi.org/10.1007/s11669-022-00970-9

If you want to cite the papers, please refer to the original.
For technical reasons we could not place the page where the original quote is coming from.

Abstract-Summary

"We demonstrated the successful application of this database in predicting phase relationships in the Co-Cr-Fe–Mn-Ni system."

"The predicted effect of each element on the stability of the σ phase was in line with experimental observations."

"By using a series of equiatomic alloys, we showed that this database can be used to understand the elemental diffusion behavior and thermal expansion in concentrated alloys in the Co-Cr-Fe-Mn-Ni system."

Introduction

"Computational tools have been developed in multiple length-scales such as first-principles calculation at nanometer scales, phase field modeling from nanometer scales to millimeter scales, and the computational thermodynamics/kinetics for bulk materials based on the CALPHAD method."

"A thermodynamic database thus enables us to calculate phase diagrams and thermodynamic properties of a multi-component system that are experimentally unavailable."

"This paper will be arranged as follows: we will first describe some details on the development of PanHEA_TH and use it to predict phase equilibria, especially the stability of the σ phase, in the popular Co-Cr-Fe-Mn-Ni subsystem; we will then demonstrate how to use PanHEA to calculate mobility and molar volume for a group of concentrated alloys in the Co-Cr-Fe-Mn-Ni family; lastly, we will discuss the extended application of the CALPHAD method by using it to develop a thermal conductivity database and an electrical resistivity database for the Co-Cr-Fe-Mn-Ni system."

Thermodynamic Database for Concentrated Alloys

"Bloomfield and others [1] also studied the effect of Fe on the phase stability of σ using a series of $CoCrFe_xMnNi$ alloys, where the atomic ratio x = 0, 0.5, and 1.5."

"The calculated phase diagram well describes the phase stability of the three alloys (11.1 at.% Fe, 20 at.% Fe, and 27.3 at.% Fe) at temperatures above 700 °C."

"Large discrepancy between the calculated and experimental observed phase stability is seen for the alloy without Fe, i.e., the equiatomic quaternary CoCrMnNi at 900 °C."

"In the Co-Cr-Fe-Mn-Ni family of alloys, the formation of the σ phase is significantly influenced by compositional variations."

Mobility Database for Concentrated Alloys

"We also developed an atomic mobility database PanHEA_MB for the transition metal high-entropy alloys."

"As the concept of HEAs was developed, Yeh and colleagues [2, 3] pointed out four core effects of this group of alloys including high entropy, sluggish diffusion, severe lattice distortion, and cocktail effects."

"According to this hypothesis, the atomic diffusion is believed to be retarded in HEAs and the retarding effect increases with the increasing number of elements in the HEAs."

"[4–6] Recently, Jin and others [7] investigated interdiffusion in seven diffusion couples varying from binary to quinary, and they reached the conclusion that elemental diffusion in HEAs is slower than those in pure and binary alloys, yet it is not necessarily true that diffusion becomes more sluggish with the increasing number of elements."

"Diffusivity is strongly influenced by the constituents of the alloy instead of the number of components."

Molar Volume Database for Concentrated Alloys

"[8, 9] By using the CALPHAD method, we have developed a molar volume database PanHEA_MV for the transition metals HEAs."

"The combined database (PanHEA [10]) is used for the design of HEAs with the desired phase stability and targeted density."

"The calculated thermal expansion coefficients agree with the experimentally measured values well for this series of alloys."

"No obvious correlation is found between the thermal expansion coefficients and the number of components, yet alloys with Mn tend to have higher thermal expansion coefficients."

"It seems that the density is lower near the Fcc + Bcc/Fcc phase boundary in both cases."

"Such contour mappings help one to select alloy compositions that meet the criteria of both phase stability and targeted density."

Thermal Conductivity

"Materials with low thermal conductivity can be used as thermal barrier coatings which are essential in preventing the key structural components of jet engines from exposure to high temperatures."

"[11] In this work, CALPHAD method is used to model the thermal conductivity of the Co-Cr-Fe-Mn-Ni system."

"If we start from CoNi binary alloy, it is found that adding the third element Fe, then the fourth element Cr, then the fifth element Mn, the thermal conductivity decreases with the increasing number of components."

"Using the thermal conductivity database for the Co-Cr-Fe-Mn-Ni system, the effect of each element on the thermal conductivity of the 5-component alloys is studied by varying one element at a time in the range of 0–1 and keep the other four elements the same ratio."

"Cr can significantly reduce the thermal conductivity of the alloys in the 5-component system."

Electrical Resistivity

"In metals and alloys free electrons are the main carriers for both thermal and electrical conduction, which make the thermal and electrical conductivities proportionally related by the Wiedemann–Franz relation: where κ is the thermal conductivity, ρ is the electrical resistivity of a metal or metal alloy, T is the temperature in Kelvin, and L is known as Lorenz number."

"Equiatomic CoNi binary alloy exhibits very low electrical resistivity."

"While the electrical resistivity increases slightly by adding Fe to CoNi to form an equiatomic CoFeNi ternary alloy, it jumps for the CoCrNi alloy."

"It is seen that the electrical resistivity tends to increase quickly when Cr is introduced to the alloys."

"An important electrical property is the temperature coefficient of resistivity (TCR), which is defined as: TCR measures the change in electrical resistance of a material per degree of temperature change."

Conclusion Remarks

"We used this database to predict the formability and phase stability of the σ phase in the Co-Cr-Fe-Mn-Ni system."

"Tracer diffusivities of Ni were calculated using the PanHEA database for a series of concentrated alloys from binary to quinary in the Ni-Co-Cr-Fe-Mn system with respect to both the absolute temperature and the homologous temperature."

"The applications of these databases were illustrated through understanding the effect of each individual element on the corresponding property for a series of equiatomic alloys in the Co-Cr-Fe-Mn-Ni system."

"Developing a database for HEAs is much more burdensome than that for traditional alloys due to the large number of constituent subsystems."

"Unlike those for the traditional alloys, the number of components should be restricted in order to develop a high-fidelity database for HEAs."

Prediction of the Phase Composition of High-Entropy Alloys Based on Cr–Nb–Ti–V–Zr Using the CALPHAD Method [208]

This is a machine-generated summary of:

Gorbachev, I. I.; Popov, V. V; Katz-Demyanetz, A.; Popov, V.; Eshed, E.: Prediction of the Phase Composition of High-Entropy Alloys Based on Cr–Nb–Ti–V–Zr Using the Calphad Method [208]

Published in: Physics of Metals and Metallography (2019)
Link to original: https://doi.org/10.1134/s0031918x19040069

If you want to cite the papers, please refer to the original.

For technical reasons we could not place the page where the original quote is coming from.

Introduction

"Upon the development of a HEA the researchers aim to obtain a single-phase structure."

"The addition of Cr in an amount comparable to the amounts of the other elements into the NbTiVZr alloy leads to the appearance of a Laves phase [12, 13], which can lead to a significant decrease in the plasticity of this alloy at room temperature."

"It transpired that the phase diagrams of all binary subsystems of the Nb–Ti–V–Zr system, except for V–Zr, demonstrate a continuous series of bcc solid solutions."

"As is noted in [14], a single-phase bcc structure can be obtained in the alloys of these elements."

"In the binary subsystems with Cr, except for Cr–V, intermetallic compounds of the Laves-phase type with the structures of the C14, C15, and C36 type (in Strukturbericht designations) can be formed."

"It is expedient to study the influence of Cr on the equilibrium phase composition of Cr–Nb–Ti–V–Zr HEAs at different temperatures."

CALPHAD Method

"The set I0 contains all such combinations of chemical elements in which only one is found in each sublattice."

"The sets of the higher order IZ correspond to different combinations of a larger number of elements from different sublattices."

"The set I1 contains all combinations of two elements; I2, of three elements; etc. Usually, upon the description of systems even with a large number of elements the interaction of no more than three elements is taken into account."

"For the sublattice model the parameters L for the set I1 are written: and for I2, when the parameter describes the interaction of three elements of one sublattice:"

Construction of the Thermodynamic Description

"The Laves phases are also not formed in the Ti–V system; therefore, the parameters for this phase can only be estimated using systems of higher order that include Ti–V. Among such systems, the work [15] can be noted."

"The parameters for the solid solutions in that work were taken from relatively old work published in 1990, and the parameters for the Laves phase with the C14 structure were estimated only based on the interaction with Fe."

"In our work, we accepted the description proposed in [16], since, first, in [16] the contemporary experimental data are considered, including those obtained by the authors of that work, and, second, the re-estimation of the Ti–V system in [16] was performed within the framework of the description of the Ti–V–Zr system, which is a subsystem of the Cr–Nb–Ti–V–Zr system."

"The same description is used in the work on the Fe–Nb–V system [17], where the parameters for the Laves phase with the structure of the C14 type were also calculated."

Results of the Calculations

"Up to the temperature of this transition, the fraction of the bcc solid solution on the basis of Nb and V grows insignificantly (except for the $Cr_{1.25}NbTiVZr$ alloy), then abruptly decreases and begins to decrease up to the complete disappearance of this phase at 686–710 °C (for the investigated alloys)."

"From 400 °C to the temperature of the disappearance of the bcc (Nb, V) solid solution there is also a decrease in the fraction of the bcc solid solution on the basis of V and Cr, and the fraction of the Laves phase grows."

"This temperature, on the contrary, the fraction of the Laves phase decreases up to the complete dissolution at 1050–1380 °C (depending on the relationship between the elements in the HEA), and the fraction of the solid solution on the basis of V and Cr grows insignificantly."

Conclusions

"On the basis of the CALPHAD method, a thermodynamic description of the Cr–Nb–Ti–V–Zr system was proposed and the influence of the content of Cr in the high-entropy $Cr_xNbTiVZr$ alloys on the temperature dependence of their phase composition was investigated."

"The calculations showed a tendency toward the formation of two and, in a certain range of compositions, three solid solutions in the thermodynamically equilibrium state."

"The realization of a single-phase structure in the equiatomic CrNbTiVZr alloy is probable only at temperatures higher than 1300 °C, but with a decrease in the amount of Cr in the alloy, the single-phase thermodynamically state of equilibrium state becomes possible also at lower temperatures."

Revealing the Selection of σ and μ Phases in $CoCrFeNiMo_x$ High Entropy Alloys by CALPHAD [209]

This is a machine-generated summary of:

Wu, Qingfeng; Wang, Zhijun; He, Feng; Li, Junjie; Wang, Jincheng: Revealing the Selection of σ and μ Phases in $CoCrFeNiMo_x$ High Entropy Alloys by CALPHAD [209]

Published in: Journal of Phase Equilibria and Diffusion (2018)
Link to original: https://doi.org/10.1007/s11669-018-0659-7

If you want to cite the papers, please refer to the original.

For technical reasons we could not place the page where the original quote is coming from.

Abstract-Summary

"The strengthening effect of σ and μ phases in $CoCrFeNiMo_x$ high entropy alloys has been confirmed, the selection of these two phases remains unclear, which restricts the further strengthening design."

"The results indicate that the selection of σ and μ phases depends on the content of Mo greatly."

"Due to the limited prediction of phase selection and transformation by the isopleth phase diagram of (CoCrFeNi)-Mo, we design a (CoFeNi)-Mo-Cr isothermal phase diagram to reveal the transformation between σ and μ phases and their selections."

Introduction

"The intermetallics of σ and μ phases are usually detrimental in traditional alloy designs [18, 19]. Surprisingly, in HEAs, Liu and others [20] reported that $CoCrFeNiMo_x$ alloys could be strengthened remarkably by these incoherent intermetallic phases without a serious embrittlement."

"Their work indicated that intermetallics σ and μ in CoCrFeNi-based HEAs could be beneficial to mechanical properties."

"Researches on the $CoCrFeNiMo_{0.85}$ alloy revealed that σ phase can exist from 600 to 900 °C and μ phase may exist at 1000 °C [21], which indicates that μ phase is more stable at high temperatures."

"Liu and others's work indicated that σ and μ phases can coexist at 1000 °C and the CALPHAD results showed that σ phase should be more stable at high temperatures in $CoCrFeNiMo_x$ alloys [20]. The limited literatures show a confusing stability of σ and μ phases in $CoCrFeNiMo_x$ HEAs."

"An isopleth phase diagram of (CoCrFeNi)-Mo is designed to reveal the thermodynamic stability of σ and μ phases based on the experimental results."

Materials and Methods

"Five multi-principal component $CoCrFeNiMo_x$ ($x = 0.1$, 0.2, 0.3, 0.4 and 0.5, in molar ratio, referred to as $Mo_{0.1}$, $Mo_{0.2}$, $Mo_{0.3}$, $Mo_{0.4}$ and $Mo_{0.5}$ respectively for convenience) alloys were prepared by arc melting under a high-purity argon atmosphere."

"Elemental Co, Cr, Fe, Ni and Mo of 99.95% purity were employed as raw materials."

"Samples cut from these ingots were annealed at 600, 750, 900 and 1000 °C for 7 days and 1080 °C for 3 days respectively, followed by water quenching."

"The (CoFeNi)-Mo-Cr isothermal phase diagram was calculated by using the Thermo-Calc software with the TTNI8 database."

Results and Discussion

"It indicates that the content of both σ and μ phases is the highest in alloys annealed at 750 °C."

"A general rule can be found that the summation of Cr and Mo content is almost fixed in both σ and μ phases for different alloys and annealed temperatures, meeting the stoichiometric ratio (CoFeNi)(MoCr) and $(CoFeNi)_7(MoCr)_6$, respectively."

"Both Cr and Mo are rich in σ and μ phases."

"The difference is that more Cr is in σ phase while more Mo in μ phase."

"Only alloys containing proper content of Cr and Mo elements are able to have the phase compositions of "fcc + σ + μ"."

"For alloys rich in Cr and lack of Mo, the transformation from μ phase to σ phase may occur with the increase of temperature."

Conclusions

"The results confirmed three phases of fcc matrix, σ phase and μ phase in these alloys annealed from 600 to 1080 °C."

"The transformation between σ and μ phases can be well explained based on a (CoFeNi)-Mo-Cr isothermal phase diagram."

"Cr is a σ phase former, while Mo is a μ phase former."

"The appearance of σ or μ phase at high temperatures depends on the content of Cr and Mo in the alloy."

Phase Formation in Equiatomic High Entropy Alloys: CALPHAD Approach and Experimental Studies [210]

This is a machine-generated summary of:

Durga, A.; Hari Kumar, K. C.; Murty, B. S.: Phase Formation in Equiatomic High Entropy Alloys: CALPHAD Approach and Experimental Studies [210]

Published in: Transactions of the Indian Institute of Metals (2012)

Link to original: https://doi.org/10.1007/s12666-012-0138-5

If you want to cite the papers, please refer to the original.

For technical reasons we could not place the page where the original quote is coming from.

Abstract-Summary

"CALPHAD approach has been used to predict the stable phases, their relative amounts and compositions in multicomponent equiatomic high entropy alloys."

Introduction

"Further, alloys with these simple structures also possess several promising properties such as high strength, wear resistance, oxidation resistance, etc. Hence, the problem of predicting solid solution formation in multicomponent equiatomic alloys is one of interest."

"Takeuchi and Inoue [22] calculated the amorphous phase forming composition range of ternary alloy systems by considering the enthalpy difference of the amorphous phase and the solid solution phase."

"Zhang and others [23] have used a plot of mixing enthalpy versus so-called δ-parameter to distinguish between multicomponent alloys that show solid solution formation vis-à-vis those that give other phases such as amorphous phase, intermetallic phase, etc. Such approaches, despite showing a reasonable match with experimental results, are limited in their utility."

"This work examines the application of the CALPHAD approach in predicting phase formation in high entropy alloys."

Calculation of Gibbs Free Energy

"The Gibbs energy of a phase φ is expressed as [24] The first term is the contribution of the pure components of the phase, the second is the ideal mixing contribution and the third is the excess Gibbs energy of mixing."

"The stable phase(s) present at equilibrium are predicted by minimisation of the total Gibbs energy of the system."

"The Gibbs energy of amorphous phase is calculated as [25] where ΔH, the enthalpy of mixing is calculated by Miedema model [26] and ΔS is given as the sum of mismatch entropy and configurational entropy."

Description of the Work

"Since high entropy alloys, when synthesized through routes like mechanical alloying, are unlikely to form phases apart from amorphous and those with simple crystal structures such as BCC (A2), FCC (A1) and HCP (A3), the computations were performed again, this time neglecting the possibility of phases other than amorphous, BCC, FCC, HCP (and also CBCC (A12) for Mn containing systems) solid solutions."

"Two equiatomic quaternary CrCuMnNi and CoCrCuNi alloys (in this paper the elements in a high entropy alloy designation are arranged in alphabetical order) were synthesized in powder form by mechanical alloying (MA) [27] in a Fritsch P-5 high-energy ball mill for 20 h. The milling was carried out with tungsten carbide milling media in toluene at a milling speed of 300 rpm and a ball to powder weight ratio of 10:1."

Results and Discussion

"We observe that the Gibbs energy of the solid solution predicted by the CALPHAD approach is lesser than that of the amorphous phase for all the systems."

"In this case also, there is a good match between the experimental results and the CALPHAD predictions, considering the kinetic constraints of phase formation during high energy ball milling."

"Even though the match between experimental results and CALPHAD predictions is not exact (in terms of the number of solid solutions present), the predominant phase as predicted by the CALPHAD approach matches with the predominant phase observed experimentally in all three cases."

"(i.e. without neglecting the possibility of formation of any phase), the predicted stable phases for the systems were a mixture of solid solutions for 13 of the 22 systems studied."

Conclusions

"The CALPHAD predictions show a good match with experimental results on phase formation for the two equiatomic high entropy alloys (CrCoCuNi and CrCuMnNi alloys) prepared by mechanical alloying, considering the kinetic constraints of the non-equilibrium processing route."

"The predictions also match well with the experimental results available in the literature."

"The CALPHAD approach does not only predict the phases possible in equiatomic multicomponent alloys, but also predicts the fraction of the phases possible and the composition of the phases."

"Among the 22 multicomponent systems (quaternary and above) studied in the Co–Cr–Cu–Fe–Mn–Ni system, 14 of them show the FCC phase to be the predominant phase, while six of them show the BCC phase to be the predominant one."

Senary Refractory High-Entropy Alloy HfNbTaTiVZr [211]

This is a machine-generated summary of:

Gao, M. C.; Zhang, B.; Yang, S.; Guo, S. M.: Senary Refractory High-Entropy Alloy HfNbTaTiVZr [211]

Published in: Metallurgical and Materials Transactions A (2015)

Link to original: https://doi.org/10.1007/s11661-015-3105-z

If you want to cite the papers, please refer to the original.

For technical reasons we could not place the page where the original quote is coming from.

Abstract-Summary

"Discovery of new single-phase high-entropy alloys (HEAs) is important to understand HEA formation mechanisms."

"The present study reports computational design and experimental validation of a senary HEA, HfNbTaTiVZr, in a body-centered cubic structure."

"Supported by both simulation and experimental results, the HEA formation rules are discussed."

Introduction

"The uniqueness of high-entropy alloys (HEAs) [2, 28] is that they contain multiple principal elements while traditional alloys are typically based on one or two principal elements (e.g., Al-, Mg-, Ti-, and Fe-based alloys)."

"The other is based on configurational entropy, and HEAs are defined as alloys having configurational entropies at a random state $\geq$ 1.5R (R is the ideal gas constant), no matter whether they are single phase or multi-phase at room temperature."

"In order to accelerate single-phase HEA design, an efficient searching strategy was proposed by Gao and Alman [29] by combining phase diagram inspection, CALPHAD (acronym of CALculation of PHAse Diagrams) modeling, and ab initio molecular dynamics (AIMD) simulations."

"They mentioned [29] that formation of isomorphous solid solution in edge binary and ternary systems will likely lead to HEA formation in higher order systems, and they suggested new single-phase HEAs based on face-centered cubic (FCC) and hexagonal close-packed (HCP) structures."

CALPHAD Modeling

"Although the BCC solid solution is found to be stable between 1258 and 1872 K (985 °C and 1599 °C), at temperatures below 1258 K (985 °C), the BCC solid solution starts to decompose into (Hf,Ti,Zr)-rich BCC#1 (0.54 mol fraction) and (Nb,Ta,V)-rich BCC#2 (0.46 mol fraction) phases."

"The simulation predicts formation of dominant BCC solid solution phase and additional V-rich BCC phase in very small fraction towards the end of the solidification."

"As the solidification starts, the liquid phase is depleted in Ta, followed by Nb, and is slightly enriched with other elements especially Zr."

"Addition of Zr causes decomposition of the BCC solid solution into two BCC phases."

AIMD Simulations

"For Ta, V, and Zr, the weakest interactions are Ta-Zr, V-Zr, and V-Zr pairs, respectively."

"For Nb, its interaction with Ta or Hf is stronger than with Nb, V, or Zr."

"For Hf, the Hf-Ti interaction is the weakest while the rest are fairly close."

"The simulated diffusion constants at 2473 K (2200 °C) in descending order are 5.35 ± 0.08, 4.96 ± 0.02, 4.07 ± 0.07, 3.96 ± 0.02, 3.86 ± 0.03, and 3.73 ± 0.06 [10^{-5} cm^2/s] for V, Ti, Zr, Nb, Hf, and Ta, respectively."

Experimental Verification

"The HfNbTaTiVZr HEA specimens were prepared using a vacuum arc melting furnace (model Edmund Bühler /MAM-1) under argon atmosphere in a water-cooled copper hearth."

"The density of the specimen was determined by measuring the samples weight and the volume."

"The hardness of the specimen was measured using a Vickers micro-hardness tester under a load of 1 kgf for 20 s."

"Pure metals of Hf, Nb, Ta, Ti, V, and Zr can be traced on the green-body disk."

"An early study on the micro-hardness of HEA HfNbTaTiZr [30] reported a value of 3.826 GPa, with a grain size around 100 to 200 μm, which is 3.8 times the estimated value predicated from the rule of mixture."

"The micro-hardnesses of BCC refractory HEAs MoNbTaW and MoNbTaWV [31] were also reported to be 4.45 and 5.25 GPa, which are 2.4 and 3.29 times of the estimated value using the rule of mixture, respectively."

Discussion

"Quantitative evaluation of entropy sources of HEAs has been provided in recent publications [32, 33]. The BCC phase has higher enthalpy than the C15 Lave phase, and their enthalpies are positive and increase with increasing temperature."

"The composition of BCC (C15) phase is 10.33 (16.96), 16.03 (0.01), 12.19 (11.43), 15.12 (2.73), 12.98 (9.08), and 33.35 (59.79) in atomic percentage for Hf, Nb, Ta, Ti, Zr, and V, respectively."

"The BCC phase has much larger total entropy than the C15 phase over the entire compositional range, but the difference becomes smaller with increasing V contents and finally diminishes at pure V. The enthalpy of the C15 phase has a minimum at 46 at."

"While the HEA community uses the ideal entropy of mixing and the enthalpy of mixing for the liquid phase to calculate Ω-parameter, it should be more meaningful if the mixing properties for the solid solution phases are used."

Conclusions

"A new senary refractory BCC HfNbTaTiVZr HEA was designed by CALPHAD modeling using TCNI7 database and was experimentally verified using XRD and SEM."

"The atomic structure and diffusion constants were predicted using AIMD simulations."

"At 2473 K (2200 °C), the predicted density is 9.076 g/cm^3, and the diffusion constants are 5.35 $\pm$ 0.08, 4.96 $\pm$ 0.02, 4.07 $\pm$ 0.07, 3.96 $\pm$ 0.02, 3.86 $\pm$ 0.03, and 3.73 $\pm$ 0.06 [10^{-5} cm^2/s] for V, Ti, Zr, Nb, Hf, and Ta, respectively."

"The BCC phase in the senary shows overall small negative excess entropy at 1273 K (1000 °C)."

"The calculated empirical parameters including atomic size difference (δ) [23], enthalpy of formation (ΔH_f) [23], electronegativity difference (Δδ) [34], and Ω-parameter [35] all favor formation of single BCC HfNbTaTiVZr HEA solid solution."

First-Principles Prediction of High-Entropy-Alloy Stability [212]

This is a machine-generated summary of:

Feng, Rui; Liaw, Peter K.; Gao, Michael C.; Widom, Michael: First-principles prediction of high-entropy-alloy stability [212]

Published in: npj Computational Materials (2017)

Link to original: https://doi.org/10.1038/s41524-017-0049-4

If you want to cite the papers, please refer to the original.

For technical reasons we could not place the page where the original quote is coming from.

Abstract-Summary

"High entropy alloys (HEAs) are multicomponent compounds whose high configurational entropy allows them to solidify into a single phase, with a simple crystal lattice structure."

"Our first-principles calculation of free energy predicts that the configurational entropy stabilizes a single body-centered cubic (BCC) phase from $T = 1700$ K up to melting, while precipitation of a complex intermetallic is favored at lower temperatures."

"We form the compound experimentally and confirm that it develops as a single BCC phase from the melt, but that it transforms reversibly at lower temperatures."

Introduction

"Trial-and-error experimental approaches to test the phase stability of HEAs are costly and time-consuming, while previously-reported phase-formation rules are empirical and susceptible to the alloy, systems [36, 37] motivating a first-principles theoretical approach to accelerate the design of HEAs through the prediction of phase stability and microstructure."

"Utilizing fully first-principles calculations of the free energies of the full quaternary and its competing binary and ternary phases, we predict that the configurational entropy stabilizes a single body-centered cubic (BCC) solid solution at high temperatures, but that it will precipitate a complex intermetallic Laves phase at lower temperatures."

"We verify the formation of the single-phase BCC solid solution from the melt, confirm the precipitation of the Laves phase at low temperatures, and observe that it reverts to a single BCC phase upon annealing at high temperatures, thus illustrating the entropic stabilization principle of HEAs."

Results

"The mixing enthalpies are relatively low in Cr–Mo and Nb–V, compared with Cr–Nb, so their estimated critical temperatures for phase separation are low (1213 and 653 K, respectively [38, 39])."

"The high mixing enthalpies that destabilize the Cr–Nb solid solution, together with the relatively-low melting temperature of 1893 K that enhances diffusion, make it possible to observe both the phase separation and the Laves-phase formation experimentally."

"Turning to the full quaternary, we will test its stability at high temperatures by calculating its free energy relative to the most-likely competing phases, namely, the coexistence of a Mo-rich BCC medium entropy alloy with a CrNbV-rich Laves phase."

"The BCC phase becomes increasingly Mo-rich, with the composition of $x_{Mo} = 1/(4 - 3\gamma)$, while the remaining elements have the composition $x_{\alpha} = (1 - \gamma)/(4 - 3\gamma)$ for α = Cr, Nb, or V. Thus, the total free energy for a two-phase mixture of Laves with a partially-transformed BCC structure is and G_{BCC} depends on x_{Mo}."

Discussion

"Notice that G_4 crosses the single-phase equiatomic HEA free energy, G_1, around $T = 800$ K, indicating a transition from the HEA to the 4-phase mixture at low temperatures."

"Because Cr_2Nb exhibits a high-temperature C14-Laves phase (Pearson type hP12), and the structure is predicted to be stable at low temperatures in NbV_2, we checked the free energy of CrNbV as a C14-Laves phase."

"By taking the ideal entropy of mixing in the BCC phase, we overestimate the mixing entropy by about 2%, which amounts to an error of 3 meV/atom at 1700 K. Repeating our phase-transition prediction, we find that this error is equivalent to underestimating the transition temperature by approximately 25 K. In contrast, our neglect of the BCC vibrational free energy composition dependence has no effect on the transition temperature, because the mole-fraction of Laves phase vanishes at the transition."

Methods

"We utilize the same series of Mo-rich configurations to evaluate electronic densities of states and electronic free energies."

"The vibrational free energy is calculated in the harmonic approximation from the vibrational DOS, $g(\nu)$."

"This value is obtained from the force-constant matrix, $\Phi_{ij} = \partial^2 U/\partial R_i \partial R_j$, which VASP determines using density functional perturbation theory at $T = 0$ K. Our calculations utilize a plane-wave basis with the energy cutoff of 400 eV, "accurate" precision to avoid wrap-around errors, and an additional support grid for augmentation charges."

"[40] We set $\Theta_D = 393$ K for the BCC HEAs, and $\Theta_D = 420$ K for the Laves phase, as appropriate values to reproduce the high temperature free energy."

"After obtaining the 1473 K-annealed samples, another annealing treatment at 1673 K for 12 h was performed using a 1473 K-annealed sample to verify the stability of the C15-Laves phase."

3.4 AI and ML in the Design of HEAs

Machine generated keywords: machine, algorithm, machine learning, learning, rule, simulation, discover, transition, model, identify, elastic, comprise, predict, modulus, useful

Machine Learning-Assisted Design of Biomedical High Entropy Alloys with Low Elastic Modulus for Orthopedic Implants [213]

This is a machine-generated summary of:

Ozdemir, H. C.; Bedir, E.; Yilmaz, R.; Yagci, M. B.; Canadinc, D.: Machine learning-assisted design of biomedical high entropy alloys with low elastic modulus for orthopedic implants [213]

Published in: Journal of Materials Science (2022)

Link to original: https://doi.org/10.1007/s10853-022-07363-w

If you want to cite the papers, please refer to the original.

For technical reasons we could not place the page where the original quote is coming from.

Abstract-Summary

"This paper focuses on finding an optimum composition for the TiTaHfNbZr quinary high entropy alloy (HEA) system with an elastic modulus close to that of bone in order to attain a better biomechanical compatibility between the bone and the implant in orthopedic applications."

"The ML algorithms utilized in this study were trained using experimental data available in the literature and then utilized to predict the optimum HEA compositions with the lowest elastic moduli."

"The $Ti_{23}Ta_{10}Hf_{27}Nb_{12}Zr_{28}$ and $Ti_{28}Ta_{10}Hf_{30}Nb_{14}Zr_{18}$ compositions were predicted as the optimum HEA compositions with elastic moduli of 83.5 ± 2.9 and 87.4 ± 2.2 GPa, respectively."

"The findings of the work reported herein demonstrate that the proposed ML model can successfully predict HEA compositions for an optimized biomechanical compatibility for orthopedic applications and warrant further biomedical research on the two new HEAs prior to their utility as orthopedic implant materials."

Introduction

"Ti-based high entropy alloys (HEAs), comprising biocompatible elements such as Ta, Hf, Nb, Zr and Mo, have been studied by researchers as potential alternative implant materials [41–44]."

"Even though some biocompatible HEAs possess lower elastic moduli than currently implemented biomedical alloys such as 316L, CoCrMo, and Ti6Al4V [43], they are still far from the elastic modulus of bone, which varies between 4 and 30 GPa [45, 46]."

"The aim of the current work is to obtain the optimum composition for $Ti_xTa_yHf_zNb_mZr_n$ biomedical HEAs with elastic moduli close to the stiffness of the bone by employing ML algorithms with the specific purpose of providing a better structural match between the implant and the bone in orthopedic applications."

"The work presented herein, thus, illustrates that the ML methods were successfully implemented to optimize $Ti_xTa_yHf_zNb_mZr_n$ biomedical HEA compositions with the aim of minimizing elastic modulus in order to ensure mechanical compatibility between the implant and bone."

Computational Methods

"One of the major difficulties in ML implementations with a high-dimensional dataset is selection of the most relevant features to train the ML algorithms."

"Too many input features result in overfitting, especially in datasets containing small number of instances, due to high degrees of freedom, they introduce into the algorithms, leading to a decrease in the generalization capability of the ML models."

"A careful selection of features which explain most of the variance in a target variable is required to discard the redundant information from a dataset, and thus, has a critical impact on the prediction accuracy and performance of a model [47]."

"Highly correlated features affect model performance and prediction accuracy due to the fact that they have the same information about the target variable, i.e., the essence of the correlated features is so similar, such that the ML algorithms cannot differentiate among them [48, 49]."

Computational Results

"The general notion in any ML algorithm is that the larger the dataset, the better the model performance will be, because a large dataset contains an extensive amount of information about the target variable."

"The aim of selecting the right features is to eliminate any redundant or duplicate information present in the dataset and decrease degrees of freedom of the models so as to reduce complexity and prevent overfitting [49]."

"To decrease the effect of high correlation in the dataset, the RF method was employed in the process of feature selection in addition to PCC."

"Since Mo and $\Delta\chi$ are highly correlated and Mo is one of the elements representing alloys in the dataset, $\Delta\chi$ was not considered in the following feature selection process."

"Various ML algorithms were trained using composition and three selected features and employed to predict novel biomedical HEAs with low elastic modulus."

Validation experiments

"The produced alloys had a button shape with an intended mass of 20 g. To increase mixing and ensure chemical homogeneity, the button-shaped samples were flipped and re-melted at least 5 times."

"The samples were exposed to immersion experiments in SBF for 28 days."

"All HEA samples were held in separate sealed tubes, where the volume of the solution in each tube was calculated in accord with the G31 ASTM standards [50]."

"To measure the amount of mass gain or loss, the weights of the samples were recorded before and after immersion experiments."

"A depth-sensing nanoindentation device (Agilent G200) equipped with a Berkovich diamond indenter tip was employed to obtain the elastic modulus values of the samples from the loading and unloading curves by using the Oliver–Pharr method [51]."

Discussion of Results

"The lattice parameter for the BCC phase of the Bio-HEA1-HT, on the other hand, was determined to be 3.452 Å. Although HEAs have a single crystal structure in as-cast condition, phase transformation and precipitation of intermetallic phases can be observed following heat treatment [52, 53]."

"This could be attributed to the formation of new phases in Bio-HEA1-HT and Bio-HEA2-HT as a result of heat treatment, which brings about a decrease in corrosion resistance due to chemical homogeneity in the matrix [54]."

"The reason that the elastic modulus is higher for Bio-HEA1-HT and Bio-HEA2-HT as opposed to Bio-HEA1 and Bio-HEA2, respectively, is associated with the change in size, total volume and distribution of intermetallic phases in the matrix upon heat treatment, leading to alterations in the mechanical properties."

Conclusions

"The results of the work presented herein demonstrate the capability of machine learning (ML) methods trained with a limited amount of experimental data to predict novel biomedical high entropy alloy (HEA) compositions with optimum elastic moduli for utility in orthopedic applications."

"The ML algorithm utilized in this work predicted two novel biomedical HEA compositions, namely $Ti_{23}Ta_{10}Hf_{27}Nb_{12}Zr_{28}$ and $Ti_{28}Ta_{10}Hf_{30}Nb_{14}Zr_{18}$, with elastic moduli of 83.5 ± 2.9 and 87.4 ± 2.2 GPa, respectively."

"The predicted elastic moduli of the new alloys were validated by nanoindentation experiments once the materials were manufactured."

"The predicted alloys were also annealed in 1000 °C for 12 h to examine the changes in microstructure, and corresponding elastic moduli and ion release behavior."

Machine-Learning-Based Intelligent Framework for Discovering Refractory High-Entropy Alloys with Improved High-Temperature Yield Strength [214]

This is a machine-generated summary of:

Giles, Stephen A.; Sengupta, Debasis; Broderick, Scott R.; Rajan, Krishna: Machine-learning-based intelligent framework for discovering refractory high-entropy alloys with improved high-temperature yield strength [214]

Published in: npj Computational Materials (2022)

Link to original: https://doi.org/10.1038/s41524-022-00926-0

If you want to cite the papers, please refer to the original.

For technical reasons we could not place the page where the original quote is coming from.

Abstract-Summary

"This work demonstrates the development of a state-of-the-art machine learning framework coupled with optimization methods to intelligently explore the vast compositional space and drive the search in a direction that improves high-temperature yield strengths."

"Upon developing and validating a robust yield strength prediction model, the coupled framework is used to discover RHEAs with superior high temperature yield strength."

"We have shown that RHEA compositions can be customized to have maximum yield strength at a specific temperature."

Introduction

"Experimental studies have shown that refractory HEAs (RHEAs) possess superior high-temperature strength compared to superalloys, making them an attractive class of alloys for further exploration [55–61] for potential use in high efficiency gas turbine engines."

"If elemental composition is varied for each base alloy, the number of HEAs becomes more than 592 billion [62]."

"An ML-based framework informed by extant literature data is still a mostly untapped route for the development of HEAs with superior mechanical properties, particularly for RHEAs which possess the potential for high strength in high-temperature applications."

"This paper, as an example, focuses on yield strengths of refractory high-entropy alloys (RHEAs) and begins with developing a comprehensive forward ML model via identifying critical descriptors selected from a set of a large number of descriptors."

"Using the ML-based model, we discovered RHEA compositions with yield strengths customized for specific temperatures."

Results and Discussion

"In the case of the 1000 °C optimal alloy, given that the value of δ is the prevailing difference between the 1000 °C optimal alloy and the base alloy, our model suggests that lower values of δ could be correlated to a lessened temperature dependence of the yield strength."

"The concept was demonstrated for three different base alloys discussed in the RHEA literature, and optimal alloy compositions with improved yield strengths, as high as 80% were predicted for both 25 °C and 1000 °C."

"The alloys optimized for yield strength at 25 °C and at 1000 °C exhibited notable differences in composition and descriptors, underscoring that the mechanisms and criteria for maximizing strength at low temperature and high temperature can be quite different, and compositions maximizing yield strength at room temperature may not improve that for high temperatures."

Methods

"We have utilized the temperature-dependent yield strength model based on edge dislocation theory formulated by Maresca and Curtin [63] to augment the experimentally available HEA dataset at temperatures between 25 °C and 600 °C."

"We used the sequential forward selection (SFS) method [64, 65] for selecting the best set of descriptors that describes the data within a given regression model."

"We then performed optimization of the hyperparameters of the regression models using the six descriptors to identify parameters that maximize R^2 (CV)."

"Once we have identified the descriptor set and optimized the hyperparameters, we performed a repeated k-fold validation."

"The repeated k-fold method provides a statistical variation of the predicted yield strength of each RHEA allowing us to compute the 95% confidence level of each yield strength data point."

"For solving the inverse problem, we used the differential evolution optimizer [66], a genetic algorithm, implemented within scipy [67] to design an atomic composition that maximizes yield strength at a specified temperature for a particular base alloy."

Impact of Lattice Relaxations on Phase Transitions in a High-Entropy Alloy Studied by Machine-Learning Potentials [215]

This is a machine-generated summary of:

Kostiuchenko, Tatiana; Körmann, Fritz; Neugebauer, Jörg; Shapeev, Alexander: Impact of lattice relaxations on phase transitions in a high-entropy alloy studied by machine-learning potentials [215]

Published in: npj Computational Materials (2019)

Link to original: https://doi.org/10.1038/s41524-019-0195-y

If you want to cite the papers, please refer to the original.

For technical reasons we could not place the page where the original quote is coming from.

Abstract-Summary

"A main challenge in identifying new HEAs is the lack of efficient approaches for exploring their huge compositional space."

"Ab initio calculations have emerged as a powerful approach that complements experiment."

"For multicomponent alloys existing approaches suffer from the chemical complexity involved."

"Our approach is based on the application of machine-learning potentials based on ab initio data in combination with Monte Carlo simulations."

"The high efficiency and performance of the approach are demonstrated on the prototype bcc NbMoTaW HEA."

"The approach is employed to study phase stability, phase transitions, and chemical short-range order."

Introduction

"For computing chemical ordering and related transitions, even for a single HEA, brute-force DFT simulations are, however, usually limited by the supercell size and number of possible configurations which can be taken into account."

"Explicit supercell calculations of a B2 ordered state as well as a disordered solid solution revealed that the order–disorder transition temperature can be significantly lower if local lattice relaxations are included in the computations."

"The CE can include implicitly local lattice relaxations and can be combined with Monte Carlo (MC) simulations to account for chemical short-range order."

"We demonstrate the power of this new approach to efficiently and accurately explore huge configuration spaces by studying the finite-temperature phase stability of bcc NbMoTaW. Based on these studies we reveal a hitherto not reported chemical ordering at ambient temperatures."

Results

"The 30% decrease in the phase transition temperatures predicted by LRP as compared to [68] might also affect other properties such as the predicted degree of chemical short-range order at elevated temperatures."

"Relaxation effects have previously been reported to stabilize the solid solution compared to the B2 ordered structure by about 200 K [69]. Here we find that the solid solution is also stabilized by about 200 K, shifting the transition temperature down to 300 K. Note that this is more than 450 K lower than predicted in ref."

"In order to understand the qualitative difference in the results of calculations with and without taking local lattice distortions into account, we compared the energies of the predicted structures before and after relaxation in DFT calculations."

"The energy of the random phase decreases by approximately 13 meV/atom, while the energies of the ordered and semiordered phases predicted in the relaxed scenario decrease by around 15 meV/atom."

Discussion

"If local lattice distortions are included in the calculations, our 1 meV/atom-accurate model predicts a sequence of transitions which significantly differ from the results by the static-lattice approach as well as previous simulation results."

"Instead of segregation of different species into sublattices we see that the solid solution persists until much lower temperatures as so far suggested, i.e., down to 300 K. Moreover, in contrast to what the previous GPM-based calculations, concentration wave method, and CE-based calculations suggested [68–75], a B2-ordering is not the most stable ordering at ambient temperatures."

"Based on our analysis of machine-learning model uncertainty we suggest that the exact phase transition temperature is in the interval of 250–350 K. As the layered structure is rather different from the originally anticipated B2 ordering, significant impact on various properties such as on the elastic tensor and mechanical anisotropy behavior could be expected."

Methods

"In the on-lattice model, LRP [76], the energy of each configuration is partitioned into contributions of the separate atomic environments of each atom as where Ω is the lattice periodically repeated in space, V is the so-called interatomic potential defining the contribution of the atom at the lattice site ξ to the total energy E, $\sigma(\xi +$

r_i) is the type of ith neighbor of the atom located at ξ, r_i is the vector connecting ξ with its neighbor, and n is the number of neighbors (including the central atom) that depends on the cutoff parameter."

"In order to find the parameters of V we minimize the following functional: where K is the number of the atomic configurations, $\sigma^{(k)}$, in the training set ($k = 1,\ldots, K$), $E(\sigma^{(k)})$ is an energy of configuration predicted by the LRP model, $E^{qm}(\sigma^{(k)})$ is the DFT reference energy."

Machine Learning Approach to Design High Entropy Alloys with Heterogeneous Grain Structures [216]

This is a machine-generated summary of:

Li, Li; Xie, Baobin; Fang, Qihong; Li, Jia: Machine Learning Approach to Design High Entropy Alloys with Heterogeneous Grain Structures [216]

Published in: Metallurgical and Materials Transactions A (2021)

Link to original: https://doi.org/10.1007/s11661-020-06099-z

If you want to cite the papers, please refer to the original.

For technical reasons we could not place the page where the original quote is coming from.

Abstract-Summary

"It is still difficult to obtain the optimized grain size in the heterogeneous-grained HEAs, which achieves their outstanding mechanical properties."

"Using a novel integration method of machine learning, a physical model and atomic simulation, the optimal grain size is designed for achieving high yield strength of heterogeneous-grained CrCoFeNi HEAs."

"Atomic simulations give the stress–strain curve, yielding strength and microstructure with the increase of small grain size."

"The physical-based strength model expands the data from the atomic simulations and obtains the transition region from the Hall–Petch to inverse Hall–Petch relationship."

"The machine learning model shows that the obvious transition point from the Hall–Petch to inverse Hall–Petch relationship occurs at the grain size of 38.4 nm for the heterogeneous-grained CrCoFeNi HEAs with the large grain size of 165 nm."

"This integrated model makes significant contributions to understanding deformation and designing the microstructure of heterogeneous-grained HEAs."

Introduction

"The nanocrystalline HEAs have been synthesized through high-pressure torsion, mechanical alloying and magnetron sputtering, the grain growth would be induced with the subsequent annealing [78–81]. Recently, the wide application of artificial intelligence, especially machine learning, has made great breakthroughs in different fields and issues, such as image recognition, speech recognition and medical diagnosis [82–84]. Machine learning contains many algorithms, such as the genetic algorithm, random forest algorithm, support vector machine algorithm and artificial neural network, ANN [85]. Moreover, machine learning has been proved to be able to mine deep connections in data."

"If there are enough training data, machine learning can learn complex nonlinear mapping from high-dimensional eigenvectors to the desired output [86]. These characteristics of machine learning make it very useful to solve new problems in the fields of physics and material science, and great progress has been made recently [87–89]. For example, the integration of machine learning and the extended Hume-Rothery rule has been successfully employed to reveal the rule of phase selection in multi-principal element alloys [90]. Furthermore, the physics-based model combined machine learning is also adopted to obtain the interatomic potential function of FeCoCrNiMn HEA with first principles accuracy and higher computational efficiency [91]. Invariably, the hierarchical distributions of microstructures, such as gradient grain structures, dual-phase structures, bimodal structures and heterogeneous lamellar structures, are commonly observed in the reported materials with high strength and ductility."

"We provide an integration method of machine learning, a physical model and atomic simulation to discover the transition from the Hall–Petch relationship to inverse Hall–Petch relationship in heterogeneous-grained CrCoFeNi HEA."

"The present research provides an innovative method to design the microstructure of heterogeneous HEAs with high strength."

Method

"The physical model then generates more yield strength data based on atomic simulation."

"The machine learning model is built based on yield strength data from the physical model to predict the optimal microstructure."

"The exploration–labeling–training (ELT) algorithm derived from active learning provides an effective solution to this problem [92]. The ELT algorithm usually proceeds with the following steps iteratively: exploration, filter the most informative data from the unlabeled dataset; labeling, label the data selected from exploration by using an experiment or simulation; training, construct a model by using a machine leaning algorithm."

"The accuracy and efficiency of the machine learning optimization system are determined by the different implementation methods of ELT algorithm."

"In the exploration stage, the grain size is randomly selected in each sub-interval, which also ensures that the selected grain size has the greatest amount of information. (2) Labeling: MD simulation and physical model."

Results and Discussion

"The first step is fitting the existing data points by using the theoretical model [93]. Then, the expanding data points around the existing data points would be predicted by adjusting the small grain size based on the theoretical model."

"The mechanical properties of the $Cr_xCo_yNi_{100-x-y}$ alloy can be accurately predicted, which makes it possible to search for all the potential alloys [94]. Hence, the trained machine learning model is employed to establish the correlation between the grain size and yielding strength of FeCoCrNi HEA with heterogeneous grain structures."

"This trend agrees well with the previous work in nanocrystalline metal materials [95, 96]. However, it seems that the relationship between grain size and yielding strength is difficult to explain clearly in various heterogeneous-grained HEAs using the classical physical model."

Conclusions

"The deformation behaviors, mechanical properties and optimal grain size in CrCoFeNi HEAs with heterogeneous grain structures are investigated by a novel approach integrating atomic simulations, physical models and machine learning."

"The atomic simulation calculates the stress–strain curves for heterogeneous CrCoFeNi HEA with various small grain sizes."

"The machine learning model uncovers that the obvious transition point from the Hall–Petch to inverse Hall–Petch relationship occurs at the grain size of 38.4 nm at a large grain size of 165 nm in CrCoFeNi HEAs."

"The present research provides an innovative method integrating atomic simulation, the physical model and machine learning model to design the microstructure (multistage grain size) of heterogeneous-grained HEAs with high strength."

Machine Learning Guided Appraisal and Exploration of Phase Design for High Entropy Alloys [217]

This is a machine-generated summary of:

Zhou, Ziqing; Zhou, Yeju; He, Quanfeng; Ding, Zhaoyi; Li, Fucheng; Yang, Yong: Machine learning guided appraisal and exploration of phase design for high entropy alloys [217]

Published in: npj Computational Materials (2019)

Link to original: https://doi.org/10.1038/s41524-019-0265-1

If you want to cite the papers, please refer to the original.

For technical reasons we could not place the page where the original quote is coming from.

Abstract-Summary

"Although many useful HEAs or CCAs were reported, the rules of phase design, if there are any, which could guide alloy screening are still an open issue."

"We explored the use of an extended set of new design parameters, which had not been considered before, for phase design in HEAs or CCAs with the ML modeling."

"To verify our ML-guided design rule, we performed various experiments and designed a series of alloys out of the Fe-Cr-Ni-Zr-Cu system."

"The outcomes of our experiments agree reasonably well with our predictions, which suggests that the ML-based techniques could be a useful tool in the future design of HEAs or CCAs."

Introduction

"[97] Based on the assumption of ideal mixing, Yeh and others [98] proposed that an equal-atomic composition maximizes the configurational entropy of an HEA, therefore, in favor of the formation of random solid solution (SS) over other phases, such as intermetallics (IM)."

"According to the Hume–Rothery rules [99], the misfit in the physical and electronic properties of constituent elements strongly affects the formation of SS in binary alloys and possibly multicomponent alloys as well, such as HEAs."

"[100–104] Despite the seemingly difference among these rules, they are in the same spirit of the classic Hume–Rothery rule, i.e., aside from increasing the number of constituent elements, one has to minimize the difference in the physical and chemical properties of the constituent elements in HEAs in order to obtain random SS."

"Alloying different types of elements can also facilitate the formation of metastable phases, such as AM [105] or metastable random SS phases."

Results

"Our ML modeling is based on three algorithms, including the artificial neural network (ANN), the one-dimensional convolutional neural network (CNN), and the support vector machine (SVM), to assess the efficiency of the existing phase design rules and to explore new ones (see Methods)."

"To be consistent with the HEA literature [23, 97, 106], we consider the existence of three general phases, i.e., random SS, IM, and AM, as the outputs or the data labels of our ML modeling."

"This could translate to the measure of sensitivity of the data label, i.e., the probability to form a certain phase, with respective to changes in the corresponding features, i.e., thermodynamic or micromechanical parameters, as designed for our ML models."

"Through the compound transformation function, the parametric sensitivity of the data labels can be assessed quantitatively, which connects the features and labels through the well-trained ML model."

Discussion

"This indicates that the formation of SS is more sensitive to the change in δ or ΔH_{mix} than that of IM and AM, which agrees with our sensitivity measure and also the recent works."

"Our ML model also indicates that low σ_T favors SS and AM, whereas high σ_T favors IM."

"Our ML modeling predicts that the HEAs should transit to IM + AM and AM when x increases, which seemingly contradicts our experimental results."

"With these ML models, we further explored the use of additional new design parameters for phase design for HEAs."

"In the meantime, our results also indicate clearly that the phases in the HEAs designed based on our ML modeling are cooling rate dependent, which is sensible as AM is a metastable phase and even SS in some HEAs, such as the cantor alloy [28], are also metastable in nature."

Methods

"The neural network contains one input layer for the 13 design parameters, one hidden layer with 20 neurons and one output layer for the label."

"The first hidden layer consists of 64 channels, each of which contains 11 neurons."

"Sixty-four channels are stacked together as the first hidden layer (convolutional layer) with different weights and biases in every of the 11 neurons."

"The second hidden layer consists of 128 channels, each containing nine neurons."

"The 128 channels, which contain different weights and biases for each of the nine neurons, are stacked together as the second hidden layer (convolutional layer)."

"The raw metals were melted and solidified to form the ingots in a laboratory-scale arc-melting furnace, which had a vacuum as high as 8×10^{-4} Pa and contained a Ti ingot to soak up any residual oxygen in the vacuum."

Materials Informatics for the Screening of Multi-principal Elements and High-Entropy Alloys [218]

This is a machine-generated summary of:

Rickman, J. M.; Chan, H. M.; Harmer, M. P.; Smeltzer, J. A.; Marvel, C. J.; Roy, A.; Balasubramanian, G.: Materials informatics for the screening of multi-principal elements and high-entropy alloys [218]

Published in: Nature Communications (2019)

Link to original: https://doi.org/10.1038/s41467-019-10533-1

If you want to cite the papers, please refer to the original.

For technical reasons we could not place the page where the original quote is coming from.

Abstract-Summary

"The field of multi-principal element or (single-phase) high-entropy (HE) alloys has recently seen exponential growth as these systems represent a paradigm shift in alloy development, in some cases exhibiting unexpected structures and superior mechanical properties."

"These tools permit the identification of promising multi-principal element alloys."

"We implement this procedure using a database for which mechanical property information exists and highlight new alloys having high hardnesses."

Introduction

"Sarker and others proposed using a new descriptor, the entropy-forming ability, to predict HE alloys having high hardnesses and validated its use by combining first-principles and experimental synthesis [107]."

"We describe here the use of data analytics to accelerate the discovery of new, useful multi-component alloys by determining maximal correlations among metrics and alloy properties in experimental databases and the exploitation of this information to search for and identify promising HE alloy candidates."

"By combining a multiple regression analysis or its generalization, a canonical correlation analysis (CCA), with a genetic algorithm (GA) optimization strategy, we will explore systematically the HE alloy chemistry/composition space to highlight those alloys that have beneficial mechanical properties (e.g., high hardness)."

"We implement this procedure using a database comprising 82 HE alloys for which reliable mechanical property information is available to highlight new alloys having potentially high hardnesses."

Results

"Having obtained canonical variates from the above analysis, the variate $V^{(1)}$ was used as a fitness function in a GA to find candidate alloys having high hardness."

"The GA produces an extremely fit population of high-hardness alloys that represent at optimality an extreme extrapolation from the $M = 82$ dataset, we conservatively select from this population a subset of alloys whose predicted hardnesses are somewhat beyond the range of the existing dataset (i.e., having fitness values that are about 10–15% greater than the maximum found with the multiple regression)."

"The relatively low hardness alloys, Ti_{39} W_{04} Nb_{31} Ta_{04} Co_{22} and Ti_{44} Ni_{02} Nb_{21} Cr_{21} Co_{12}, were found to be quite ductile (i.e., $\gtrsim$95% of the indentations without cracks), which is generally consistent with observations that Group IV element additions, such as Ti, are often associated with increased ductility [108]."

"With this methodology we have identified several alloys that are both hard and relatively ductile."

Discussion

"In this context, the CCA can also be generalized to identify alloys having other useful thermomechanical and kinetic properties (e.g., yield strength, electrical conductivity, corrosion resistance) or, as described above, to Pareto-optimize multiple properties simultaneously using a multi-objective GA."

"To illustrate the use of the CCA to examine multiple output variables, we consider the following set of predictor (x) and output (y) variables, respectively, for the HE alloy example, where fcc, bcc, and IM denote the presence of a fcc solid solution, a bcc solid solution, or intermediate phase(s), respectively."

"The remaining variates primarily reveal associations between the alloy metrics and the phase behavior variables with only a weak dependence on hardness."

"This general CCA links the alloy metrics with both properties and phase information and is especially useful to workers who distinguish between HE alloys and multi-principal element alloys based on whether the system is a solid solution."

Methods

"This is accomplished by constructing from products of blocks of the correlation (or covariance) matrix an operator whose eigenvalues (λ_i, where i runs from 1 to the number of variates pairs, nv) capture the degree of correlation between combinations of predictor and outcome variables and whose eigenvectors determined the relative weights of the variables [109–111]."

"A GA is a fitness function optimizer that begins with a random distribution of chromosomes (i.e., alloys) represented as fixed-length bit strings and, via evolutionary processes, produces successive generations that are better trial solutions to the problem (i.e., having good mechanical properties), as determined from the fitness function [112–114]."

"Secondary electron imaging and X-ray EDS were performed on a FEI Scios Dual Beam FIB/SEM operated at an accelerating voltage of 20 kV and beam current of 13 nA. Secondary electron images were collected using an Everhart-Thornley detector."

3.5 Fabrication with Casting

Machine generated keywords: solidification, dendrite, cocrfeniti, liquid, nps, sample, cast, copper, gradient, first time, crystal, pct, casting, synthesis, bodycenteredcubic

Morphology Transition from Dendrites to Equiaxed Grains for AlCoCrFeNi High-Entropy Alloys by Copper Mold Casting and Bridgman Solidification [219]

This is a machine-generated summary of:

Zhang, Y.; Ma, S. G.; Qiao, J. W.: Morphology Transition from Dendrites to Equiaxed Grains for AlCoCrFeNi High-Entropy Alloys by Copper Mold Casting and Bridgman Solidification [219]

Published in: Metallurgical and Materials Transactions A (2011)

Link to original: https://doi.org/10.1007/s11661-011-0981-8

If you want to cite the papers, please refer to the original.

For technical reasons we could not place the page where the original quote is coming from.

Abstract-Summary

"The AlCoCrFeNi high-entropy alloys (HEAs) were prepared by the copper mold casting and Bridgman solidification."

"Compared to the as-cast sample, the plasticity of alloys synthesized by Bridgman solidification was improved by a maximum of 35 pct."

Introduction

"The alloys with high entropy of mixing mainly form face-centered-cubic or body-centered-cubic (bcc) solid solutions [115, 116]. At present, the arc melting and copper mold casting are usually used for the preparation of HEAs [117, 118], during which the solidification process cannot be easily controlled, which leads to a changing microstructure from the surface to the center of the samples, e.g., inhomogeneous distribution of the dendrites."

"Compared to the copper mold casting, the direction of the thermal conduction and extraction for rod-shape samples obtained by Bridgman solidification is mainly along the longitudinal direction, and the temperature gradient and growth rates can be precisely controlled by adjusting the heating power and withdrawal velocities."

"It was reported that dendrites/bulk-metallic-glass (BMG) composites were successfully tailored by Bridgman solidification, during which the dendrites are homogeneously distributed within the BMG matrix, and the dendrite spanning length has a strong dependence on the temperature gradient and growth rates [119, 120]. In this study, the microstructure and mechanical properties of the AlCoCrFeNi HEAs were investigated by these two solidification techniques."

Experimental Procedure

"One part was remelted and then cast into cylindrical rods with diameters of 3 and 5 mm."

"The cross sections of the samples were etched with the aqua regia solution, and the microstructures were investigated by metallographic microscope."

"The uniaxial compressive tests were performed on 6-mm-long sections cut from the 3-mm-diameter cylinders by using an MTS 809 materials testing machine at room temperature with a strain rate of 2×10^{-4} s^{-1}."

"The microstructure morphologies of the longitudinal sections were investigated by the metallographic microscope and the scanning electron microscope (SEM) equipped with an energy-dispersive X-ray spectrometer."

Results

"The dendrites are intensively embedded on the grain matrix and surrounded by the grain boundary."

"Most of the primary dendrite trunks, inclined to an angle of about 10 to 20 deg with respect to the vertical, are aligned along a specific crystallographic direction."

"The side arms of the dendrites that developed on the left side of the primary dendrite trunks usually have an angle of about 60 deg with respect to the trunks."

"These nanoscale precipitations may originate from the spinodal decomposition, leading to a modulated structure composed of ordered (B2) and disordered (A2) phases, respectively."

Discussion

"For Bridgman solidification, the temperature gradient in the liquid/solid interface front is almost a constant of ~ 70 K/mm, and the growth rate can be roughly considered to be the withdrawal velocities of 200 to 1800 μm/s (generally, the growth rate is always lower than the withdrawal velocity), which is much lower than that of the as cast."

"For a solid solution, to keep the stability of the grain size, a qualitative expression can be as follows: in which A, B, and C are the angle of the intersection of two adjacent grain boundaries, respectively; σ_A, σ_B, and σ_C are the interface tensions on the grain boundaries, respectively; and sinA, sinB, and sinC are the sine values of the corresponding angle, respectively."

Conclusions

"The AlCoCrFeNi HEAs are mainly composed of a bcc solid solution phase by copper mold casting and Bridgman solidification."

"The morphology changes from dendrites to equiaxed grains in this alloy after Bridgman solidification, which may be due to the lower heating temperature, longer holding time, and higher G/V values for the Bridgman solidification."

"The plasticity of alloys was improved to some extent by Bridgman solidification."

AlNiCrFeMn Equiatomic High Entropy Alloy: A Further Insight in Its Microstructural Evolution, Mechanical and Surface Degradation Response [220]

This is a machine-generated summary of:

Ananiadis, E.; Lentzaris, K.; Georgatis, E.; Mathiou, C.; Poulia, A.; Karantzalis, A. E.: AlNiCrFeMn Equiatomic High Entropy Alloy: A Further Insight in Its Microstructural Evolution, Mechanical and Surface Degradation Response [220]

Published in: Metals and Materials International (2019)

Link to original: https://doi.org/10.1007/s12540-019-00401-4

If you want to cite the papers, please refer to the original.

For technical reasons we could not place the page where the original quote is coming from.

Abstract-Summary

"The microstructure consists of two phases (NiAl-rich B2 and FeCr-rich A2) featuring various eutectic morphologies, ranging from regular to anomalous development."

"Solidification phenomena (undercooling and recalescense) are used in order to assess the various morphological features along with theories of anomalous eutectic growth."

"High temperature oxidation tests show multiple oxidation phenomena of different kinetics, formed oxide phases and material deterioration phenomena."

Introduction

"Though, first evidence revealed that HEAs exhibited intriguing features, such as (1) simple solid solution phases; (2) high hardness and wear resistance; (3) wide range of promising mechanical properties and (4) superior corrosion/oxidation resistance [121–146]."

"As for the tribological properties, Wu and others [127] studied the adhesive wear behavior of $Al_xCoCrCuFeNi$ HEAs as a function of Al content, pointing out that, for higher Al content, the worn surface is smooth and yields to fine oxidized debris, providing in this way a large wear resistance improvement."

"In terms of the oxidation behavior, it is well established that in conventional alloys, oxidation resistance can be largely improved by adding suitable amounts of Al, Cr, and Si, due to these elements dense, stable and coherent oxide layer formation ability [147]."

"The reduced diffusivity and high concentration of elements in HEAs have been assumed to improve oxidation resistance [140]."

Materials and Methods

"The aforementioned composition (nominal and measured) fulfills the solid solution formation criteria of HEAs, concerning atomic size difference (δ), enthalpy of mixing (ΔH_{mix}), entropy of mixing (ΔS_{mix}) while the Valence electron concentration (VEC) criterion suggests a bcc crystal lattice."

"The testing parameters were: load of 2 N and linear speed of 10 cm/s. A 100Cr6 steel ball of 6 mm diameter was used as counter body material for a sliding distance of 1000 m. Specimens were acetone cleaned and weighted before testing."

"The samples were compressed using a strain rate of 5×10^{-3} mm/s. At least three different specimens were tested."

"Tests were carried out to 150 h. Before the oxidation tests, rectangular coupons ($4 \times 4 \times 3$ mm) of the examined HEA were ground to a 1200 grit surface finish using SiC paper, cleaned with acetone, and placed into annealed alumina boats."

Results and Discussion

"As a final result the microstructure consists of a "ghost" dendrite phase decomposed in A2 and B2, a halo perimetrically to the "ghost" dendrite comprising mainly Fe and Cr, regular eutectic developed radially to the halos and conjunction regular eutectics parting lines rich in Fe and Al."

"According to the Ellingham diagrams [148] the order of the possible stable formed oxides in the case of the examined HEA is: Oxides of Al > Oxides of Mn > Oxides of Cr > Oxides of Fe > Oxides of Ni."

"At 9 h (initiation of Stage I.b.) a coherent and stable oxide scale (probably Al based according to Ellingham diagram) is formed on the HEA surface which protects the exposured material for further degradation."

"It can be assumed that a uniform and compact Al-rich scale has been formed during the initial stages of the oxidation (up to 75 h) that has obstructed O_2 to further access the HEA substrate."

Conclusions

"A refined approach on the solidification behavior of AlCrMnNiFe high entropy alloy was followed The microstructure consists of two phases (a rich in Ni–Al B2 and a rich in Fe–Cr A2 phase respectively with Mn showing a neutral behavior) The two phases are associated in different modes, comprising a eutectic microstructure which is, however, varied from regular to anomalous as a function of cooling rate, recalescence and nucleation phenomena Sliding wear response is mainly characterized by a surface oxidation and delamination material removal mode, partially associated with abrasive mode due to the presence of the hard oxide- based formed debris Compression testing showed acceptably high values for modulus of elasticity and yield point and a satisfactory ductility."

"The fracture mode is of brittle nature with all the characteristic features of brittle fracture surface High temperature oxidation is characterized of various steps of oxide formation and of different kinetics, with the early steps being governed by the formation of a protective Al-based oxide film and the late stages showing a scale spallation-reformation mechanism associated with the presence of Mn based oxides."

Solidification Behaviour of Ti–Cu–Fe–Co–Ni High Entropy Alloys [221]

This is a machine-generated summary of:

Mishra, Ajit Kumar; Samal, Sumanta; Biswas, Krishanu: Solidification Behaviour of Ti–Cu–Fe–Co–Ni High Entropy Alloys [221]

Published in: Transactions of the Indian Institute of Metals (2012)

Link to original: https://doi.org/10.1007/s12666-012-0206-x

If you want to cite the papers, please refer to the original.

For technical reasons we could not place the page where the original quote is coming from.

Abstract-Summary

"Ti–Cu–Fe–Co–Ni multicomponent alloys with varying Ti/Cu (x) molar ratio (x = 1/3, 3/7, 3/5, 9/11, 1, 11/9 and 3/2) have been prepared through the tailoring of microstructure to get understanding of the phase formation and the microstructural evolution of these multicomponent HEAs."

"X-ray diffraction and scanning electron microscopy coupled with energy dispersive spectroscopic results confirm the presence of $(Cu)_{ss}$, $(Co)_{ss}$ and $(\beta\text{-}Ti)_{ss}$ dendrites with ultrafine eutectic between cubic $(Cu)_{ss}$ and Laves phase (Ti_2Co type)."

"For x = 9/11, 1, 11/9 and 3/2; firstly, $(\beta\text{-}Ti)_{ss}$ dendrite is formed from the liquid, followed by the liquid phase separation between the cobalt-rich solid solution $(Co)_{ss}$ and copper-rich solid solution $(Cu)_{ss}$ and finally, the remaining liquid undergoes eutectic reaction between copper solid solution $(Cu)_{ss}$ and the Laves phase (Ti_2Co Type), whereas for x = 1/3, 3/7 and 3/5; $(\beta\text{-}Ti)_{ss}$ dendrite is formed first from the liquid and then remaining liquid undergoes the liquid phase separation resulting two different dendrites of $(Cu)_{ss}$ and $(Co)_{ss}$ phases."

Introduction

"Yeh and others [2], for the first time, adopted a brand new alloy design strategy, popularly known as multicomponent high entropy alloys (HEAs) consisting of at least 5 principal metallic elements in equimolar or near equimolar ratio whose concentration lies between 0.05 and 0.35 mol fraction."

"The study of the microstructural evolution during solidification of the multicomponent HEAs is scarce and has been less understood in terms of experiment as well as theory."

"The primary objectives of the present investigation are to synthesize novel Ti–Cu–Fe–Co–Ni multicomponent high entropy alloys with systematic variation of (Ti/Cu) molar ratio (Ti/Cu: 1/3, 3/7, 3/5, 9/11, 1, 11/9 and 3/2) as well as to achieve proper understanding of the phase selection and the microstructural evolution of these newly designed HEAs."

Experimental

"The multicomponent Ti–Cu–Fe–Co–Ni high entropy alloy ingots were synthesized by vacuum arc melting of pure elements Ti, Cu, Fe, Co, and Ni ($\geq$99.99% mass purity, Sigma-Aldrich, USA) on a water-cooled copper hearth using non consumable tungsten electrode under high purity argon atmosphere."

"Phase identification in different samples was carried out using BRUKER, D8 FOCUS X-ray diffractometer (XRD) in the θ–2θ configuration with Ni-filtered Cu K_α ($\lambda = 0.154056$ nm) radiation."

"The peaks in the diffraction patterns in each specimen were identified using International Committee for Diffraction Data (ICDD) database."

"The phase distribution and compositional analysis of all the polished samples was carried out using a Carl Zeiss EVO 50, scanning electron microscope (SEM) coupled with an energy dispersive spectrometer (EDS, INCA PENTA FET X3), operated at 20 kV."

Results and Discussion

"The solid solution phases, $(Cu)_{ss}$, $(Co)_{ss}$ and $(\beta\text{-}Ti)_{ss}$ are predominately present in the as-cast sample."

"The EDS analysis (not shown here) of the multicomponent Ti–Cu–Fe–Co–Ni HEAs shows the presence of different phases i.e. Ti–rich solid solution $(\beta\text{-}Ti)_{ss}$ (black contrast), Cu-rich solid solution $(Cu)_{ss}$ (white contrast), Co-rich solid solution $(Co)_{ss}$ (light-gray contrast) and the Laves phase Ti_2Co type (dark gray contrast)."

"After the formation of $(\beta\text{-}Ti)_{ss}$ phase, the remaining liquid is observed to undergo liquid phase separation to form the cobalt-rich solid solution $(Co)_{ss}$ and Cu-rich solid solution $(Cu)_{ss}$ phase."

"This causes phase separation between Cu-rich liquid and Co-rich liquid for high Cu-concentration, i.e. (Ti/Cu) molar ratio = 1/3, 3/7 and 3/5."

"We observed two solid solution dendritic phases i.e. $(Cu)_{ss}$ and $(Co)_{ss}$ in the microstructure of these alloys."

Thermodynamic Stability of Multicomponent Ti–Cu–Fe–Co–Ni HEAs

"Zhang and others [23, 149] proposed that the formation of stable solid solution in the multicomponent HEAs is possible due to small atomic size differences (<6.5%) and small values of the absolute enthalpy of mixing ($-15 \leq \Delta H_{mix} \leq 5$ kJ mol^{-1})."

"Although it has been observed that the absolute value of enthalpy of mixing in some binary alloys is higher, however the multicomponent equimolar or near equimolar alloys forms the stable solid solution because of high value of configurational entropy."

"It is also envisaged that the formation of stable solid solution in multicomponent high entropy alloys can be attributed to two thermodynamic parameters i.e. (i) a parameter Ω should be ≥ 1.1 and (ii) atomic size difference (δ) should be $\leq 6.6\%$ [101]."

Conclusions

"Multicomponent Ti–Cu–Fe–Co–Ni high entropy alloys have been synthesized successfully by vacuum arc melting technique."

"The sequence of phase evolution has been monitored with the aid of XRD and SEM during the solidification of these multicomponent HEAs."

"It has been experimentally observed that it is possible to form more than two solid solution phases during solidification."

"Detailed thermodynamic calculations have been carried to rationalize the formation of stable solid solution phases which show the thermodynamic parameters Ω and δ play significant role in stabilizing the solid solution phases."

On the Solidification and Phase Stability of a Co-Cr-Fe-Ni-Ti High-Entropy Alloy [222]

This is a machine-generated summary of:

Yeh, An-Chou; Chang, Yao-Jen; Tsai, Che-Wei; Wang, Yen-Chun; Yeh, Jien-Wei; Kuo, Chen-Ming: On the Solidification and Phase Stability of a Co-Cr-Fe-Ni-Ti High-Entropy Alloy [222]

Published in: Metallurgical and Materials Transactions A (2013)

Link to original: https://doi.org/10.1007/s11661-013-2097-9

If you want to cite the papers, please refer to the original.

For technical reasons we could not place the page where the original quote is coming from.

Abstract-Summary

"This material is shown to possess mainly a face-centered cubic (FCC) structure; the η phase is present at the interdendritic region in the as-cast condition, and it is stable between 1073 and 1273 K (800 °C and 1000 °C); γ' particles are found throughout the microstructures below 1073 K (800 °C)."

"Results show that Co, Cr, and Fe partition toward the dendritic region, while Ni and Ti partition toward the interdendritic areas."

"Scheil analysis indicates that the solid–liquid partitioning ratio of each element is very similar to those in typical single crystal superalloys."

Introduction

"The casting process was carried out with withdrawal velocities of 200 to1800 μm/s through a temperature gradient of ~ 70 K/mm into the water-cooled Ga-In-Sn liquid alloys; it was found that the casting microstructure changed from dendrites to equiaxed grains in this alloy [151]. Until recently, a two-step Bridgeman process with Ga-In-Sn liquid metal cooling was employed successfully to cast a FCC $Al_{0.3}CoCrFeNi$ alloy into a single crystal with 3 mm in diameter and about 50 mm in length [152]. In one of our previous studies [153], $Co_{1.5}CrFeNi_{1.5}Ti_{0.5}$ high-entropy alloy has been shown to possess mainly the face-centered cubic (FCC) phase, and the hardness of the as-cast state has achieved HV 400."

"Two aspects of research have been conducted, high-temperature phase stability and the possibility to cast this alloy into a single crystal sample by the conventional Bridgeman process."

"The typical thermal gradient of the liquid metal cooling casting process is more than double that of the conventional process [150], and so far there has been no reported study to cast high-entropy alloy into a single crystal by the conventional Bridgeman process."

Experimental Procedures

"The thermal properties of $Co_{1.5}CrFeNi_{1.5}Ti_{0.5}$ high-entropy alloy were analyzed with a differential thermal analyzer (DTA) with a heating rate of 10 K/minute and a cooling rate of 20 K/minute; the sample was heat treated at 1423 K (1150 °C) for 6 h prior to analysis."

"For phase stability analysis, as-cast samples were first heat treated at 1423 K (1150 °C) for 6 h, which was determined to be the solvus temperature of the η phase in the next section, in order to homogenize the microstructure."

"Samples for transmission electron microscopy (TEM) were prepared by twin-jet polishing in a solution of $HClO_4 + 9CH_3OH$ and then ion milled; samples were observed by a JEOL JEM-2010 TEM under 200 kV. In the first attempt, a single crystal (SC) bar 20 mm in height and 9 mm in diameter was successfully fabricated by the conventional Bridgeman furnace with a single crystal seed material of Rene 142 obtained from in-house casting."

Results and Analysis

"As-homogenized samples were then isothermal aged at 873 K, 973 K, 1073 K, 1173 K, and 1273 K (600 °C, 700 °C, 800 °C, 900 °C, and 1000 °C) for various durations."

"Microstructural observations revealed that the size of γ' increased as aging temperature increased from 923 to 1073 K (650 °C to 800 °C)."

"The elemental distributions were fitted with the Scheil equation to determine the partition coefficients for the various alloying elements [154], where C_s is the element concentration, k is the partition coefficient, C_0 is the mean composition, and f_s is the volume fraction of solid."

"The partition coefficient of each element was calculated from the gradient of the best-fit line of the previous equation."

Discussion

"Although $Co_{1.5}CrFeNi_{1.5}Ti_{0.5}$ high-entropy alloy can be cast into single crystal by the conventional Bridgeman process with a thermal gradient ~ 30 K/mm, casting segregation has resulted in the formation of the as-cast η phase, which is present in the interdendritic region, where Ti and Ni are enriched, and solution heat treatment is required to eliminate the η phase."

"For future alloy design to improve its phase stability, the Ti content can be substituted by some Al content, as Ni3(Al, Ti) type of γ' are more stable than that of

Ni3Ti γ' [150]. Although $Co_{1.5}CrFeNi_{1.5}Ti_{0.5}$ high-entropy alloy may not be able to compete with single crystal superalloys at temperatures above 1273 K (1000 °C), its behaviors of microstructural evolution do suggest some potentials to be applied at an intermediate temperature 873 K ~ 1073 K (600 °C ~ 800 °C), as a heat-resistant polycrystalline material."

Conclusions

"$Co_{1.5}CrFeNi_{1.5}Ti_{0.5}$ high-entropy alloy can be cast into a single crystal by a conventional Bridgeman technique."

"The segregation behavior of $Co_{1.5}CrFeNi_{1.5}Ti_{0.5}$ high-entropy alloy is comparable with that of conventional superalloy; Fe, Cr, and Co partition to the dendrite region; Ni and Ti partition to the interdendritic region."

"The γ-γ' microstructure of $Co_{1.5}CrFeNi_{1.5}Ti_{0.5}$ high-entropy alloy is stable up to 1073 K (800 °C)."

A Successful Synthesis of the $CoCrFeNiAl_{0.3}$ Single-Crystal, High-Entropy Alloy by Bridgman Solidification [223]

This is a machine-generated summary of:

Ma, S. G.; Zhang, S. F.; Gao, M. C.; Liaw, P. K.; Zhang, Y.: A Successful Synthesis of the $CoCrFeNiAl_{0.3}$ Single-Crystal, High-Entropy Alloy by Bridgman Solidification [223]

Published in: JOM (2013)

Link to original: https://doi.org/10.1007/s11837-013-0733-x

If you want to cite the papers, please refer to the original.

For technical reasons we could not place the page where the original quote is coming from.

Abstract-Summary

"A face-centered-cubic, single-crystal $CoCrFeNiAl_{0.3}$ (designated as Al0.3), high-entropy alloy (HEA) was successfully synthesized by the Bridgman solidification (BS) method, at an extremely low withdrawal velocity through a constant temperature gradient, for which it underwent two BS steps."

"At the equiaxed-grain region, some visible annealing twins were observed, which indicates a low stacking fault energy of the Al0.3 alloy."

"A similar morphological transition from dendrites to equiaxed grains occurred at the equiaxed-grain region in Al1 alloy, but the annealing twins were not observed probably because a higher Al addition leads to a higher stacking fault energy for this alloy."

Introduction

"The traditional alloy-design strategy was based on one or two principal metal elements that dictated the primary phase, while minor alloying elements that promote the formation of the secondary phase(s) were added in order to tailor the microstructures and properties."

"This multiprincipal elements alloy usually contains more than five elements and was named high-entropy alloys (HEAs) by Yeh and others2 on the basis of the high configurational entropy of mixing."

"155 These impressive properties make HEAs promising engineering materials especially for high-temperature applications, such as the turbine blades in aircraft engines, which are made of expensive nickel-based superalloys."

"A single-crystal FCC HEA was prepared successfully for the first time, and this report details the synthesis process and microstructure evolution of the single-crystal HEA."

Experimental Procedures

"Two alloy ingots of nominal compositions of $CoCrFeNiAl_x$ (x value in a molar ratio, and $x = 0.3$ and 1, denoted by Al0.3 and Al1, respectively) were synthesized by arc-melting pure elements with a purity higher than 99.95 wt.% under a high-purity argon atmosphere on a water-cooled Cu hearth."

"The synthesized alloys were crashed into pieces and placed in an alumina tube with an internal diameter of 3 mm and a wall thickness of ~ 1 mm."

"The samples were inductively heated to the completely melting condition by adjusting the heating power, holding for 15 min."

"Two BS steps were conducted, which means that after the first BS step, the same sample was turned by 180° lengthwise, and the BS was repeated under the same conditions as the first BS process."

Results and Discussion

"This feature differs from the Al0.3 alloy as the Al element generally has high stacking fault energies of around 160 mJ/m–200 mJ/m [120, 156] in which twinning crystals are difficult to observe."

"As addressed by the analyses above, the FCC Al0.3 alloy underwent a microstructural transition from dendrites to equiaxed grains, and then to columnar grains, and last to a single-crystal structure."

"The BCC Al1 alloy went through almost the same mcirostructural transition process, but the final-stage solidification structure consists of columnar crystals not of a single crystal under the same conditions."

"The Al addition can effectively tailor the stacking fault energy or potential twinning-crystal density (under appropriate conditions) of the alloy system, for which it has been successfully applied to improve the tensile ductility of metallic alloys (e.g., Fe-6.5 wt.% Si alloy [157] or bulk metallic glass composites [158])."

Conclusions

"Using two BS steps, we first successfully prepared the FCC-structured, single-crystal Al0.3 HEA, and the BCC-structured, columnar-crystal Al1 HEA, with instinctively preferential crystallographic orientations, under an extremely slow growth velocity."

"The microstructure of the Al0.3 alloy started with the dendrites, and then underwent equiaxed grains and columnar crystals, and lastly ended with the single crystal, while for the Al1 alloy, only the last transition was not obtained."

"In the equiaxed-grain region of the $Al_{0.3}$ alloy, some annealing-twin-like crystals with a twin-boundary spacing of about several tens of micrometers were observed."

"The ready-to-wear growth models for annealing twins indicate that the $Al_{0.3}$ alloy may have a low stacking fault energy."

Kinetic Ways of Tailoring Phases in High Entropy Alloys [224]

This is a machine-generated summary of:

He, Feng; Wang, Zhijun; Li, Yiyan; Wu, Qingfeng; Li, Junjie; Wang, Jincheng; Liu, C. T.: Kinetic ways of tailoring phases in high entropy alloys [224]

Published in: Scientific Reports (2016)

Link to original: https://doi.org/10.1038/srep34628

If you want to cite the papers, please refer to the original.

For technical reasons we could not place the page where the original quote is coming from.

Abstract-Summary

"Up to now, investigations of the phase selection in HEAs mainly focused on the thermodynamic equilibrium phase, while kinetic ways of tailoring the phases in HEAs are seldom considered."

"The kinetic effect in $CoCrFeNiTi_{0.4}$ HEAs was investigated to reveal the possibility of controlling phase selection via kinetic ways for HEAs."

"The theoretical analyses revealed the kinetic effect on phase selection."

"The method proposed here, tailoring the phases with different kinetic ways, could be used to prepare promising HEAs with very rich composition design."

Introduction

"These previous efforts are helpful to search for single phase HEAs, however, it was found that only several alloys consist of single solid solution while most HEAs are with multi-phases."

"In spite of that the kinetics of phase transformation is well known, in HEAs, it deserves more attentions."

"The phases in HEAs could be very complex because of their concentrated multi-components while phase compositions in traditional alloys are much simpler."

"As a new class of alloys, the kinetic effect in the phase selection of HEAs is of critical importance in both scientific studies and practical applications."

"Some reported works also showed that the processing routes have great influence on the phase compositions of HEAs, which actually reflected that the kinetics acted on the phase selection of HEAs [159–162]."

"Inspired by kinetics, tailoring phases in HEAs is extremely attractive and promising."

Experimental

"The $CoCrFeNiTi_{0.4}$ HEA was prepared by arc melting and then cooled in the arc furnace and alloys in this state were noted as as-cast (AC) HEA."

"The laser was held on the as-cast $CoCrFeNiTi_{0.4}$ base for 0.3 s forming a laser remelted zone (noted as LR zone) where rapid solidification occurred after removing the laser spot."

"Two different cooling ways, rapid quenching in water and furnace cooling, were used."

"The rapidly quenched samples are noted as Q1 (annealed at 1080 °C) and Q2 (annealed at 900 °C), and the samples that were cooled in furnace are noted as FC1 (annealed at 1080 °C) and FC2 (annealed at 900 °C)."

Phase selection of $CoCrFeNiTi_{0.4}$ HEA

"Besides the rapid solidification, the effects of solid-state transformation kinetics on the phase selection were investigated via different cooling rate after annealing the $CoCrFeNiTi_{0.4}$ HEA at 900 °C and 1080 °C."

"The cooling rate of LR HEA is the highest, thus R, σ and γ' phases are all inhibited in this condition; with the decrease of cooling rate, the R phase shows up in an AC sample."

"As for the solid-state transformation process, the cooling rate of the Q1 HEA is quite high, and σ phase was restrained under the rapid quenching condition."

"The R and γ' phases will also be inhibited when the cooling rate goes up further and only FCC and R phases can be found when the cooling rate reaches up around 80 °C/s. Consequently, the σ phase in Q1 HEA is restrained because of the high cooling rate when quenched in water."

Conclusions

"The kinetic effect of phase transformation and selection in HEAs are investigated by taking the $CoCrFeNiTi_{0.4}$ HEA as example."

"The phase selection of $CoCrFeNiTi_{0.4}$ HEA under different process routines was determined."

"Extremely rapid solidification process inhibited the formation of σ, γ' and R phases in the $CoCrFeNiTi_{0.4}$ HEA."

"The significant influence of the kinetic effect on the phase selection of $CoCrFeNiTi_{0.4}$ HEA indicates that it is possible and important to take the solidification and solid-state phase transformation kinetics into consideration in controlling the phase selection of HEAs for a better performance."

Effects of Solidification Conditions on Microstructure and Properties of High-Entropy Alloys from the CoCrFeMnNi Family [225]

This is a machine-generated summary of:

Stasiak, Tomasz; Oleszak, Dariusz; Fraczkiewicz, Anna: Effects of Solidification Conditions on Microstructure and Properties of High-Entropy Alloys from the CoCrFeMnNi Family [225]

Published in: JOM (2022)

Link to original: https://doi.org/10.1007/s11837-022-05543-2

If you want to cite the papers, please refer to the original.

For technical reasons we could not place the page where the original quote is coming from.

Abstract-Summary

"Alloys from the CoCrFeMnNi family remain the most studied austenitic high-entropy alloys."

"Four alloys, i.e., Cantor alloy, A3S (modified nonequiatomic Cantor composition), both "pure" or doped with carbon (200 wt ppm) and niobium (1.3 wt.%), were investigated."

"The effects of the solidification rate and the presence of carbon and niobium on the microstructure and hardness were studied."

"The lattice parameter of the fcc phase increases with the increasing cooling rate, and with the addition of niobium and carbon, which confirms at least a partial presence of these elements in solid solution."

"The higher cooling rate led to a small decrease in the grain size and a slight increase in hardness."

"The hardness of doped alloys can be further improved by annealing (500 °C for 24 h) through NbC precipitation."

Introduction

"[163] The first discovered HEA, equiatomic CoCrFeMnNi alloy, usually called Cantor alloy, is still the most studied."

"[164, 165] The most characteristic properties of the Cantor alloy are an increase in yield strength and ductility with decreasing temperature, from room down to liquid nitrogen temperature, while the fracture toughness remains almost unchanged."

"[166] Melt-spinning is a rapid solidification technique in which a material is solidified from a liquid state by pouring it onto a moving refrigerated wheel; the obtained temperature gradient enables the reaching of a cooling rate of 10^5–10^7 K s^{-1}."

"Despite many articles dealing with HEAs prepared by melt-spinning, there is a lack of papers discussing the effects of the cooling rate and the addition of carbon on the microstructure and mechanical properties of materials elaborated by this technique."

"The effects of the cooling rate and the addition of niobium and carbon on the microstructure and hardness of equiatomic and non-equiatomic Cantor alloys were studied."

Experimental

"The structure, microstructure, and microhardness of the melt-spun ribbons were investigated in this study."

"The scanning step was 0.02° from $2\theta = 10°$ to 100° with a time step of 5 s. For scanning electron microscopy (SEM) and electron backscatter diffraction (EBSD) analyses, the samples were cold-mounted, then manually polished with SiC grinding papers (P320–P4000), and finally polished by Struers LectroPol-5, using an electropolishing bath of 6% perchloric acid ($HClO_4$) and ethanol (C_2H_5OH) with an applied voltage between 18 and 25 V for a few seconds."

"Transmission electron microscope (TEM) observations were carried out using a Philips CM200 TEM equipped with an EDX detector operated at 200 kV. The TEM samples were prepared by conventional twinjet electropolishing."

"The samples for microhardness analysis were prepared similarly to those for the SEM investigation except for the electropolishing."

Results and Discussion

"Similar to other studies [167, 168], the addition of niobium and carbon causes an increase in lattice parameters in both alloys."

"The addition of carbon in this study was 0.02 wt.% (0.1 at.%); however, the increase of the lattice parameter is about 0.0025 Å for E and 0.0042 Å for N alloy."

"Alloys doped with niobium and carbon present slightly larger FWHMs than pure alloys, which could result from a smaller crystallite size and higher lattice strain in alloys with the addition of niobium and carbon."

"Chattopadhyay and others [169] found a slightly smaller grain size (about 1 μm) in the melt-spun ribbons (40 m/s linear wheel speed) of the Cantor alloy."

"Ribbons cooled at 70 R doped with niobium and carbon (especially E-D) present a significantly smaller grain size than pure alloys."

Conclusion

"A higher cooling rate caused a slight increase in the lattice parameter."

"The grain size is about a few micrometers and is decreases slightly with an increasing cooling rate."

"The microhardness slightly increases for the higher cooling rate, which could be mainly related to the smaller grain size."

"As-spun ribbons with the addition of carbon and niobium presented a slightly higher microhardness than the pure alloy, which is the effect of the nanometric NbC precipitates."

"Further increases of microhardness of the doped alloys could be achieved through annealing at 500 °C for 24 h. Such heat treatment causes precipitation from the solid solution of more NbC precipitates."

"It was shown that the cooling rates obtained in the melt-spinning process are high enough to retain niobium atoms in solid solution in the matrix, and make them available for precipitation hardening in selected thermal conditions of annealing."

General Synthesis of High-Entropy Alloy and Ceramic Nanoparticles in Nanoseconds [226]

This is a machine-generated summary of:

Wang, Bing; Wang, Cheng; Yu, Xiwen; Cao, Yuan; Gao, Linfeng; Wu, Congping; Yao, Yingfang; Lin, Zhiqun; Zou, Zhigang: General synthesis of high-entropy alloy and ceramic nanoparticles in nanoseconds [226]

Published in: Nature Synthesis (2022)

Link to original: https://doi.org/10.1038/s44160-021-00004-1

If you want to cite the papers, please refer to the original.

For technical reasons we could not place the page where the original quote is coming from.

Abstract-Summary

"High-entropy materials, which include high-entropy alloys and high-entropy ceramics, show promise for their use in many fields, yet a robust synthesis strategy is lacking."

"We present a simple and general approach, laser scanning ablation, to synthesize a library of high-entropy alloy and ceramic nanoparticles."

"As a laser pulse confines energy to the desired microregions, the laser scanning ablation method renders a high-entropy material nanoparticle loading on various substrates, which include thermally sensitive substrates."

"Applied as electrocatalysts for overall water splitting, the as-prepared high-entropy material nanoparticles can achieve an overpotential of 185 mV @ 10 mA cm^{-2}."

Main

"High-entropy materials (HEMs), which include high-entropy alloys (HEA) and high-entropy ceramics (HEC), are a class of materials that contain at least five near-equimolar principal metal atoms in an amorphous structure or a solid-solution phase [170]."

"For synthetic methods that use mild conditions [171, 172], only a small variety of HEA NPs can be produced because the synthesis either requires source materials with the same composition as the NPs (e.g., if we need to synthesize PtIrCuNiCr HEA NPs, bulk source materials of PtIrCuNiCr must be provided) or results in the formation of amorphous NPs."

"It is of great significance to develop a general yet robust route to synthesizing both HEA and HEC NPs under mild conditions over a diversity of substrates."

"We developed a laser scanning ablation (LSA) approach using a pulsed nanosecond laser to create a library of HEA and HEC NPs at atmospheric temperature and pressure."

"As the laser pulse precisely confines the energy to desired microregions, HEM NPs can form on various substrates, even thermally sensitive ones, such as metals and glass."

Results and Discussion

"A large amount of energy is released alongside the formation of the HEA NPs."

"HEA NPs formed on these substrate surfaces, and each element was evenly distributed in the particles, which demonstrates the broad substrate applicability of LSA."

"The effects of lattice distortions, a high mixing entropy and sluggish diffusion contribute to the formation of amorphous HEC NPs."

"Both effects (that is, high entropy and sluggish diffusion) synergistically impart the formation of amorphous HEC NPs."

"The derived metal–metal bond length in HEA NPs is clearly different from that in pure foils, which indicates the formation of an alloy structure."

"The coordination numbers of the Pt and Ir atoms in HEA NPs are much lower than those in the pure metal foils."

"The unsaturated coordination around the Pt and Ir sites can increase the initial potential energy of the HEA NPs, which leads to a greatly decreased activation energy barrier [173]."

Conclusions

"A simple, general and scalable LSA route to creating HEA and HEC NPs under mild conditions with a broad substrate applicability."

"The superior electrocatalytic performance of HEM NPs can be attributed to the severe lattice distortion and unsaturated coordination of the materials."

"Going beyond electrocatalysis, we envision that by judiciously designing HEM morphology and screening their compositions, HEM NPs may provide tremendous potential for applications in biomedicine, catalysis, energy storage and sensors."

Methods

"HEA NPs on the surface of substrates were prepared by pulsed laser ablation of the corresponding precursor-loaded substrates (for example, CNFs, wood, glass or copper foam)."

"To synthesize the HES, the salt-precursor-loaded substrates were dipped into a CS_2 solution with the sulfur at 0.1 M to allow a uniform sulfur layer to deposit on the surface of the substrate before laser irradiation."

"To synthesize HEOs, the salt-precursor-loaded substrates were treated by a 0.1 M NaOH solution to convert the metal chlorides into metal hydroxides on the substrate, followed by washing with distilled water to remove the by-product NaCl prior to laser irradiation."

"To synthesize high-entropy borides and nitrides, the salt-precursor-loaded substrates were dipped into 0.1 M sodium borohydride and 0.1 M ammonium chloride aqueous solutions, respectively, before the laser irradiation."

"To synthesize high-entropy phosphides, the salt-precursor-loaded substrates were dipped into a 0.1 M triphenyl phosphine ethanol solution before the laser irradiation."

3.6 Additively Manufactured HEA

Machine generated keywords: manufacture, manufacturing, additive, coating, powder, substrate, laser, additive manufacturing, selective, equiatomic, deformation behavior, alcocrfeni, radius, deformation, ehea

Mechanism Controlling Elevated Temperature Deformation in Additively Manufactured Eutectic High-Entropy Alloy [227]

This is a machine-generated summary of:

Vikram, R. J.; Verma, S. K.; Dash, K.; Fabijanic, D.; Murty, B. S.; Suwas, Satyam: Mechanism Controlling Elevated Temperature Deformation in Additively Manufactured Eutectic High-Entropy Alloy [227]

Published in: Metallurgical and Materials Transactions A (2022)

Link to original: https://doi.org/10.1007/s11661-022-06777-0

If you want to cite the papers, please refer to the original.

For technical reasons we could not place the page where the original quote is coming from.

Abstract-Summary

"High-temperature compression studies show that the yield strength tends to increase with the increase in temperature up to 400 °C and then starts declining till 800 °C."

"The yield strength was found to be almost similar at room temperature and 600 °C."

"To understand the anomaly in the temperature dependence of yield strength, first-principles density functional theory (DFT) calculations were performed using the diffuse multi-layer fault model (DMLF) to estimate the planar fault energies associated with the $L1_2$ crystal structure."

"Further, a time-dependent deformation study at different constant loads was performed at an elevated temperature where the anomaly was observed."

Introduction

"High-entropy alloys (HEAs) are viewed as a potential substitute to nickel-based superalloys for high-temperature applications."

"Because of its high-temperature strength and ductility, this EHEA has received much attention."

"[174] The primary opportunities for metal additive manufacturing are in aerospace applications, which seek significant cost savings, reduced lead times, lightweight designs, and a major focus on high-temperature materials that can be customized and provide high thermal efficiency by eliminating traditional joining processes."

"A positive yield stress dependence can be meritorious by inheriting high strength at elevated temperatures for high-temperature structural application aspects."

"The focus was placed on the $L1_2$ phase as this phase governs yield strength anomaly or high-temperature deformation in superalloys."

"[175] This work aims to estimate the PFE of high-entropy alloys by modeling the $L1_2$ compound by giving site preference to individual elements that give the crystal's lowest energy."

Experimental Methods

"The EHEA ($AlCoCrFeNi_{2.1}$) with a composition of Al, Co, Cr, Fe 16.39 at.pct, and Ni 34.49 at.pct was manufactured using a laser engineered net shaping (LENS) (Optomec MR7) process, which is a powder-blown direct energy deposition (DED) technique."

"The initial spherical gas-atomized powder size was sieved between 45 and 150 μm, and the deposition was built on a 316L stainless steel plate substrate using processing parameters developed in a prior study [176]; 900 W power, 35 inches/min scanning speed, and 30 g/min powder feed."

"32, 32, 64, and 64 atoms supercells were used to model the structures γ_a', $D0_{23}$, $D0_{24}$, and χ, respectively, in the DMLF model."

"APT measurements were taken using a local electrode atom probe (LEAP, 5000 XR, Cameca Instruments) in laser pulse mode at a repetition rate of 200 kHz and pulse energy of 0.4 nJ. The data reconstruction analysis was performed using the software application IVAS™ 3.8.2."

Results and Discussions

"Because the $L1_2$ phase is harder than the bcc phase in the present alloy (DFT elastic constant calculations), the primary deformation process is solely dependent on the $L1_2$ phase."

"Thermally activated cross slip from (111) to (100) is commonly observed in the deformation of the $L1_2$ phase at high temperatures [177], which is attributed to anisotropy of APB energy and elastic constants."

"Four major parameters were used to determine the nature of such anomaly: $\Delta\omega$, ΔE, W_c, and ΔH [177], where $\Delta\omega$ is the energy difference between two modes of dislocation splitting (APB (111) vs SISF (111)), ΔE is the driving force for the cross slip of dislocations from octahedral to cube plane, W_c is the constriction energy required to join the two Shockley partials surrounding the CSF (111), and ΔH is the activation energy barrier for the cross slip from the octahedral to the cube plane."

Conclusion

"The alloy displayed yield anisotropy at room temperature along (XY) and across (YZ) the build-up plane and yield anomaly in the XY plane at high temperatures."

"Since the $L1_2$ phase was present, the high-temperature yield strength anomaly is mostly dominated by the $L1_2$ phase due to its underlying planar defects."

"Since Peierls–Nabarro stress for dislocation glide is high for {001} slip plane, the dislocations get pinned and enhance the strength at high temperatures."

"Thermodynamics of observed yield strength anomaly was established for the $L1_2$ phase and it was found that APB {111} coupled dissociation of perfect dislocation is favorable, with a positive driving force ($\Delta E = 2.42 \times 10^{-2}$ eV/Å) and an activation energy barrier ($\Delta H = 0.97$ eV) for the cross slip."

Superior Temperature-Dependent Mechanical Properties and Deformation Behavior of Equiatomic CoCrFeMnNi High-Entropy Alloy Additively Manufactured by Selective Laser Melting [228]

This is a machine-generated summary of:

Kim, Young-Kyun; Yang, Sangsun; Lee, Kee-Ahn: Superior Temperature-Dependent Mechanical Properties and Deformation Behavior of Equiatomic CoCr-FeMnNi High-Entropy Alloy Additively Manufactured by Selective Laser Melting [228]

Published in: Scientific Reports (2020)

Link to original: https://doi.org/10.1038/s41598-020-65073-2

If you want to cite the papers, please refer to the original.

For technical reasons we could not place the page where the original quote is coming from.

Abstract-Summary

"The microstructure, temperature-dependent mechanical properties and deformation behaviors of equiatomic CoCrFeMnNi high-entropy alloy (HEA) additively manufactured by selective laser melting (SLM) were investigated."

"Room- and high-temperature compression tests confirmed that SLM-built HEA has outstanding mechanical properties in all temperature ranges compared to equiatomic CoCrFeMnNi HEAs reported up to the present."

"The excellent mechanical properties of SLM-built HEA were achieved with fine grains, high dislocation density and fine precipitates at low temperatures (25 °C to 600 °C), and by high dislocation density and fine precipitates at high temperatures (700 °C or higher)."

"Findings, this study also discusses correlations among the microstructure, superior mechanical properties and deformation mechanisms of SLM-built equiatomic CoCrFeMnNi HEA."

Introduction

"[178], we confirmed that SLM-built equiatomic CoCrFeMnNi HEA has outstanding room temperature mechanical properties."

"Equiatomic CoCrFeMnNi HEA is known to have high phase stability, which will allow it to be used in high temperatures [179]."

"Despite the possibility of its use as a structural material in high temperature or extreme environments, studies conducted on AM-processed equiatomic CoCrFeMnNi HEA have been limited to fundamental level investigations such as manufacturing, microstructural analysis and room temperature tensile properties [180–185]."

"There have been no studies that investigated the influence of the unique microstructure of SLM-built HEA on high-temperature mechanical properties and deformation mechanism."

"Equiatomic CoCrFeMnNi HEA, which has outstanding phase stability, was manufactured using SLM, and its temperature-dependent mechanical properties were investigated."

"This study attempted to identify the room- and high-temperature deformation mechanisms of SLM-built equiatomic CoCrFeMnNi HEA in relation to its unique microstructure."

Results

"Equiatomic CoCrFeMnNi HEA is known to form a Cr-Mn-based oxide or σ-phase of a few μm in size [186, 187]."

"SLM-built HEA did not form a Cr-Mn-based oxide or σ-phase, which can partially hinder the mechanical properties of the alloy, and only Mn_2O_3 tens of nm in size was identified."

"The main reason why SLM-built HEA has outstanding mechanical properties is suspected to be due to the dislocation network, fine grain size and nano-sized Mn_2O_3 that were observed in the initial microstructure."

"Compared to CoCrFeMnNi HEAs manufactured with other manufacturing processes, the SLM-built HEA has higher mechanical properties in room temperature as well as high temperature ranges [188, 189]."

"Grains of hundreds of nm up to a few μm were observed at the grain boundary at 700 °C or higher, the temperature where yield strength drop occurred."

Discussion

"When comparing mechanical properties with EBM-built Ti-48Al-2Cr-2Nb gamma alloy, which is attracting great attention as a high temperature material, SLM-built HEA showed higher yield strength up to ~ 800 °C [190]."

"It is confirmed that the outstanding room temperature mechanical properties of SLM-built HEA can be attributed to fine grain sizes, dislocation networks and nano-sized oxides, and based on this, the authors presented their interpretation of a multiple strengthening mechanism [178]."

"In the range of temperatures above 700 °C, the deformed samples had a clear cellular and columnar structure composed of dislocation networks as well as nano-sized oxides, and this confirmed that the slip and partial RX observed in EBSD are the main deformation mechanisms."

"The reason why a dramatic decrease in flow stress did not occur in SLM-built equiatomic CoCrFeMnNi HEA deformations up to 900 °C is suspected to be due to full recrystallization not occurring easily."

Conclusions

"Correlations among the microstructure, temperature-dependent mechanical properties, and deformation behavior of selective laser-melted equiatomic CoCrFeMnNi high-entropy alloy were investigated."

"SLM-built equiatomic CoCrFeMnNi HEA was confirmed to have outstanding mechanical properties in room to high temperature ranges."

"Fine grain size, dislocation network and nano-sized oxides are suggested to be contributing factors, and in high temperature conditions, dislocation forest hardening and Orowan strengthening were confirmed to be the major strengthening mechanisms."

"Compressive deformation microstructure observation confirmed that slip and DT were the dominant behaviors from room temperature to 600 °C, and slip and DRX were confirmed to be the major deformation mechanisms above 700 °C."

"An extreme flow stress decrease does not occur in SLM-built HEA up to a condition of ~ 900 °C, and this is suspected be the result of a decrease in driving force of recrystallization caused by nano-sized oxides and migrated grains."

Methods

"To observe the initial and compressive deformed microstructures of SLM-built equiatomic CoCrFeMnNi HEA, samples were ground using silicon carbide papers with #400 ~ #1200 grit, and 1 μm diamond suspension, and then mirror polished more using a colloidal silica (CS) suspension."

"To analyze nano-sized precipitates in SLM-built equiatomic CoCrFeMnNi, a Cs-corrected scanning transmission electron microscope (STEM, JEM-ARM200F, Japan) and field emission TEM (FE-TEM, JEM-2100F, Japan) were used."

"To evaluate the temperature-dependent mechanical properties of SLM-built equiatomic CoCrFeMnNi, a cylindrical specimen with a size of Φ 4 mm × height 6 mm was prepared by using electrical discharge machining according to scanning direction (SD)."

"Before the temperature-dependent compressive tests, the SLM-built equiatomic CoCrFeMnNi HEAs were preheated in an electric resistance furnace at a heating rate of 10 °C/min."

Additive Manufacturing of High-Entropy Alloys by Laser Processing [229]

This is a machine-generated summary of:

Ocelík, V.; Janssen, N.; Smith, S. N.; De Hosson, J. Th. M.: Additive Manufacturing of High-Entropy Alloys by Laser Processing [229]

Published in: JOM (2016)

Link to original: https://doi.org/10.1007/s11837-016-1888-z

If you want to cite the papers, please refer to the original.

For technical reasons we could not place the page where the original quote is coming from.

Abstract-Summary

"This contribution concentrates on the possibilities of additive manufacturing of high-entropy clad layers by laser processing."

"The effects of the laser surface processing parameters on the microstructure and hardness of high-entropy alloys (HEAs) were examined."

"AlCoCrFeNi alloys with different amounts of aluminum prepared by arc melting were investigated and compared with the laser beam remelted HEAs with the same composition."

"It is concluded that a high-power laser is a versatile tool to synthesize interesting HEAs with additive manufacturing processing."

Introduction

"[191] We have noticed also disadvantages of these direct deposition methods of HEAs, namely, the evaporation of certain elements, different rates of intake of the powders, as well as dilution of the coating materials by the substrate during laser cladding."

"In the case when pre-alloyed HEA powders are used [192], a wider range of scanning speeds could be used to study the influence of solidification rates on the microstructure and properties of laser-deposited HEAs."

"Comparative study of arc melted and laser beam fabricated $Al_xCoCrFeNi$ ($x = 0.3, 0.6$, and 0.85) HEAs [193] shows similar microstructures for low and high Al content alloys with a strong ⟨001⟩ fiber texture along the direction of solidification."

"Direct laser deposition of HEA coatings from elemental powders mixture should prove an ability of this method for fast and versatile HEA surface coatings fabrication."

Experimental Procedure

"Energy dispersive X-ray spectroscopy (EDS) analysis performed after sample preparation reveals different compositions in comparison with the nominal compositions."

"The second set of specimens was prepared by surface laser remelting of the samples prepared as mentioned earlier."

"A laser beam spot with a diameter of 1.5 mm (Gaussian distribution) was used to scan the surface with speeds of 5 and 10 mm/s and laser power of 300 and 450 W, respectively."

"A third group of samples was prepared by direct laser beam deposition on the SS 305 steel bar substrate using the laser beam scanning speed of 5 mm/s and laser power between 600 and 650 W. A premixture of pure elemental powders (particles size between 5 μm and 120 μm) with equimolar nominal compositions of AlCoCrFeNi and AlCrFeNiTa was fed by a Sulzer Metco Twin10C powder feeder using a side cladding nozzle and forming a single layer coating with a thickness of about 500–800 μm."

Results and Discussion

"The composition of the fcc phase in Al_0.7 and Al_1.0 is also very close to each other, but the composition of the bcc phase in these two is different."

"To investigate the microstructure and properties of HEAs with the same compositions, but faster solidification, remelted tracks were made on the original arc-melted samples using a moving laser beam."

"The phase fractions and distribution of fcc and bcc phases inside the remelted zones differ substantially, as well as the elemental compositions of fcc and bcc phases."

"The amount of the bcc phase in the remelted zone has increased to 7.5% in comparison with only 3% in the nonremelted Al_0.7 sample."

"This difference with the phases present in the original vacuum arc-melted sample (bcc phase fraction of 35%) finds its origin in the high solidification rate after laser remelting."

Conclusion

"From the laser remelting experiment, we see that the solidification rate is a very important factor for determination of the presence and distribution of phases in AlCoCrFeNi HEAs."

"The high thermal stability of AlCoCrFeNi base HEA has been confirmed by the absence of a heat-affected zone in the vicinity of the remelting track."

"It goes without saying that the success of direct laser deposition of HEAs coatings using a feeding of mixture of elemental powders is affected by several materials and processing parameters, e.g., in-take of particular powders, melting temperatures, laser beam velocity, and power density."

Synthesis of $CoCrFeNiMnW_{0.25}$ High-Entropy Alloy Powders by Mechanical Alloying and Plasma Spheroidization Processes for Additive Manufacturing [230]

This is a machine-generated summary of:

Makhmutov, Tagir; Razumov, Nikolay; Kim, Artem; Ozerskoy, Nikolay; Mazeeva, Alina; Popovich, Anatoliy: Synthesis of $CoCrFeNiMnW_{0.25}$ High-Entropy Alloy Powders by Mechanical Alloying and Plasma Spheroidization Processes for Additive Manufacturing [230]

Published in: Metals and Materials International (2020)

Link to original: https://doi.org/10.1007/s12540-020-00747-0

If you want to cite the papers, please refer to the original.

For technical reasons we could not place the page where the original quote is coming from.

Abstract-Summary

"Samples obtained from low-energy MA-powders have an inhomogeneous microstructure and phase composition."

"Samples obtained from MA-powders with homogeneous structures have a high degree of sphericity and chemical homogeneity; the face-centered cubic solid solution represents the phase composition of the powders."

Introduction

"The variability of the compositions allows to obtain HEAs with unique properties such as high-temperature strength, corrosion resistance, wear-resistance, and fracture-resistance at cryogenic temperatures [155, 194–196]."

"The equiatomic FeCrMnNiCo alloy was one of the first HEAs obtained by Cantor in his advanced work devoted to equiatomic multicomponent alloys [28]."

"Many variations of this alloy with various alloying elements and produced by various manufacturing methods were obtained [197]."

"Not all existing methods for producing powders for AM make it possible to obtain powders doped with refractory elements because of their high melting point."

"Mechanical alloying in high-energy mills and attritors makes it possible to obtain alloys bypassing the melting processes [198], and the following plasma spheroidization (PS) provides a high rate of the powders sphericity and makes them suitable for use in AM [199]."

"Work, the CoCrFeNiMn alloy with W addition was chosen as a promising one for the manufacturing of corrosion-resistant units with high strength and plasticity."

Materials and Methods

"The calculation of the optimal alloy composition was carried out using the CALPHAD method."

"Mechanical alloying was carried out in Fritsch Pulverisette 4 planetary mill in argon atmosphere at the following modes: MA duration: 5–20 h; the main disk rotation speed: 200–400 rpm; the bowl rotation speed: 400–800 rpm (rotation opposite to the main disk); every 15 min the milling was stopped for 15 min to prevent overheating."

"Steel balls of 10 mm in diameter were used as the milling media with the ball-to-powder ratio of 20:1."

"The MA-powders were sieved using a 160 μm sieve and spheroidized using a Tekna TEK-15 (15 kW) plant with inductively coupled plasma."

"The phase composition was analyzed with Bruker D8 Advance X-ray diffractometer (XRD) using Cu-Kα ($\lambda = 1.5418$ Å) irradiation."

Results and Discussion

"The powders obtained by MA have an irregular shape typical for milling processes."

"The samples obtained at low-energy mode have an inhomogeneous structure with CoCrFeNiMn matrix and irregularly distributed inclusions of W up to several microns in size."

"A homogeneous structure with uniformly distributed submicron inclusions of W is typical for powders obtained at high-energy conditions."

"It may take place due to the irregular distribution of W over the volume of the MA-powders, where there are W inclusions of various sizes, that have a higher (compared to the CoCrFeNiMn alloy) melting point."

Conclusions

"Spherical $CoCrFeNiMnW_{0.25}$ HEA powders obtained using mechanical alloying (MA), followed by plasma spheroidization (PS), have different phase compositions and microstructure depending on the MA-modes."

"A CoCrFeNiMn matrix represents the microstructure and phase composition of spherical powders after PS with inclusions of a solid solution based on W. High-energy MA followed by PS allowed to obtain spherical powders with a homogeneous chemical and phase composition."

"The phase composition of these powders is represented by a solid solution with the FCC lattice; and the lattice parameter of the final powder is 3.605 Å. Time-varying of the MA-process allowed to identify the time and doping mode appropriate to obtain homogeneous powders after PS."

Investigation of the Structure and Properties of a Coating from a High-Entropy FeCoCrNiMn Alloy Obtained by a Wire Arc Additive Manufacturing [231]

This is a machine-generated summary of:

Gromov, V. E.; Konovalov, S. V.; Ivanov, Yu. F.; Efimov, M. O.; Panchenko, I. A.; Vorob'ev, S. V.: Investigation of the Structure and Properties of a Coating from a High-Entropy FeCoCrNiMn Alloy Obtained by a Wire Arc Additive Manufacturing [231]

Published in: Physics of the Solid State (2022)

Link to original: https://doi.org/10.1134/s1063783422070034

If you want to cite the papers, please refer to the original.

For technical reasons we could not place the page where the original quote is coming from.

Abstract-Summary

"A coating of high-entropy Cantor alloy FeCoCrNiMn of nonequiatomic composition was formed on a 5083 aluminum alloy substrate by wire-arc additive manufacturing (WAAM)."

"The deposition of the FeCoCrNiMn high-entropy coating on the 5083 alloy surface is accompanied by the formation of microhardness and elemental composition gradients."

"A transition layer up to 450 μm thick is formed at the interface between the coating and the substrate."

"We analyzed the elemental composition gradient of the transition layer and noted a high level of chemical homogeneity of the coating."

"The found doping of the coating with substrate elements (aluminum) leads to the formation of a FeC-oCrNiMnAl high-entropy coating, causing a lamellar structure at the interface between the transition layer and the substrate."

Introduction

"The attention of researchers in the field of physical materials science has been drawn to the study of high-entropy alloys (HEAs)."

"Articles, analytical reviews, and monographs consider the original results in the field of HEAs in detail, discuss the microstructure, properties, and thermodynamics of HEAs, assess the methods for simulating HEA structure, and offer new methods for obtaining multicomponent alloys [200, 201]."

"One of the first high-entropy alloys under study is the FeCoCrNiMn alloy, which can maintain a face-centered cubic structure in a wide temperature range and has a good balance of strength and ductility [202]."

"One of the aspects of technical progress is the development and application of new HEAs with high mechanical, cryo- and heat-resistant, corrosion-resistant, special magnetic and electrical properties, as well as their compounds (carbides, nitrides, oxides, borides, silicides) [203, 204]."

"This work analyzes the structure and properties of the Fe-CoCrNiMn Kantor HEA coating deposited on the 5083 alloy."

Experimental

"Samples of the coating–substrate system were used as the objects of research."

"The coating was a high-entropy alloy of nonequiatomic elemental composition FeCoCrNiMn, formed on the substrate using the wire electric arc additive manufacturing (WAAM) [203, 205, 206]."

"The structure and elemental composition of the coating and the substrate layer adjacent to the coating were studied by scanning electron microscopy (SEM) using an LEO EV50 microscope (Carl Zeiss) with an energy-dispersive analyzer."

"The properties of the coating and substrate were characterized by microhardness using an HV-1000 instrument at an indenter load of 0.5 N and wear resistance using a TRIBOtester instrument with factors: normal load 2 N, rotation speed 25 mm/s, friction path length 100 m, the radius of the friction track is 2 mm; the counterbody was a tungsten carbide ball with a diameter of 6 mm; tests were performed by dry friction in air at room temperature."

Results and Discussion

"Tribological tests performed on the longitudinal section of the coating (section parallel to the coating/substrate interface) in its middle part showed that the wear factor of the coating is 2.3×10^{-4} mm^3/N m, and its friction coefficient is 0.7."

"The study of the cross section of the coating revealed micropores and microcracks."

"Depending on the degree of etching, the coating can be divided into three sublayers."

"The difference in the degree of etching indicates the inhomogeneity of the elemental composition of the coating."

"Significant concentration changes were revealed only for aluminum, which is an element of the substrate and got into the coating by diffusion during the formation of the coating–substrate system."

Conclusions

"Using scanning electron microscopy, we studied the structure, elemental composition, mechanical, and tribological properties of the HEA coating–5083 alloy substrate system formed by WAAM by determining its microhardness and wear resistance."

"The following main results have been obtained: —Deposition of HEA on the 5083 alloy surface is accompanied by the formation of a gradient structure characterized by a regular change in microhardness, elemental, and phase composition; —A stepwise nature of the change in the microhardness of the HEA coating–5083 alloy substrate system was revealed; —The formation of a transition layer up to 450 μm thick, located at the interface between the coating and the substrate, was found; —Dopping of the coating with substrate elements (aluminum) was observed, leading to the formation of a HEA of the chemical composition Al–Mn–Fe–Cr–Co–Ni; —The formation of a lamellar structure in the substrate layer adjacent to the coating was revealed."

References

1. M.E. Bloomfield, K.A. Christofidou, F. Monni, Q. Yang, M. Hang, and N.G. Jones, The Influence of Fe Variations on the Phase Stability of CrMnFexCoNi Alloys Following Long-Duration Exposures at Intermediate Temperatures, Intermetallics, 2021, 131, p 107108.
2. J.W. Yeh, S.K. Chen, S.J. Lin, J.Y. Gan, T.S. Chin, T.T. Shun, C.H. Tsau, and S.Y. Chang, Adv. Eng. Mater. 6, 299 (2004).
3. J.W. Yeh, Ann. Chim. Sci. Mater. 31, 633 (2006).
4. K. Jin, B.C. Sales, G.M. Stocks, G.D. Samolyuk, M. Daene, W.J. Weber, Y. Zhang, and H. Bei, Tailoring the Physical Properties of Ni-Based Single-Phase Equiatomic Alloys by Modifying the Chemical Complexity, Sci. Rep., 2016, 6, p 20159.
5. Z. Wu, H. Bei, G.M. Pharr, and E.P. George, Temperature Dependence of the Mechanical Properties of Equiatomic Solid Solution Alloys with Face-Centered Cubic Crystal Structures, Acta Mater., 2014, 81, p 428–441.
6. B. Gludovatz, A. Hohenwarter, K.V.S. Thurston, H.B. Bei, Z.G. Wu, E.P. George, and R.O. Ritchie, Exceptional Damage-Tolerance of a Medium-Entropy Alloy CrCoNi at Cryogenic Temperatures, Nat. Commun., 2016, 7, p 10602.
7. K. Jin, C. Zhang, F. Zhang, H.B. Bei, Influence of compositional complexity on interdiffusion in Ni-containing concentrated solid-solution alloys. Mater. Res. Lett. 6(5), 293–299 (2018)
8. X.G. Lu, M. Selleby, and B. Sundman, Assessments of Molar Volume and Thermal Expansion for Selected bcc, fcc and hcp Metallic Elements, Calphad, 2005, 29, p 68–89.

9. D.D. Huang, S.H. Liu, and Y. Du, Modeling on the Molar Volume of the Al-Cu-Mg-Si System, Calphad, 2020, 68, p 101693.
10. PanHEA, Database for Multi-component High Entropy Alloys, www.computherm.com .
11. L. Huang, S. Liu, Y. Du, and C. Zhang, Thermal Conductivity of the Mg–Al–Zn Alloys: Experimental Measurement and CALPHAD Modeling, Calphad, 2018, 62, p 99–108.
12. O. N. Senkov, C. Woodward, and D. B. Miracle, "Microstructure of aluminum-containing refractory high-entropy alloys," JOM 66, 2030–2042 (2014).
13. E. Eshed, N. Larianovsky, A. Kovalevsky, V. Popov, Jr., I. Gorbachev, V. Popov, and A. Katz-Demyanetz, "Microstructural evolution and phase formation in 2nd-generation refractory-based high entropy alloys," Materials 11, 175–187 (2018).
14. F. Zhang, C. Zhang, S. L. Chen, J. Zhu, W. S. Cao, and U. R. Kattner, "An understanding of high entropy alloys from phase diagram calculations, "CALPHAD 45, 1–10 (2014).
15. C. Guo, C. Li, X. Zheng, and Zh. Du, "Thermodynamic modeling of the Fe–Ti–V system," CALPHAD 38, 155–160 (2012).
16. J. Cui, C. Guo, L. Zou, Ch. Li, and Zh. Du, "Experimental investigation and thermodynamic modeling of the Ti–V–Zr system," CALPHAD 55 (Part 2), 189–198 (2016).
17. A. V. Khvan, K. Chang, and B. Hallstedt, "Thermodynamic assessment of the Fe–Nb–V system," CALPHAD 43, 143–148 (2013).
18. Y.S. Na, N.K. Park, and R.C. Reed, Sigma Morphology and Precipitation Mechanism in Udimet 720Li, Scr. Mater., 2000, 43(7), p 585–590
19. M. Simonetti and P. Caron, Role and Behaviour of μ Phase During Deformation of a Nickel-Based Single Crystal Superalloy, Mater. Sci. Eng., A, 1998, 254(1), p 1–12
20. W.H. Liu, Z.P. Lu, J.Y. He, J.H. Luan, Z.J. Wang, B. Liu, Y. Liu, M.W. Chen, and C.T. Liu, Ductile CoCrFeNiMox High Entropy Alloys Strengthened by Hard Intermetallic Phases, Acta Mater., 2016, 116, p 332–342
21. T.-T. Shun, L.-Y. Chang, and M.-H. Shiu, Age-Hardening of the CoCrFeNiMo0.85 High-Entropy Alloy, Mater. Charact., 2013, 81, p 92–96
22. Takeuchi A ,and Inoue A, Mater Trans 42 (2001) 1435.
23. Y. Zhang, Y.J. Zhou, J.P. Lin, G.L. Chen, and P.K. Liaw, Adv. Eng. Mater. 10, 534 (2008).
24. H.L. Lukas, S.G. Fries, and B. Sundman, Computational Thermodynamics: The CALPHAD Method. Cambridge University Press, Cambridge, 2007.
25. Srinivasa Rao B, Bhatt J, and Murty B S, Mater Sci Eng A 449–451 (2007) 211.
26. Niessen A K, de Boer F R, Boom R, de Chatel P F, Mattens W C M, and Miedema A R, CALPHAD 7 (1983) 51.
27. Murty B S, and Ranganathan S, Int Mater Rev 43 (1998) 101.
28. B. Cantor, I. T. H. Chang, P. Knight, et al., "Microstructural development in equiatomic multicomponent alloys," Mater. Sci. Eng. A, 375, 213–218 (2004).
29. M. C. Gao and D. E. Alman, "Searching for next single-phase high-entropy alloy composition," Entropy, 15, No. 10, 4504–4519 (2013).
30. O.N. Senkov, J.M. Scott, S.V. Senkova, D.B. Miracle, and C.F. Woodward, J. Alloys Compd. 509, 6043 (2011).
31. O.N. Senkov, G.B. Wilks, D.B. Miracle, C.P. Chuang, and P.K. Liaw, Intermetallics 18, 1758 (2010).
32. M.C. Gao, C. Niu, C. Jiang, and D.L. Irving: in High-Entropy Alloys: Fundamentals and Applications, M.C. Gao, J.W. Yeh, P.K. Liaw, and Y. Zhang, eds., Springer, Berlin, 2015.
33. M. Widom: in High-Entropy Alloys: Fundamentals and Applications, M.C. Gao, J.W. Yeh, P.K. Liaw, and Y. Zhang, eds., Springer, Cham, 2015.
34. M.G. Poletti and L. Battezzati: Acta Mater., 2014, vol. 75, pp. 297–306.
35. Y. Zhang, Z.P. Lu, S.G. Ma, P.K. Liaw, Z. Tang, Y.Q. Cheng and M.C. Gao: MRS Comm., 2014, vol. 4, pp. 57–62.
36. Feng, R. et al. Design of light-weight high-entropy alloys. Entropy 18, 333 (2016).
37. Gao, M. C. et al. Thermodynamics of concentrated solid solution alloys. Curr. Opin. Solid State Mater. Sci. 21, 238–251 (2017).

38. Turchi, P., Kaufman, L. & Liu, Z. Modelling of Ni-Cr-Mo based alloys: part i-phase stability. CALPHAD 30, 70–87 (2006).
39. Molokanov, V. V., Chernov, D. B. & Budberg, P. B. Calculations of the phase diagrams for the vanadium-chromium, vanadium-niobium, and chromium-niobium systems in terms of the regular solution approximation. Russ. J. Phys. Chem. 51, 1181–1183 (1977).
40. Moruzzi, V. L., Janak, J. F. & Schwarz, K. Calculated thermal properties of metals. Phys. Rev. B. 37, 790–799 (1988).
41. Gurel S, Yagci MB, Bal B, Canadinc D (2020) Corrosion behavior of novel Titanium-based high entropy alloys designed for medical implants. Mater Chem Phys 254:123377
42. Gurel S, Yagci MB, Canadinc D et al (2020) Fracture behavior of novel biomedical Ti-based high entropy alloys under impact loading. Mater Sci Eng A 140456. https://doi.org/10.1016/j.msea.2020.140456
43. Motallebzadeh A, Peighambardoust NS, Sheikh S et al (2019) Microstructural, mechanical and electrochemical characterization of TiZrTaHfNb and Ti1.5ZrTa0.5Hf0.5Nb0.5 refractory high-entropy alloys for biomedical applications. Intermetallics 113:106572. https://doi.org/10.1016/j.intermet.2019.106572
44. Gurel S, Nazarahari A, Canadinc D et al (2021) Assessment of biocompatibility of novel TiTaHf-based high entropy alloys for utility in orthopedic implants. Mater Chem Phys 266:124573. https://doi.org/10.1016/j.matchemphys.2021.124573
45. Geetha M, Singh AK, Asokamani R, Gogia AK (2009) Ti based biomaterials, the ultimate choice for orthopaedic implants—a review. Prog Mater Sci 54:397–425. https://doi.org/10.1016/j.pmatsci.2008.06.004
46. Song Y, Xu DS, Yang R et al (1999) Theoretical study of the effects of alloying elements on the strength and modulus of β-type bio-titanium alloys. Mater Sci Eng A 260:269–274. https://doi.org/10.1016/s0921-5093(98)00886-7
47. Diao Y, Yan L, Gao K (2021) Improvement of the machine learning-based corrosion rate prediction model through the optimization of input features. Mater Des 198:109326. https://doi.org/10.1016/j.matdes.2020.109326
48. Schmidt J, Marques MALMRG, Botti S, Marques MALMRG (2019) Recent advances and applications of machine learning in solid-state materials science. npj Comput Mater 5:1–36. https://doi.org/10.1038/s41524-019-0221-0
49. Gregorutti B, Michel B, Saint-Pierre P (2017) Correlation and variable importance in random forests. Stat Comput 27:659–678. https://doi.org/10.1007/s11222-016-9646-1
50. Astm G (2004) G 31–72: Standard Practice for Laboratory Immersion Corrosion Testing of Metals. Annu B Stand
51. Oliver WC, Pharr GM (1992) An improved technique for determining hardness and elastic modulus using load and displacement sensing indentation experiments. J Mater Res 7:1564–1583
52. Wang M, Ma ZL, Xu ZQ, Cheng XW (2021) Designing VxNbMoTa refractory high-entropy alloys with improved properties for high-temperature applications. Scr Mater 191:131–136. https://doi.org/10.1016/j.scriptamat.2020.09.027
53. Chen SY, Tong Y, Tseng KK et al (2019) Phase transformations of HfNbTaTiZr high-entropy alloy at intermediate temperatures. Scr Mater 158:50–56. https://doi.org/10.1016/j.scriptamat.2018.08.032
54. Masemola K, Popoola P, Malatji N (2020) The effect of annealing temperature on the microstructure, mechanical and electrochemical properties of arc-melted AlCrFeMnNi equi-atomic High entropy alloy. J Mater Res Technol 9:5241–5251. https://doi.org/10.1016/j.jmrt.2020.03.050
55. Senkov, O. N., Senkova, S. V., Miracle, D. B. & Woodward, C. Mechanical properties of low-density, refractory multi-principal element alloys of the Cr-Nb-Ti-V-Zr system. Mater. Sci. Eng. A 565, 51–62 (2013).
56. Kang, B., Lee, J., Ryu, H. J. & Hong, S. H. Ultra-high strength WNbMoTaV high-entropy alloys with fine grain structure fabricated by powder metallurgical process. Mater. Sci. Eng. A 712, 616–624 (2018).

57. Liu, Y. et al. Microstructure and mechanical properties of refractory HfMo0.5NbTiV0.5Sixhigh-entropy composites. J. Alloy. Compd. 694, 869–876 (2017).
58. Maiti, S. & Steurer, W. Structural-disorder and its effect on mechanical properties in single-phase TaNbHfZr high-entropy alloy. Acta Mater. 106, 87–97 (2016).
59. Senkov, O. N., Isheim, D., Seidman, D. N. & Pilchak, A. L. Development of a refractory high entropy superalloy. Entropy https://doi.org/10.3390/e18030102 (2016).
60. Zhang, M., Zhou, X. & Li, J. Microstructure and mechanical properties of a refractory CoCrMoNbTi high-entropy alloy. J. Mater. Eng. Perform. 26, 3657–3665 (2017).
61. Zhang, Y., Yang, X. & Liaw, P. K. Alloy design and properties optimization of high-entropy alloys. JOM 64, 830–838 (2012).
62. Miracle, D. B. High entropy alloys as a bold step forward in alloy development. Nat. Commun. 10, 1–3 (2019).
63. Maresca, F. & Curtin, W. A. Mechanistic origin of high strength in refractory BCC high entropy alloys up to 1900K. Acta Mater. 182, 235–249 (2020).
64. Pudil, P., Novovičová, J. & Kittler, J. Floating search methods in feature selection. Pattern Recognit. Lett. 15, 1119–1125 (1994).
65. Ferri, F. J., Pudil, P., Hatef, M. & Kittler, J. Comparative study of techniques for large-scale feature selection. Pattern Recognit. Pract. IV 16, 403–413 (1994).
66. Storn, R. & Price, K. Differential evolution—a simple and efficient heuristic for global optimization over continuous spaces. J. Glob. Optim. 11, 341–359 (1997).
67. Virtanen, P. et al. SciPy 1.0: fundamental algorithms for scientific computing in Python. Nat. Methods 17, 261–272 (2020).
68. Körmann, F., Ruban, A. V. & Sluiter, M. H. Long-ranged interactions in bcc NbMoTaW high-entropy alloys. Mater. Res. Lett. 5, 35–40 (2017).
69. Körmann, F. & Sluiter, M. H. Interplay between lattice distortions, vibrations and phase stability in NbMoTaW high entropy alloys. Entropy 18, 403 (2016).
70. Singh, P., Smirnov, A. V. & Johnson, D. D. Ta-Nb-Mo-W refractory high-entropy alloys: anomalous ordering behavior and its intriguing electronic origin. Phys. Rev. Mater. 2, 055004 (2018).
71. Wang, Y. et al. Computation of entropies and phase equilibria in refractory V-Nb-Mo-Ta-W high-entropy alloys. Acta Mater. 143, 88–101 (2018).
72. Widom, M., Huhn, W. P., Maiti, S. & Steurer, W. Hybrid Monte Carlo/molecular dynamics simulation of a refractory metal high entropy alloy. Mat. Met. Trans. A. 45, 196–200 (2013).
73. Widom, M. Prediction of structure and phase transformations. In High-Entropy Alloys, (eds. Gao, M. C., Yeh, J.-W., Liaw, P. K. & Zhang, Y.) 267–298 (Springer, Cham, 2016).
74. Huhn, W. P. & Widom, M. Prediction of A2 to B2 phase transition in the high-entropy alloy Mo-Nb-Ta-W. JOM 65, 1772–1779 (2013).
75. Huhn, W. P. Thermodynamics From First Principles: Prediction of Phase Diagrams and Materials Properties Using Density Functional Theory. Ph.D. thesis. http://repository.cmu.edu/dissertations/369. (2014).
76. Shapeev, A. Accurate representation of formation energies of crystalline alloys with many components. Comput. Mater. Sci. 139. 26–30 (2017).
77. E. Alpaydin: MIT Press, 2020.
78. B. Schuh, F. Mendez-Martin, B. Völker, E.P. George, H. Clemens, R. Pippan, and A. Hohenwarter: Acta Mater., 2015, vol. 96, pp. 258–68.
79. D.H. Lee, J.A. Lee, Y. Zhao, Z. Lu, J.Y. Suh, J.Y. Kim, and J.I. Jang: Acta Mater., 2017, vol. 140, pp. 443–51.
80. S. Varalakshmi, M. Kamaraj, and B.S. Murty: Mater. Sci. Eng. A., 2010, vol. 527, pp. 1027–30.
81. W.H. Liu, Y. Wu, J.Y. He, T.G. Nieh, and Z.P. Lu: Scr. Mater., 2013, vol. 68, pp. 526–29.
82. J. Schmidhuber: Neural Netw., 2015, vol. 61, pp. 85–117.
83. Y. Zhang, S. Yang, and J. Evans: Acta Mater., 2008, vol. 56. pp. 1094–105.
84. M. Rupp, A. Tkatchenko, K.R. Müller, and O.A.V. Lilienfeld: Phys. Rev. Lett., 2012, vol. 108, pp. 058301.
85. B. Meredig, and C. Wolverton: Nat. Mater., 2013, vol. 12, pp. 123.

86. L. Ward, A. Agrawal, A. Choudhary, and C. Wolvertonv: NPJ Comput. Mater., 2016, vol. 2, p. 16028.
87. Y. Liu, J. Zhang, and J. Zhong: Comput. Mater. Sci., 2008, vol. 43, pp. 752–58.
88. H. Bhadeshiam: Encycl. Mater. Sci. Technol., 2008, vol. 10, pp. 1–5.
89. Sozerdem M, Kolukisa S (2009) Mater Des 30:764–69
90. N. Islam, W.J. Huang, and H.L.L. Zhuang: Comput. Mater. Sci., 2018, vol. 150, pp. 230–35.
91. Wang H, Zhang LF, Han JQ, Weinan E (2018) Comput. Phys Commun 228:178–84
92. Zhang LF, Wang H, Weinan E (2018) J Chem Phys 148:124113
93. L. Li, H. Chen, Q. Fang, J. Li, F. Liu, Y. Liu, and P.K. Liaw: Intermetallics, 2020, vol. 120, pp. 106741.
94. J. Li, B. B. Xie, Q. H. Fang, B. Liu, Y. Liu, and P. K. Liaw: J. Mater. Sci. Technol., 2021, vol. 68, pp. 70–5.
95. J. Li, L. Li, C. Jiang, Q. Fang, F. Liu, Y. Liu, and P.K. Liaw: J. Mater. Sci. Technol., 2020, vol. 57, pp. 85–91
96. T. H. Fang, W. L. Li, N. R. Tao, and K. Lu: Science, 2011, vol. 6024, pp. 1587.
97. Miracle, D. B. & Senkov, O. N. A critical review of high entropy alloys and related concepts. Acta Mater. 122, 448–511 (2017).
98. MacDonald, B. E. et al. Rerch. JOM 69, 2024–2031 (2017).
99. Mizutani, U. Hume-Rothery Rules for Structurally Complex Alloy Phases. (CRC Press: Boca Raton, 2016). 323–329.
100. Guo, S. & Liu, C. T. Phase stability in high entropy alloys: Formation of solid-solution phase or amorphous phase. Prog. Nat. Sci.: Mater. Int. 21, 433–446 (2011).
101. X. Yang and Y. Zhang, Mater. Chem. Phys. 132, 233 (2012).
102. Guo, S., Hu, Q., Ng, C. & Liu, C. T. More than entropy in high-entropy alloys: Forming solid solutions or amorphous phase. Intermetallics 41, 96–103 (2013).
103. Rajan, K. Materials informatics: an introduction. Inform. Mater. Sci. Eng. Data-Driven Discov. Accel. Exp. Appl. 8, 1–16 (2013).
104. Senkov, O. N., Miller, J. D., Miracle, D. B. & Woodward, C. Accelerated exploration of multi-principal element alloys for structural applications. Calphad Comput. Coupling Phase Diagr. Thermochem. 50, 32–48 (2015).
105. Inoue, A. Stabilization of metallic supercooled liquid and bulk amorphous alloys. Acta Materialia 48, 279–306 (2000).
106. Guo S (2015) Phase selection rules for cast high entropy alloys: an overview. Mater Sci Technol (United Kingdom) 31:1223–1230. https://doi.org/10.1179/1743284715Y.0000000018
107. Sarker, P. et al. High-entropy high-hardness metal carbides discovered by entropy descriptors. Nat. Commun. 9, 4980 (2018).
108. Senkov, O. N., Miracle, D. B., Chaput, K. J. & Couzinie, J.-P. Development and exploration of refractory high entropy alloys—A review. J. Mater. Res. 33, 3092–3128 (2018).
109. Knapp, T. R. Canonical correlation analysis: a general parametric significance-testing system. Psychol. Bull. 85, 410–416 (1978).
110. Rickman, J. M., Wang, Y., Rollett, A. D., Harmer, M. P. & Compson, C. Data analytics using canonical correlation analysis and Monte Carlo simulation. npj Comput. Mater. 3, 26 (2017).
111. Jobson, J. D. Applied Multivariate Data Analysis, Vol. II (Springer-Verlag, New York, 1992).
112. Whitley, D. A genetic algorithm tutorial. Stat. Comput. 4, 65–85 (Springer, New York, NY, 1994).
113. Michalewicz, Z. Genetic Algorithms+Data Structures=Evolution Programs, 3rd edn (Springer, New York, 1996).
114. Sivanandam, S. N. & Deepa, S. N. Introduction to Genetic Algorithms (Springer-Verlag, Berlin, 2008).
115. G.L. Chen and C.T. Liu: Int. Mater. Rev., 2001, vol. 46, pp. 253–60.
116. X.F. Wang, Y. Zhang, Y. Qiao, and G.L. Chen, Intermetallics 15, 357 (2007).
117. Y.J. Zhou, Y. Zhang, Y.L. Wang, and G.L. Chen: Appl. Phys. Lett., 2007, vol. 90, pp. 181904-1-3.
118. C. Li, J.C. Li, M. Zhao, and Q. Jiang: J. Alloy Compd., 2010, vol. 504S, pp. S516–18.

119. J.W. Qiao, Y. Zhang, and P.K. Liaw: Adv. Eng. Mater., 2008, vol. 10, pp. 1039–42.
120. J.W. Qiao, A.C. Sun, E.W. Huang, Y. Zhang, P.K. Liaw, and C.P. Chuang: Acta Mater., 2011, vol. 59, pp. 4126–37.
121. B.S. Murty, J.W. Yeh, S. Ranganathan, High-Entropy Alloys (Butterworth-Heinemann, Oxford, 2014)
122. J.W. Yeh, S.K. Chen, S.J. Lin, J.Y. Gan, T.S. Chin, T.T. Shun, C.H. Tsau, S.Y. Chang, Nanostructured high-entropy alloys with multiple principal elements: novel alloy design concepts and outcomes. Adv. Eng. Mater. 6, 299 (2004)
123. C.M. Lin, H.L. Tsai, H.Y. Bor, Effect of aging treatment on microstructure and properties of high-entropy Cu0.5CoCrFeNi alloy. Intermetallics 18, 1244 (2010)
124. S.T. Chen, W.Y. Tang, Y.F. Kuo, S.Y. Chen, C.H. Tsau, T.T. Shun, J.W. Yeh, Microstructure and properties of age-hardenable AlxCrFe1.5MnNi0.5 alloys. Mater. Sci. Eng. A 527, 5818 (2010)
125. C.Y. Hsu, C.C. Juan, T.S. Sheu, S.K. Chen, J.W. Yeh, Effect of aluminum content on microstructure and mechanical properties of AlxCoCrFeMo0.5Ni high-entropy alloys. J. Met. 65, 1840 (2013)
126. C.Y. Hsu, T.S. Sheu, J.W. Yeh, S.K. Chen, Effect of iron content on wear behavior of AlCoCrFexMo0.5Ni high-entropy alloys. Wear 268, 653 (2010)
127. J.M. Wu, S.J. Lin, J.W. Yeh, S.K. Chen, Y.S. Huang, H.C. Chen, Adhesive wear behavior of AlxCoCrCuFeNi high-entropy alloys as a function of aluminum content. Wear 261, 513 (2006)
128. M.H. Chuang, M.H. Tsai, W.R. Wang, S.J. Lin, J.W. Yeh, Microstructure and wear behavior of AlxCo1.5CrFeNi1.5Tiy high-entropy alloys. Acta Mater. 59, 6308 (2011)
129. C.Y. Hsu, J.W. Yeh, S.K. Chen, T.T. Shun, Wear resistance and high-temperature compression strength of FCC CuCoNiCrAl0.5Fe alloy with boron addition. Metall. Mater. Trans. A 35, 1465 (2004)
130. C.P. Lee, C.C. Chang, Y.Y. Chen, J.W. Yeh, H.C. Shih, Effect of the aluminium content AlxCrFe1.5MnNi0.5 high-entropy alloys on the corrosion behavior in aqueous environments. Corros. Sci. 50, 2053 (2008)
131. P.K. Liaw, Z. Tang, L. Huang, W. He, Alloying and processing effects on the aqueous corrosion behavior of high-entropy alloys. MDPI-Entropy 16, 895 (2014)
132. S. Frankel, Introduction to Metallurgically Influenced Corrosion, ASM Handbook, Corrosion Fundamentals, Testing, and Protection, vol. 13 (ASM International, Novelty, 2003), p. 257
133. M.J. Yao, K.G. Pradeep, C.C. Tasan, D. Raabe, A novel, single phase, non- equiatomic FeMnNiCoCr high-entropy alloy with exceptional phase stability and tensile ductility. Scripta Mater. 72, 5 (2014)
134. M.S. Lucas, G.B. Wilks, L. Mauger, J.A. Munoz, O.N. Senkov, E. Michel, J. Horwath, S.L. Semiatin, M.B. Stone, D.L. Abernathy, E. Karapetrova, Absence of long-range chemical ordering in equimolar FeCoCrNi. Appl. Phys. Lett. 100, 251907 (2012)
135. Y.F. Kao, T.D. Lee, S.K. Chen, Y.S. Chang, Electrochemical passive properties of AlxCoCrFeNi ($x = 0, 0.25, 0.50, 1.00$) alloys in sulfuric acids. Corros. Sci. 52, 1026 (2010)
136. K. Sieradzki, R. Newman, Percolation model for passivation in stainless steels. J. Electrochem. Soc. 133, 1979 (1986)
137. Y.L. Chou, Y.C. Wang, J.W. Yeh, H.C. Shih, Pitting corrosion of the high entropy alloy $Co_{1.5}CrFeNi_{1.5}Ti_{0.5}Mo_{0.1}$ in chloride-containing sulphate solutions. Corros. Sci. 52, 3481 (2010)
138. Y. Zhang, T.T. Zuo, Z. Tang, M.C. Gao, K.A. Dahmen, P.K. Liaw, Z.P. Lu, Microstructures and properties of high-entropy alloys. Prog. Mater Sci. 61, 1 (2014)
139. Y.F. Kao, T.J. Chen, S.K. Chen, J.W. Yeh, Microstructure and mechanical property of as-cast, -homogenized, and -deformed AlxCoCrFeNi ($0 \leq x \leq 2$) high- entropy alloys. J. Alloys Compd. 488, 57 (2009)
140. G.R. Holcomb, J. Tylczak, C. Carney, Oxidation of CoCrFeMnNi high entropy alloys. JOM 67, 2326 (2015)

141. Y.K. Kim, Y.A. Joo, H.S. Kim, K.A. Lee, High temperature oxidation behavior of Cr-Mn-Fe-Co-Ni high entropy alloy. Intermetallics 98, 45 (2018)
142. W. Kai, C.C. Li, F.P. Cheng, K.P. Chu, R.T. Huang, L.W. Tsay, J.J. Kai, Air-oxidation of FeCoNiCr-based quinary high-entropy alloys at 700–900∘C. Corros. Sci. 121, 116 (2017)
143. O.N. Senkov, S.V. Senkova, D.M. Dimiduk, C. Woodward, D.B. Miracle, Oxidation behaviour of a refractory NbCrMo0.5Ta0.5TiZr alloy. J. Mater. Sci. 47, 6522–6534 (2012)
144. C.M. Liu, H.M. Wang, S.Q. Zhang, H.B. Tang, A.L. Zhang, Microstructure and oxidation behaviour of new refractory high entropy alloys. J. Alloys Compd. 583, 162169 (2014)
145. G. Zhu, Y. Liu, J. Ye, Early high-temperature oxidation behavior of Ti (C, N)- based cermets with multi-component AlCoCrFeNi high-entropy alloy binder. Int. J. Refract. Met. Hard Mater. 44, 3541 (2014)
146. Y. Zhang, High Entropy Materials-A Brief Introduction. 1st edn. (Springer, 2019), pp. 77–89
147. C.T. Sims, W.C. Hagel, The Superalloys (Wiley, New York, 1972)
148. A. Takeuchi, A. Inoue, Classification of bulk metallic glasses by atomic size difference, heat of mixing and period of constituent elements and its applications to characterization of the main alloying element. Mater. Trans. 46, 2817 (2005)
149. Zhang Y and Zhou Y J, Mater Sci Forum 561–565 (2007) 1337.
150. R.C. Reed: The Superalloys: Fundamentals and Applications, Cambridge University Press, New York, NY, 2006.
151. Y. Zhang, S.G. Ma, and J.W. Qiao: Metall. Mater. Trans. A, vol. 43A, pp. 2012–25.
152. S.G. Ma, S.F. Zhang, M.C. Gao, P.K. Liaw, and Y. Zhang: JOM, 2013, https://doi.org/10.1007/s11837-013-0733-x .
153. Chuang, M. H., Tsai, M. H., Wang, W. R., Lin, S. J. & Yeh, J. W. Microstructure and wear behavior of AlxCo1.5CrFeNi1.5Tiy high-entropy alloys. Acta Materialia 59, 6308–6317 (2011).
154. M.N. Gungor: Metall. Trans. A, 1989, vol. 20, pp. 2529–33.
155. O.N. Senkov, G.B. Wilks, J.M. Scott, and D.B. Miracle, Intermetallics 19, 698 (2011).
156. L.E. Murr, Interfacial Phenomena in Metals and Alloys (Reading: Addison-Wesley Publishig Co, 1975).
157. J.X. Xie, H.D. Fu, Z.H. Zhang, and Y.B. Jiang, Intermetallics 23, 20 (2012).
158. Y. Wu, D.Q. Zhou, W.L. Song, H. Wang, Z.Y. Zhang, D. Ma, X.L. Wang, and Z.P. Lu, Phys. Rev. Lett. 109, 245506 (2012).
159. Wang, Z., Guo, S. & Liu, C. T. Phase Selection in High-Entropy Alloys: From Nonequilibrium to Equilibrium. JOM. 1–7 (2014).
160. Li, J. et al. Enhanced mechanical properties of a CoCrFeNi high entropy alloy by supercooling method. Materials & Design 95, 183–187 (2016).
161. Liu, N. et al. Rapid solidification and liquid-phase separation of undercooled CoCrCuFexNi high-entropy alloys. Intermetallics 72, 44–52 (2016).
162. Syed Ghazi, S. & Ravi, K. R. Phase-evolution in high entropy alloys: Role of synthesis route. Intermetallics 73, 40–42 (2016).
163. Y. Shi, B. Yang, X. Xie, J. Brechtl, K.A. Dahmen, and P.K. Liaw, Corros. Sci. 119, 33 (2017).
164. F. Otto, A. Dlouhý, K.G. Pradeep, M. Kuběnová, D. Raabe, G. Eggeler, and E.P. George, Acta Mater. 112, 40 (2016).
165. E.J. Pickering, R. Muñoz-Moreno, H.J. Stone, and N.G. Jones, Scr. Mater. 113, 106 (2016).
166. M. Mroz, Design and Structural Optimization of a High Entropy Alloy (HEA) of the CoCrFeMnNi Family with High Mechanical Resistance, Ph.D. thesis, Ecole des Mines de Saint Etienne, France (2018).
167. E. Abbasi, and K. Dehghani, J. Alloys Compd. 783, 292 (2019).
168. E. Abbasi, and K. Dehghani, Mater. Sci. Eng. A 772, 138771 (2020). https://doi.org/10.1016/j.msea.2019.138771
169. K. Biswas, J.W. Yeh, P.P. Bhattacharjee, and J.T.M. DeHosson, Scr. Mater. 188, 54 (2020).
170. Ding, Q. et al. Tuning element distribution, structure and properties by composition in high-entropy alloys. Nature 574, 223–227 (2019).

171. Waag, F. et al. Kinetically-controlled laser-synthesis of colloidal high-entropy alloy nanoparticles. RSC Adv. 9, 18547–18558 (2019).
172. Glasscott, M. W. et al. Electrosynthesis of high-entropy metallic glass nanoparticles for designer, multi-functional electrocatalysis. Nat. Commun. 10, 2650 (2019).
173. Li, X. Y., Rong, H. P., Zhang, J. T., Wang, D. S. & Li, Y. D. Modulating the local coordination environment of single-atom catalysts for enhanced catalytic performance. Nano Res. 13, 1842–1855 (2020).
174. E. MacDonald and R. Wicker: Science, 2016, vol. 353, p. 6307.
175. K.V. Vamsi and S. Karthikeyan: Acta Mater., 2018, vol. 145, pp. 532–42.
176. R.J. Vikram, B.S. Murty, D. Fabijanic, and S. Suwas: J. Alloys Compd., 2020, vol. 827, 154034.
177. K.V. Vamsi and S. Karthikeyan: Scripta Mater., 2017, vol. 130, pp. 269–73.
178. Kim, Y.-K., Choe, J. & Lee, K.-A. Selective laser melted equiatomic CoCrFeMnNi high-entropy alloy: microstructure, anisotropic mechanical response, and multiple strengthening mechanism. J. Alloy. Compd. 805, 680–691 (2019).
179. TCHEA4: TCS High Entropy Alloy Database, http://www.thermocalc.com/products-services/databses/mobility (2020).
180. Guo, J. et al. On the machining of selective laser melting CoCrFeMnNi high-entropy alloy. Mater. Des. 153, 211–220 (2018).
181. Chew, Y. et al. Microstructure and enhanced strength of laser aided additive manufactured CoCrFeMnNi high entropy alloy. Mater. Sci. Eng. A 744, 137–144 (2019).
182. Xiang, S. et al. Effects of process parameters on microstructures and tensile properties of laser melting deposited CrMnFeCoNi high entropy alloys. Mater. Sci. Eng. A 743, 412–417 (2019).
183. Li, R. et al. Selective laser melting of an equiatomic CoCrFeMnNi high-entropy alloy: Processability, non-equilibrium microstructure and mechanical property. J. Alloy. Comp. 746, 125–134 (2018).
184. Xiang, S. et al. Microstructures and mechanical properties of CrMnFeCoNi high entropy alloys fabricated by using laser metal deposition technique. J. Alloy. Compd. 773, 387–392 (2019).
185. Zhu, Z. G. et al. Hierarchical microstructure and strengthening mechanism of a CoCrFeMnNi high entropy alloy additively manufactured by selective laser melting. Scr. Mater. 154, 20–24 (2018).
186. Zhou, W. et al. Deformation stimulated precipitation of a single-phase CoCrFeMnNi high entropy alloy. Intermetallics 85, 90–97 (2017).
187. Kim, Y.-K., Ham, G.-S., Kim, H. S. & Lee, K.-A. High-cycle fatigue and tensile deformation behaviors of coarse-grained equiatomic CoCrFeMnNi high-entropy alloy and unexpected hardening behavior during cyclic loading. Intermetallics 111, 106486 (2019).
188. Jang, M. J. et al. High-temperature tensile deformation behavior of hot rolled CrMnFeCoNi high-entropy alloy. J. Alloy. Compd. 730, 242–248 (2018).
189. Otto, F. et al. The influences of temperature and microstructure on the tensile properties of a CoCrFeMnNi high-entropy alloy. Acta Mater. 61, 5743–5755 (2013).
190. Kim, Y.-K., Youn, S.-J., Kim, S.-W., Hong, J. & Lee, K.-A. High-temperature creep behavior of gamma Ti-48Al-2Cr-2Nb alloy additively manufactured by electron beam melting. Mater. Sci. Eng. A 763, 138138 (2019).
191. C. Huang, Y. Zhang, J. Shen, and R. Vilar, Surf. Coat. Technol. 206, 1389 (2011).
192. I. Kunce, M. Polanski, K. Karczewski, T. Plocinski, and K.J. Kurzydlowski, J. Alloy. Compd. 648, 751 (2015).
193. J. Joseph, T. Jarvis, X. Wu, N. Stanford, P. Hodgson, and D.M. Fabijanic, Mater. Sci. Eng. A 633, 184 (2015).
194. J.W. Yeh, S.K. Chen, H.C. Shih, Y. Zhang, T.T. Zuo, Functional properties. High-Entropy Alloys Fundam. Appl. (2016). https://doi.org/10.1007/978-3-319-27013-5_7
195. S.A. Firstov, V.F. Gorban, N.A. Krapivka, M.V. Karpets, A.D. Kostenko, Wear resistance of high-entropy alloys. Powder Metall. Met. Ceram. (2017). https://doi.org/10.1007/s11106-017-9882-8

196. F. Otto, A. Dlouhý, C. Somsen, et al., "The influences of temperature and microstructure on the tensile properties of a CoCrFeMnNi high-entropy alloy," Acta Mater., 61, No. 15, 5743–5755 (2013).
197. Z. Li, S. Zhao, R.O. Ritchie, M.A. Meyers, Mechanical properties of high-entropy alloys with emphasis on face-centered cubic alloys. Prog. Mater. Sci. (2019). https://doi.org/10.1016/j.pmatsci.2018.12.003
198. A.A. Popovich, N.G. Razumov, Dissolution of alloying elements and phase formation in powder materials Fe–18Cr–8Ni–12Mn − xN during mechanical alloying. Adv. Mater. Lett. 5, 683–687 (2014). https://doi.org/10.5185/amlett.2014.6585
199. I.S. Goncharov, N.G. Razumov, A.O. Silin, N.E. Ozerskoi, A.I. Shamshurin, A. Kim, Q.S. Wang, A.A. Popovich, Synthesis of Nb-based powder alloy by mechanical alloying and plasma spheroidization processes for additive manufacturing. Mater. Lett. (2019). https://doi.org/10.1016/j.matlet.2019.03.014
200. George, E. P., Curtin, W. A. & Tasan, C. C. High entropy alloys: a focused review of mechanical properties and deformation mechanisms. Acta Materialia 188, 435–474 (2020).
201. Y. Zhang, X. Chen, S. Jayalakshmi, R. A. Singh, V. B. Deev, and E. S. Prusov, J. Alloys Compd. 857, 157625 (2021). https://doi.org/10.1016/j.jallcom.2020.157625
202. T. Zhang, L. Xin, F. Wu, Xiang J. Chen, S. Jiang, Y. Huang, and S. Chen, J. Mater. Sci. Technol. 35, 2331 (2019).
203. V. E. Gromov, S. V. Konovalov, Yu. F. Ivanov, and K. A. Osintsev, Structure and Properties of High-Entropy Alloys, Vol. 107 of Advanced Structured Materials (Springer, New York, 2021).
204. A. D. Pogrebnjak, A. A. Bagdasaryan, I. V. Yakushchenko, and V. M. Beresnev, Russ. Chem. Rev. 83, 1027 (2014).
205. Y. F. Ivanov, K. A. Osintsev, V. E. Gromov, S. V. Konovalov, and I. A. Panchenko, Steel Transl. 51 (1), 27 (2021). https://doi.org/10.3103/S0967091221010046
206. K. Osintsev, S. Konovalov, V. Gromov, I. Panchenko, and X. Chen, Mater. Today: Proc. 46, 961 (2021). https://doi.org/10.1016/j.matpr.2021.01.079
207. Zhang, C.; Zhu, J.; Cao, W. S.; Chen, S. L.; Liang, S. M.; Zhang, F. Development of Phase-Based Databases via CALPHAD Method for the Design of High-Entropy Alloys. Journal of Phase Equilibria and Diffusion (2022). https://doi.org/10.1007/s11669-022-00970-9
208. Gorbachev, I. I.; Popov, V. V; Katz-Demyanetz, A.; Popov, V.; Eshed, E. Prediction of the Phase Composition of High-Entropy Alloys Based on Cr–Nb–Ti–V–Zr Using the Calphad Method. Physics of Metals and Metallography (2019). https://doi.org/10.1134/s0031918x19040069
209. Wu, Qingfeng; Wang, Zhijun; He, Feng; Li, Junjie; Wang, Jincheng Revealing the Selection of σ and μ Phases in $CoCrFeNiMo_x$ High Entropy Alloys by CALPHAD. Journal of Phase Equilibria and Diffusion (2018). https://doi.org/10.1007/s11669-018-0659-7
210. Durga, A.; Hari Kumar, K. C.; Murty, B. S. Phase Formation in Equiatomic High Entropy Alloys: CALPHAD Approach and Experimental Studies. Transactions of the Indian Institute of Metals (2012). https://doi.org/10.1007/s12666-012-0138-5
211. Gao, M. C.; Zhang, B.; Yang, S.; Guo, S. M. Senary Refractory High-Entropy Alloy HfNbTaTiVZr. Metallurgical and Materials Transactions A (2015). https://doi.org/10.1007/s11661-015-3105-z
212. Feng, Rui; Liaw, Peter K.; Gao, Michael C.; Widom, Michael First-principles prediction of high-entropy-alloy stability. npj Computational Materials (2017). https://doi.org/10.1038/s41524-017-0049-4
213. Ozdemir, H. C.; Bedir, E.; Yilmaz, R.; Yagci, M. B.; Canadinc, D. Machine learning-assisted design of biomedical high entropy alloys with low elastic modulus for orthopedic implants. Journal of Materials Science (2022). https://doi.org/10.1007/s10853-022-07363-w
214. Giles, Stephen A.; Sengupta, Debasis; Broderick, Scott R.; Rajan, Krishna Machine-learning-based intelligent framework for discovering refractory high-entropy alloys with improved high-temperature yield strength. npj Computational Materials (2022). https://doi.org/10.1038/s41524-022-00926-0

215. Kostiuchenko, Tatiana; Körmann, Fritz; Neugebauer, Jörg; Shapeev, Alexander Impact of lattice relaxations on phase transitions in a high-entropy alloy studied by machine-learning potentials. npj Computational Materials (2019). https://doi.org/10.1038/s41524-019-0195-y
216. Li, Li; Xie, Baobin; Fang, Qihong; Li, Jia Machine Learning Approach to Design High Entropy Alloys with Heterogeneous Grain Structures. Metallurgical and Materials Transactions A (2021). https://doi.org/10.1007/s11661-020-06099-z
217. Zhou, Ziqing; Zhou, Yeju; He, Quanfeng; Ding, Zhaoyi; Li, Fucheng; Yang, Yong Machine learning guided appraisal and exploration of phase design for high entropy alloys. npj Computational Materials (2019). https://doi.org/10.1038/s41524-019-0265-1
218. Rickman, J. M.; Chan, H. M.; Harmer, M. P.; Smeltzer, J. A.; Marvel, C. J.; Roy, A.; Balasubramanian, G. Materials informatics for the screening of multi-principal elements and high-entropy alloys. Nature Communications (2019). https://doi.org/10.1038/s41467-019-10533-1
219. Zhang, Y.; Ma, S. G.; Qiao, J. W. Morphology Transition from Dendrites to Equiaxed Grains for AlCoCrFeNi High-Entropy Alloys by Copper Mold Casting and Bridgman Solidification. Metallurgical and Materials Transactions A (2011). https://doi.org/10.1007/s11661-011-0981-8
220. Ananiadis, E.; Lentzaris, K.; Georgatis, E.; Mathiou, C.; Poulia, A.; Karantzalis, A. E. AlNiCrFeMn Equiatomic High Entropy Alloy: A Further Insight in Its Microstructural Evolution, Mechanical and Surface Degradation Response. Metals and Materials International (2019). https://doi.org/10.1007/s12540-019-00401-4
221. Mishra, Ajit Kumar; Samal, Sumanta; Biswas, Krishanu Solidification Behaviour of Ti–Cu–Fe–Co–Ni High Entropy Alloys. Transactions of the Indian Institute of Metals (2012). https://doi.org/10.1007/s12666-012-0206-x
222. Yeh, An-Chou; Chang, Yao-Jen; Tsai, Che-Wei; Wang, Yen-Chun; Yeh, Jien-Wei; Kuo, Chen-Ming On the Solidification and Phase Stability of a Co-Cr-Fe-Ni-Ti High-Entropy Alloy. Metallurgical and Materials Transactions A (2013). https://doi.org/10.1007/s11661-013-2097-9
223. Ma, S. G.; Zhang, S. F.; Gao, M. C.; Liaw, P. K.; Zhang, Y. A Successful Synthesis of the $CoCrFeNiAl_{0.3}$ Single-Crystal, High-Entropy Alloy by Bridgman Solidification. JOM (2013). https://doi.org/10.1007/s11837-013-0733-x
224. He, Feng; Wang, Zhijun; Li, Yiyan; Wu, Qingfeng; Li, Junjie; Wang, Jincheng; Liu, C. T. Kinetic ways of tailoring phases in high entropy alloys. Scientific Reports (2016). https://doi.org/10.1038/srep34628
225. Stasiak, Tomasz; Oleszak, Dariusz; Fraczkiewicz, Anna Effects of Solidification Conditions on Microstructure and Properties of High-Entropy Alloys from the CoCrFeMnNi Family. JOM (2022). https://doi.org/10.1007/s11837-022-05543-2
226. Wang, Bing; Wang, Cheng; Yu, Xiwen; Cao, Yuan; Gao, Linfeng; Wu, Congping; Yao, Yingfang; Lin, Zhiqun; Zou, Zhigang General synthesis of high-entropy alloy and ceramic nanoparticles in nanoseconds. Nature Synthesis (2022). https://doi.org/10.1038/s44160-021-00004-1
227. Vikram, R. J.; Verma, S. K.; Dash, K.; Fabijanic, D.; Murty, B. S.; Suwas, Satyam Mechanism Controlling Elevated Temperature Deformation in Additively Manufactured Eutectic High-Entropy Alloy. Metallurgical and Materials Transactions A (2022). https://doi.org/10.1007/s11661-022-06777-0
228. Kim, Young-Kyun; Yang, Sangsun; Lee, Kee-Ahn Superior Temperature-Dependent Mechanical Properties and Deformation Behavior of Equiatomic CoCrFeMnNi High-Entropy Alloy Additively Manufactured by Selective Laser Melting. Scientific Reports (2020). https://doi.org/10.1038/s41598-020-65073-2
229. Ocelík, V.; Janssen, N.; Smith, S. N.; De Hosson, J. Th. M. Additive Manufacturing of High-Entropy Alloys by Laser Processing. JOM (2016). https://doi.org/10.1007/s11837-016-1888-z

230. Makhmutov, Tagir; Razumov, Nikolay; Kim, Artem; Ozerskoy, Nikolay; Mazeeva, Alina; Popovich, Anatoliy Synthesis of $CoCrFeNiMnW_{0.25}$ High-Entropy Alloy Powders by Mechanical Alloying and Plasma Spheroidization Processes for Additive Manufacturing. Metals and Materials International (2020). https://doi.org/10.1007/s12540-020-00747-0
231. Gromov, V. E.; Konovalov, S. V.; Ivanov, Yu. F.; Efimov, M. O.; Panchenko, I. A.; Vorob'ev, S. V. Investigation of the Structure and Properties of a Coating from a High-Entropy FeCoCrNiMn Alloy Obtained by a Wire Arc Additive Manufacturing. Physics of the Solid State (2022). https://doi.org/10.1134/s1063783422070034

Chapter 4
Microstructure and Mechanical Properties of High Entropy Alloys

4.1 Introduction by Editor

The complex reciprocation between the composition, microstructure, and mechanical behavior of HEAs has prompted researchers to delve deeply into the underlying mechanisms governing their extraordinary properties. Understanding the microstructural evolution of these alloys and their impact on mechanical properties has become critical as they depart from the conventional material-design approach. Furthermore, in HEAs, multiple strengthening mechanisms such as solid solution, precipitation, work hardening, stacking fault assisted strengthening, transformation/twin induced deformation and grain boundary strengthening found to be active individually or concurrently. In this context, this chapter provides in-depth knowledge of the captivating realm of HEAs, presenting the microstructure and mechanical properties of eutectic HEAs, metastable HEAs, refractory HEAs, and lightweight HEAs along with discussion on deformation mechanisms activated in them.

The origin of HEAs begins with a curiosity to understand the effect of a multi-element matrix as of five equiatomic elements combined in an alloy commonly known as Cantor alloy. This alloy specifically constituted from the 3d-series elements like CoCrFeMnNi and was proficient enough to describe the effective behavior that these kinds of alloy may show, i.e. the core effects namely, high entropy effect, severe lattice distortion, sluggish diffusion, and cocktail effects. However, limiting up to this alloy alone wasn't sufficient from the material's evolution point of view. This led to an exciting pathway in searching for alloys, first closely related to this alloying system, followed by further enhancement in properties like high-temperature deformation, fatigue, corrosion, and a wide range of functional properties.

Beginning with some modification in Cantor alloys, introducing Al in this system showed excellent features by improving the strength-ductility and high-temperature properties simultaneously with reduced density. The involvement of Al in conventional Cantor alloy first became an exciting thing in terms of the crystal structure. The contribution of high Al content to form a b.c.c. phase over f.c.c. phase made this

S. S. Nene (ed.), *High Entropy Alloys*, https://doi.org/10.1007/978-981-99-7173-2_4

alloy system exciting to study ultimately contributing to understanding the behavior of alloying addition in HEAs. This gave a notion of developing HEAs that varying alloying content over a traditional methodology of developing an equiatomic HEA. Further, switching from the equiatomic to non-equiatomic HEAs has successively helped in getting other phases that include eutectics and spinodals.

With time, the evolution in the definition of HEAs changed spatially and is associated with the development of non-equiatomic HEAs which demonstrated further enhancement in the mechanical properties while broadening the dimensional space for HEA design. This formed a basis for understanding the microstructure evolution and subsequent mechanical properties correlating it more to the physical metallurgy in HEAs. Not only the properties in as-cast condition, a desired heat treatment or an optimized thermomechanical treatment too can eventually play a crucial role in developing the microstructure leading to a targeted properties. The alloying composition, the phase distribution, and the grain morphology together dictated the properties of HEAs.

Many studies on other d-series elements like refractory alloys or introducing p-block elements also led to the formation of lightweight HEAs. These alloys which prominently exhibit b.c.c. phase found to improve the material performance in comparison to the pre-existing conventional alloys of the same category. Refractory HEAs primarily showed excellent high-temperature strength with reduced density in comparison with conventional refractory alloys. From a light-weight alloy point of view, introducing multi-principal elements effectively has shown promising results from alloy mechanical properties. The complexity of the phases overall has played a crucial role in developing the understanding or successfully setting a notion providing flexibility in the material properties and behavior. Bringing in the concept of metastability in HEAs again proved to be a breakthrough in understanding the concepts and linking them closely with conventional metals.

The relation of the microstructural evolution of these alloys and their impact on mechanical properties has become critical as they depart from the conventional material-design approach. Furthermore, in HEAs, multiple strengthening mechanisms such as solid solution, precipitation, twinning/transformation induced deformation, stacking fault strengthening, and grain boundary strengthening are found active individually or concurrently based on not only the microstructure of the alloy but also by the selection of the alloy system. Conventional refractory alloy demonstrates limitations like high densities and limited room temperature specific strength which are being significantly overcome by multi-principal elemental alloy design approaches. As a result, this makes a successful contribution in the metal industry in developing alloys for modern structural applications leading to alloy development from applications point of view. The work on lightweight alloys is prominently focused on improving the specific strength of the alloy maintaining the strength-ductility synergy.

In this context, this chapter provides a summary of recent work in broad range of categories of alloys in HEAs specific for structural applications. This chapter is designed with specific research articles primarily focusing on the captivating realm of HEAs, presenting microstructure and mechanical properties of eutectic HEAs, metastable HEAs, refractory HEAs, and lightweight HEAs along with the discussion on deformation mechanisms activated in them. The articles were selectively chosen to give a brief idea about the evolution occurring in every category of alloys with their subsequent characteristics. The microstructure, phases, and property correlations forms the base for these alloys.

4.2 Machine Generated Summaries

Disclaimer: The summaries in this chapter were generated from Springer Nature publications using extractive AI auto-summarization: An extraction-based summarizer aims to identify the most important sentences of a text using an algorithm and uses those original sentences to create the auto-summary (unlike generative AI). As the constituted sentences are machine selected, they may not fully reflect the body of the work, so we strongly advise that the original content is read and cited. The auto generated summaries were curated by the editor to meet Springer Nature publication standards. To cite this content, please refer to the original papers.

Machine generated keywords: ehea, refractory, eutectic, mpa, rhea, strain, bcc, yield strength, dislocation, hcp phase, entropy alloy, strain rate, yield, hcp, deformation

4.3 CoCrFeMnNi High Entropy Alloy

Machine generated keywords: cocrfemnni, dislocation, ref, strain rate, high strain, dynamic, strain, rate, cantor alloy, strengthen, cantor, plasma sinter, spark, spark plasma, motion

The Origin of Jerky Dislocation Motion in High-Entropy Alloys [203]

This is a machine-generated summary of:

Utt, Daniel; Lee, Subin; Xing, Yaolong; Jeong, Hyejin; Stukowski, Alexander; Oh, Sang Ho; Dehm, Gerhard; Albe, Karsten: The origin of jerky dislocation motion in high-entropy alloys [203]

Published in: Nature Communications (2022)

Link to original: https://doi.org/10.1038/s41467-022-32134-1

If you want to cite the papers, please refer to the original.

For technical reasons we could not place the page where the original quote is coming from.

Abstract-Summary

"Dislocations in single-phase concentrated random alloys, including high-entropy alloys (HEAs), repeatedly encounter pinning during glide, resulting in jerky dislocation motion."

"While solute-dislocation interaction is well understood in conventional alloys, the origin of individual pinning points in concentrated random alloys is a matter of debate."

"Atomistic simulations reproduce the jerky dislocation motion and link the repeated pinning to local fluctuations in the Peierls friction."

Introduction

"Dislocation pinning and the resulting strengthening are also observed in ideally random alloys [1]."

"The unexpected wide range of stacking fault (SF) energies observed for HEAs could provide an alternative explanation for the pinning experienced by dislocations [2, 3]."

"Such a mechanism would only exist in HEAs, but not in conventional dilute solid solutions as their local fluctuations in the SF energy remain comparatively small."

"We investigate the repeated local pinning along subsequent dislocations in the face-centered cubic (FCC) CoCrFeMnNi HEA by combining in-situ TEM deformation, atomic resolution analytical scanning transmission electron microscopy (STEM), and atomistic simulations."

"The underlying mechanism is directly linked to local fluctuations in Peierls (force/friction) barriers, which lead to dislocation pinning points in concentrated random alloys."

"While not a full solid solution strengthening model, the atomic fluctuations in frictional forces provide predictive design guidelines to tailor the dislocation pinning in HEAs."

Results

"This value is proportional to the strength of this atom acting as a pinning point for the dislocation line and is denoted as F_P."

"Our simulations have found an atomic scale descriptor for dislocation pinning points that is in line with the established theory of the Peierls model, we test this descriptor for a wide range of HEAs and search if an increased number of high F_P lattice sites occurs."

"We investigate the effect of the pinning point spectrum on the critical force required to initiate dislocation glide as well as their mobility."

"A clear linear relationship between the pinning point density and the critical force can be seen within the small and large sample series confirming the expected correlation, where an increased density of strong pinning points leads to an increased force required to move a dislocation."

"The critical force required for a dislocation to move correlates with the density of high strength pinning points identified from the atomic scale resolved GSF curves."

Discussion

"The identification of dislocation pinning points allows us to sample their density for different binary and ternary alloys and compare those to the CoCrFeMnNi HEA."

"The intuitive trends expected from the densities of strong pinning points, i.e., increasing resistance against dislocation glide and reduced mobility as the fraction of strong pinning points in a material increases are confirmed by our single dislocation shear simulations."

"In the alloy samples investigated, these changes are highly non-linear and the interaction of Co and Cr leads to the highest density of strong dislocation pinning points."

"This understanding of the local Peierls stress landscape and the resulting pinning forces acting on dislocations gives access to a different parameter that can be tuned in the design of alloys with increased resistance against the onset of dislocation glide and overall reduced dislocation mobility."

Methods

"The atomic-resolution STEM EDS chemical mapping was carried out on a JEM-ARM 200F (JEOL) equipped with a spherical aberration corrector (ASCOR, CEOS) and energy dispersive X-ray spectrometer (JED-EDS, JEOL)."

"Multi-frame EDS maps were averaged by correcting the frame-to-frame displacement caused by sample drift during acquisition."

"During this annealing and the subsequent shear simulations the surface layers were thermostatted using a Langevin thermostat [4], while the atoms in the center of the sample were integrated in the NVE ensemble."

"Applying the conventional simulation methodology prevents the atomic relaxation during rigid displacement of the two crystallites within the GSF plane [5], however, the relaxation of the atoms normal to the GSF remains unconstrained."

"While this effect averages out in the global GSF calculation, the resulting atomic energy landscapes are not smooth but include discontinuities when an atom relaxes into such a metastable state."

Effect of Strain Rate on the Tensile Behavior of CoCrFeNi and CoCrFeMnNi High-Entropy Alloys [204]

This is a machine-generated summary of:

Shabani, Mitra; Indeck, Joseph; Hazeli, Kavan; Jablonski, Paul D.; Pataky, Garrett J.: Effect of Strain Rate on the Tensile Behavior of CoCrFeNi and CoCrFeMnNi High-Entropy Alloys [204]

Published in: Journal of Materials Engineering and Performance (2019)

Link to original: https://doi.org/10.1007/s11665-019-04176-y

If you want to cite the papers, please refer to the original.

For technical reasons we could not place the page where the original quote is coming from.

Abstract-Summary

"Tensile experiments at quasi-static to dynamic strain rates (10^{-4}–10^{3} s^{-1}) were performed on two single-phase face-centered cubic HEAs, CoCrFeNi and CoCrFeMnNi."

"The dominant deformation mechanisms were dislocation slip at quasi-static strain rates with the addition of deformation nano-twins at dynamic strain rates."

"Ultimate dynamic tensile strength and ductility improved with the increase in strain rate, which can be attributed to the activation of deformation nano-twins in HEAs."

"CoCrFeNi and CoCrFeMnNi both have low stacking fault energies, which could promote twinning at high strain rates to accommodate plastic deformation."

"The strain rate sensitivity of the flow stress increased with increasing strain rate, beginning with negligible strain rate sensitivity in the quasi-static range to high strain rate sensitivity in the dynamic range."

"CoCrFeNi, with less configurational entropy, had higher mechanical properties and strain-hardening rates compared to CoCrFeMnNi, which was attributed to the weakening effect of the addition of Mn on the solid solution hardening."

Introduction

"High-entropy alloys (HEAs) provide a viable replacement to conventional metal alloys (e.g., stainless steel and nickel-base superalloy) in critical load-bearing applications because of their high strength, ductility, and fracture resistance, some even down to cryogenic temperatures (Refs. [6–12])."

"Mechanical properties of HEAs such as yield and ultimate strength, ductility, hardness and fracture toughness were found to be in the same range or better than conventional metal alloys such as stainless steel and nickel-base superalloy (Refs. [7, 8, 12–18])."

"While there have been studies on its mechanical behavior and microstructure (Refs. [7, 8, 11, 19, 20]), less research has been focused on the CoCrFeNi HEA compare to the Cantor alloy."

"Huo and others studied the tensile quasi-static strain rate effect of CoCrFeNi and found an increase in the yield strength, ultimate tensile strength, and ductility with rising the strain rate (Ref. [11])."

"Previous studies have examined the Cantor alloy's mechanical properties and microstructure under dynamic compression (Refs. [21, 22])."

Materials and Methods

"For the 1×10^{-1} s^{-1} strain rate, strains were measured using an MTS extensometer with a 12-mm gauge length due to camera limitations."

"Dynamic tension experiments, at strain rates of 1.0×10^2–1.0×10^3 s^{-1}, were performed on dogbone tensile specimens using a modified Kolsky bar (Split-Hopkinson Pressure Bar)."

"Impact velocities ranged between 3.5 and 24 m/s. A piece of 14 gauge copper wire was wrapped around the incident bar to act as a pulse shaper for the high strain rate experiments."

"Strain data from DIC were synchronized with stress data using software from REL Inc. A constant strain rate was not reached in the dynamic tests until approximately 30 μs due to the rise time of the loading wave."

"The maximum length of the striker bar able to fit inside the launch chamber was 610 mm, limiting the tensile loading pulse duration."

Results and Discussion

"Strain rate sensitivity (SRS) values were calculated to further investigate the effect of strain rate on the mechanical properties of HEA-1 and HEA-2 for both quasi-static and dynamic strain rates."

"High SRS at dynamic strain rates was believed to be related to the distorted lattice of the HEAs (Ref. [23]) and/or thermally activated mechanisms (Ref. [24])."

"Higher SRS is observed with the increase in strain rate (Ref. [25])."

"HEA-1 SHRs showed a fluctuating value in the quasi-static range, initially increasing 28% then subsequently decreasing for the strain rate of 1×10^{-2} s^{-1}."

"Although the global tensile stress on the sample reached a magnitude that presumably would satisfy the CRSS for twinning being reached, no twinning was observed in the postmortem TEM micrographs of HEA-2 specimens under quasi-static strain rates."

"No twinning was observed in the HEA-1 TEM micrographs of the specimens loaded at quasi-static strain rates."

Conclusions

"The strain rate effect on the tensile behavior of CoCrFeNi (HEA-1) and CoCrFeMnNi (HEA-2) was studied through tensile experiments at strain rates ranging from quasi-static, 10^{-4} s^{-1}, to dynamic, 10^3 s^{-1}."

"The SRS component was found to be low for quasi-static strain rates, showing negligible SRS, and higher for dynamic strain rates, showing positive SRS for both HEAs."

"Both alloys showed promising possible application as structural materials due to their strength, ductility, SRS and SHR both in the quasi-static and dynamic strain rate range."

A Micromechanical Study on the Effects of Precipitation on the Mechanical Properties of CoCrFeMnNi High Entropy Alloys with Various Annealing Temperatures [205]

This is a machine-generated summary of:

Huang, Chang-Wei; Su, Pei-Ying; Yu, Chi-Hua; Wang, Chia-Ling; Lo, Yu-Chieh; Jang, Jason Shian-Ching; Hu, Hsuan-Teh: A micromechanical study on the effects of precipitation on the mechanical properties of CoCrFeMnNi high entropy alloys with various annealing temperatures [205]

Published in: Scientific Reports (2023)

Link to original: https://doi.org/10.1038/s41598-023-30508-z

If you want to cite the papers, please refer to the original.

For technical reasons we could not place the page where the original quote is coming from.

Abstract-Summary

"The application of CoCrFeMnNi is limited because of the relatively low tensile strength."

"We proposed a micromechanical model which adopted from the theory of dislocation density to investigate the strengthening mechanisms of precipitation of chromium-rich non-equiatomic CoCrFeMnNi alloy."

"The proposed framework is validated by comparing simulations with experiments of uniaxial tensile tests on the CoCrFeMnNi alloys under different annealing temperatures."

"The numerical results indicate that HEAs with a precipitate distribution closer to a normal distribution and with smaller average size will tend to have higher strength and ductility."

Introduction

"Several relevant factors influence its yield strength and post-yield behavior, such as the components of the HEA, lattice distortion, the grain distribution, and the presence of defects (e.g., doping, precipitation, and dislocation)."

"These mechanisms can be expressed as the intrinsic strength, solid solution strengthening, grain boundary strengthening, and precipitation strengthening of an HEA."

"Recent studies have shown that in CoCrFeNiMn alloys, second-phase precipitates can effectively hinder grain growth and form fine-grained structures [26–28]."

"The precipitation strengthening effect produced by the σ-phase can effectively improve the yield strength, tensile strength, and strain hardening rate [29, 30]."

"The purpose of this study is to conduct a comprehensive investigation of the mechanical properties of CoCrFeMnNi high-entropy alloys at different annealing temperatures."

"The alloys used in the study were tested for tensile strength at room temperature and the deformation mechanism of the alloy was simulated with controlled slip differentials."

Materials and Methods

"The SEM-BSE images show a fully recrystallized coarse-grained microstructure without the σ-phase after annealing at a high temperature of 1373 K; however, after annealing at the lower temperature of 973 K, the σ-phase was observed to precipitate, while the recrystallization of the face-centered crystal (FCC) structure matrix was not complete."

"P_s and ρ_g represent the statistically stored and the geometrically necessary dislocation densities, respectively, a is a constant and is set to be 0.3 in this study, and G and b are the shear modulus at room temperature and the Burgers vector of the CoCrFeMnNi alloy, respectively."

"SEM-BSE images of CoCrFeMnNi alloys with three different annealing temperatures (from 1073 to 1273 K) were analyzed using the ImageJ software to obtain the precipitate data."

"The corresponding grain size distributions of the CoCrFeMnNi specimens with the three different annealing temperatures were obtained by analyzing the SEM-BSE images using the LineCut method [31]."

Results

"The grain boundary strengthening mechanism dominated the initial yield strength for all annealing temperatures, since the mean grain size decreases from 4.9 to 1.41 μm as the annealing temperature decreases."

"The contribution of precipitation strengthening increases with decreasing annealing temperature."

"The statistically stored dislocation density ρ_s increases with increasing annealing temperature."

"In the case of the 1273 K annealing temperature, $\rho_s < \rho_g$ during the early deformation stage."

"The mean value of the precipitate radius, r_m, and the mean value of the grain size, d_g, can be described by (unit: μm): where T is the annealing temperature in terms of the absolute temperature."

Discussion

"We can observe that the lower the annealing temperature, the larger the standard deviation of the precipitate size distribution, allowing the experimental data to be matched more accurately."

"The geometrically necessary dislocation density decreases with an increase in the standard deviation of the precipitate size distribution, which reduces the strain hardening in the stress–strain curve."

"The lower the annealing temperature, the lower the statistical storage dislocation density and the higher the geometrically necessary dislocation density, which resulted in more obvious strain hardening stress–strain curves."

"Although the model enables us to study the mechanisms behind precipitates, there are some limitations that can be addressed in future work by expanding the experimental data to include more alloy systems with more annealing temperatures."

"Our presentation of dislocation density-based CPFEM will help inform the design of mechanically robust HEAs at desired annealing temperatures, especially given the vast design space of complex metal elements."

Conclusions

"A computational framework that integrated a theoretical model with the dislocation-based crystal plasticity finite element method was proposed to investigate the effects of precipitate distribution on the mechanical properties of CoCrFeMnNi alloys."

"SEM-BSE images of CoCrFeMnNi specimens with different annealing temperatures were analyzed using the ImageJ software to obtain the real precipitate distributions."

"Numerical simulations indicated that the precipitate distribution played an important role in the strain hardening behavior of the CoCrFeMnNi samples."

"The normal distribution, which had a standard deviation of 0.4 times the mean radius of the precipitate, provided an approximate precipitate distribution to efficiently predict the mechanical behavior of the CoCrFeMnNi specimens."

Mechanical Properties and Microstructure of the CoCrFeMnNi High Entropy Alloy Under High Strain Rate Compression [206]

This is a machine-generated summary of:

Wang, Bingfeng; Fu, Ao; Huang, Xiaoxia; Liu, Bin; Liu, Yong; Li, Zezhou; Zan, Xiang: Mechanical Properties and Microstructure of the CoCrFeMnNi High Entropy Alloy Under High Strain Rate Compression [206]

Published in: Journal of Materials Engineering and Performance (2016)

Link to original: https://doi.org/10.1007/s11665-016-2105-5

If you want to cite the papers, please refer to the original.

For technical reasons we could not place the page where the original quote is coming from.

Abstract-Summary

"Dynamic compressive tests of the CoCrFeMnNi high entropy alloy were deformed at varying strain rates ranging from 1×10^3 to 3×10^3 s^{-1} using a split-Hopkinson pressure bar (SHPB) system."

"The dynamic yield strength of the CoCrFeMnNi high entropy alloy increases with increasing strain rate."

"The mechanism for serration behavior of the alloy deformed at high strain rate is proposed."

Introduction

"Different techniques have been explored to synthesize high entropy alloys, such as arc-melting (Ref. [32]), laser cladding (Ref. [33]), mechanical alloying, and spark plasma sintering (Refs. [34, 35]), etc. Among these methods, the arc melt casting process is widely used in preparing high entropy alloys."

"The influence of composition (Ref. [36]) or grain size (Ref. [37]) on microstructure and mechanical properties of the high entropy alloy are investigated."

"The microstructural evolution and related serration behavior of the equiatomic CoCrFeMnNi high entropy alloy under high strain rate deformation are also seldom reported."

"We focus on the dynamic mechanical behavior of the equiatomic CoCrFeMnNi high entropy alloy prepared by spark plasma sintering."

"The aims of this paper are: (1) to report the mechanical properties and the microstructure of the CoCrFeMnNi high entropy alloy under high strain rate deformation; (2) to obtain the Zerilli-Armstrong plastic model of the alloy; (3) to discuss the microstructural mechanism for serration behavior of the alloy deformed at high strain rate."

Experimental and Procedures

"The melt was dropped through a ceramic tube and atomized by high purity Ar with the atomization pressure of 4 MPa."

"Cylindrical specimens with diameters of 4 or 6 mm were machined from the sintered specimens with the impact axes parallel to the longitudinal direction."

"The impact tests were performed at varying strain rates ranging from 1×10^3 to 3×10^3 s^{-1} and ambient temperature, using a split-Hopkinson pressure bar system."

"When the cylindrical specimens were loaded by a split-Hopkinson pressure bar, the force applied to the deformation region was calculated from data collected by the strain gages on the incident and transmitted bars."

"E_0 and C_0 are elastic modulus and elastic wave speed in the split-Hopkinson pressure bar; A_0 is the cross-sectional area of the bar; A_s and L_s are the cross-sectional area and the length of the cylindrical specimens; $\varepsilon_r(t)$ and $\varepsilon_t(t)$ are the experimentally measured strain of incident and transmitted stress pulse on the split-Hopkinson pressure bars, respectively."

Results and Discussion

"These stress–strain curves demonstrate the deformation process of the CoCrFeMnNi high entropy alloy."

"The Zerilli-Armstrong plastic model which is used for predicting the strain rate flow behavior of the CoCrFeMnNi high entropy alloy with face-centered cubic (FCC) structure can be represented as follows (Ref. [38]): where C_1, C_2, C_3, C_4, and P are material constants."

"The local hotspots and thermal softening caused by high strain rate deformation are important reasons causing the collapse of the grains and the serration behavior of the CoCrFeMnNi high entropy alloy at high strain rate."

"The above analysis on the microstructure and the thermodynamic calculation results of the CoCrFeMnNi high entropy alloy specimens deformed at high strain rates suggest that the collapse of the grains in the deformation zone and the thermal softening are the essential reasons for the related serration behavior of the CoCrFeMnNi high entropy alloy at high strain rate."

Conclusion

"The Zerilli-Armstrong plastic model of the CoCrFeMnNi high entropy is obtained as follows: The serration behavior of the CoCrFeMnNi high entropy alloy at high strain rate is sensitive to the strain rate."

"Microvoids form on the boundaries of the grains in the specimen deformed at high strain rate."

"With increasing of the strain rate, the shear bands with the width about 2 μ are found in the specimens."

"The collapse of the grains in the deformation zone and the thermal softening are the essential reasons for the related serration behavior of the CoCrFeMnNi high entropy alloy under high strain rate deformation."

Composition Dependence of Phase Stability, Deformation Mechanisms, and Mechanical Properties of the CoCrFeMnNi High-Entropy Alloy System [207]

This is a machine-generated summary of:

Tasan, C. C.; Deng, Y.; Pradeep, K. G.; Yao, M. J.; Springer, H.; Raabe, D.: Composition Dependence of Phase Stability, Deformation Mechanisms, and Mechanical Properties of the CoCrFeMnNi High-Entropy Alloy System [207]

Published in: JOM (2014)

Link to original: https://doi.org/10.1007/s11837-014-1133-6

Copyright of the summarized publication:

If you want to cite the papers, please refer to the original.

For technical reasons we could not place the page where the original quote is coming from.

Abstract-Summary

"The proposal of configurational entropy maximization to produce massive solid-solution (SS)-strengthened, single-phase high-entropy alloy (HEA) systems has gained much scientific interest."

"Although most of this interest focuses on the basic role of configurational entropy in SS formability, setting future research directions also requires the overall property benefits of massive SS strengthening to be carefully investigated."

"A comparative assessment of the results with respect to room temperature behavior of binary Fe-Mn alloys reveals only limited benefits of massive SS formation."

Introduction

"High-entropy alloys (HEAs) are newly developed, multicomponent alloys wherein the configurational entropy (CE) is higher than the fusion entropy of most common metals."

"[16] These alloys draw increasing scientific interest, especially regarding the fundamental aspects of massive solid-solution (SS) phases and the basic role of CE therein."

"[39, 40] These efforts are well motivated, as the HEA concept was originally based on CE maximization (i.e., by increasing the number of equiatomically proportioned alloying elements)."

"In what follows, taking the reported weak CE dependence of the $Co_{20}Cr_{20}Fe_{20}Mn_{20}Ni_{20}$ alloy as the starting point, we investigate alloys of different compositions and/or different number of alloying elements to assess compositional effects on single-phase SS formation, deformation mechanisms, and resulting mechanical properties."

"The possibilities of expanding the limits of the property space of this alloy system will be investigated at the expense of decreasing configurational entropy."

Methodology

"These two systems were selected as they enable the investigation of the influence of several parameters (e.g., configurational entropy, Ni content, and stacking fault energy) on the overall properties."

"A more comprehensive investigation of the property trends in these systems will be treated elsewhere."

"To investigate the influence of cold working on the mechanical properties, some of the hot-rolled sheets were further cold rolled to 64% engineering thickness reduction."

"The resulting microstructures were investigated in the as-cast, homogenized, and cold-rolled conditions using various tools and techniques."

"The deformation and damage mechanisms were also investigated in the latter microscope, employing the electron channelling contrast imaging (ECCI) technique."

"[41] To investigate the mechanical performance of the two alloy systems, tensile tests were carried out employing a miniaturized deformation stage (Kammrath-Weiss GmbH, Dortmund, Germany) and Aramis digital image correlation (DIC) system (GOM GmbH, Braunschweig, Germany)."

Results

"This alloy develops a microstructure that consists of a single fcc phase at all processing states, i.e., including the hot rolled and cold rolled states, the images of which are not shown here."

"While the five-component, equiatomic $Co_{20}Cr_{20}Fe_{20}Mn_{20}Ni_{20}$ and nonequiatomic $Fe_{40}Mn_{27}Ni_{26}Co_5Cr_2$ alloys show a stable microstructure consisting of a single fcc phase, the microstructure characterization of the four-component equiatomic $Co_{25}Cr_{25}Fe_{25}Mn_{25}$ alloy reveals a contrasting picture."

"These observations suggest that the four-component, nonequiatomic $Fe_{37}Mn_{45}Co_9Cr_9$ alloy develops microstructures that consist of a dominant fcc phase for the investigated processing states."

"This comparison reveals that (I) the mechanical properties of all three alloys are comparable with high-Mn binary Fe-Mn alloys (although grain refinement typically leads to improved properties) and (II) all three alloys have comparable mechanical response with respect to each other, although the four-component $Fe_{37}Mn_{45}Co_9Cr_9$ alloy has relatively lower strength and ductility values."

Discussion

"It can be observed that a ~ 23% decrease in the CE from $Co_{20}Cr_{20}Fe_{20}Mn_{20}Ni_{20}$ to $Fe_{40}Mn_{27}Ni_{26}Co_5Cr_2$ alloy does not lead to any change in the stability of the massive SS fcc phase, whereas a decrease of ~ 20% in the CE from the $Co_{25}Cr_{25}Fe_{25}Mn_{25}$ to the $Fe_{37}Mn_{45}Co_9Cr_9$ alloy even favors the formation of a single-phase SS."

"From the perspective of mechanical properties, although it is rather contrary to what is expected in massively SS strengthened alloys (e.g., other HEA), the excellent tensile ductility and low strength combinations reported for this alloy system correlate well with the presented single fcc phase microstructures and the deformation mechanisms."

"Given the first indications of wide phase stability observed in this article for different compositions, compositional changes can be considered in future work to decrease the stacking fault energy and to introduce a mechanically induced twinning mechanism at room temperature."

Conclusion

"The heavy alloying requirements and the relatively small gains in mechanical performance (even with respect to some binary Fe-Mn systems) limit the room- or high-temperature applications of the most successful HEA $Co_{20}Cr_{20}Fe_{20}Mn_{20}Ni_{20}$."

"The presented investigations reveal that the compositional requirements to stabilize the massive solid solution fcc phase in this alloy system are more relaxed what was originally proposed."

"As expected from the similar deformation mechanisms observed (e.g., planar slip, nanotwinning, etc.), nonequiatomic HEA demonstrate mechanical performances comparable with those of the equiatomic alloy."

"Well-controlled compositional variations can now be considered not only to decrease the processing and alloying costs but also to explore improved mechanical property combination, e.g., by the introduction of a mechanically induced twinning mechanism at room temperature."

4.4 AlXCoCrFeNi High Entropy Alloys

Machine generated keywords: band, strain rate, disorder, shear, deformation, strain, alcocrfeni hea, order, behavior highentropy, rate, configurational, configurational entropy, enthalpy, principal element, powder

Microstructure, Mechanical and Corrosion Properties of AlCoCrFeNi High-Entropy Alloy Prepared by Spark Plasma Sintering [208]

This is a machine-generated summary of:

Zhou, P. F.; Xiao, D. H.; Yuan, T. C.: Microstructure, Mechanical and Corrosion Properties of AlCoCrFeNi High-Entropy Alloy Prepared by Spark Plasma Sintering [208]

Published in: Acta Metallurgica Sinica (English Letters) (2019)

Link to original: https://doi.org/10.1007/s40195-019-00962-8

If you want to cite the papers, please refer to the original.

For technical reasons we could not place the page where the original quote is coming from.

Abstract-Summary

"The sintered samples were annealed at 700, 800 and 900 °C, and the effect of annealing temperature on the microstructure, mechanical and corrosion properties was studied."

"The size and quantity of the nanoscale precipitates increase with increasing annealing temperature."

"The annealing temperature has an obvious effect on the mechanical properties and corrosion resistance of the spark plasma sintered AlCoCrFeNi HEA."

"When the annealing temperature is 700 °C, the hardness, yield strength and fracture strength reach the maximum with the value of 545 HV, 1430 MPa and 2230 MPa, respectively."

"The compressive ratio reaches the maximum of 17.2%, with the annealing temperature increasing to 800 °C."

"The corrosion resistance of the samples decreases with increasing the annealing temperature."

Introduction

"During PM, the AlCoCrFeNi HEA powders is mainly synthesized by mechanical alloying (MA) during which the pre-alloy powders are prepared by mixing raw elemental materials powders through high-energy ball milling [42–44]."

"The other way to prepare AlCoCrFeNi HEA powders is through gas atomization during which the raw materials are melted first and then gas-atomized to the pre-alloy powders [45–47]."

"In the reported literature, heat treatment especially annealing has an obvious effect on the phases, microstructure and mechanical properties of the AlCoCrFeNi HEA [48–50]."

"Both the preparation of AlCoCrFeNi HEA with gas-atomized powders combined with SPS technique and the effect of annealing temperature on the spark plasma sintered AlCoCrFeNi HEA have been rarely reported."

"The AlCoCrFeNi HEA was fabricated using gas-atomized pre-alloy powders combined with the SPS technique."

"The effect of annealing temperature on the phases, microstructure, mechanical and corrosion properties of the spark plasma sintered AlCoCrFeNi HEA was investigated."

Experimental

"The AlCoCrFeNi HEA pre-alloy powders(− 400 mesh) were sintered by SPS (FCT, HP D25/3) at 1200 °C for 10 min with a diameter of 40 mm in Ar atmosphere under the pressure of 30 MPa and heating rate of 50 °C/min."

"The phases of the sintered and annealed samples were characterized under X-ray diffractometer (XRD, SIEMENS D500) with Cu Kα radiation, through 2θ ranging from 20° to 90° with a scan rate of 4°/min."

"The hardness of the sintered and annealed samples was tested on the Vickers hardness tester (Micromet 5104) at a load of 100 g. Five random indents were performed for each sample."

"Compression tests were performed on the samples with a dimension of φ 4 mm × 8 mm using the Instron 3369 testing machine at a strain rate of 1×10^{-4} s^{-1} under room temperature."

Results and Discussion

"The Al content changes both in FCC and BCC/B2 phases, due to part of BCC/B2 phase transforms to FCC and σ phases during annealing process, which results in the peak shift of the FCC and BCC/B2 phases."

"It can be inferred that most of the Cr element in Cr-rich BCC phase diffuses into nanoscale precipitates, FCC and B2 matrix phases, and the remaining Cr element forms Cr-, Fe-rich σ phase during annealing."

"Cr-rich BCC phase disappears and σ phase forms during annealing process, which results in a reduction in corrosion resistance after annealing."

"In the microstructure, FCC phase increases and the Cr-rich BCC phase transforms to other phases, resulting in the decrease in the corrosion resistance with increasing the annealing temperature."

Conclusion

"AlCoCrFeNi HEA has been prepared using gas atomization combined with SPS technique."

"The effect of annealing temperature on the microstructure, mechanical and corrosion properties of the spark plasma sintered AlCoCrFeNi HEA has been studied, and the results are as follows: (1) There is phase transformation during annealing process with the B2 + BCC + FCC phases in the spark plasma sintered sample transforming to B2 + BCC + FCC + σ phases in the annealed samples; (2) except for the σ phase and nanoscale precipitates, the annealed twin is also observed after annealing at 900 °C; (3) annealing temperature has an obvious effect on the mechanical properties of the samples, with the highest hardness, yield strength, fracture strength and compressive ratio 545 HV, 1420 MPa, 2236 MPa and 17.2%, respectively; (4) the corrosion resistance of the spark plasma sintered AlCoCrFeNi HEA decreases with increasing annealing temperature."

Excellent Ballistic Impact Resistance of $Al_{0.3}$CoCrFeNi Multi-principal Element Alloy with Unique Bimodal Microstructure [209]

This is a machine-generated summary of:

Muskeri, Saideep; Gwalani, Bharat; Jha, Shristy; Yu, Anqi; Jannotti, Philip A.; Haridas, Ravi Sankar; Schuster, Brian E.; Lloyd, Jeffrey T.; Mishra, Rajiv S.; Mukherjee, Sundeep: Excellent ballistic impact resistance of $Al_{0.3}$CoCrFeNi multi-principal element alloy with unique bimodal microstructure [209]

Published in: Scientific Reports (2021)

Link to original: https://doi.org/10.1038/s41598-021-02209-y

If you want to cite the papers, please refer to the original.

For technical reasons we could not place the page where the original quote is coming from.

Abstract-Summary

"The mechanical response of $Al_{0.3}$CoCrFeNi alloy, with unique bimodal microstructure, was evaluated at quasistatic, dynamic, and ballistic strain rates."

"High density of deformation bands was observed at dynamic strain rates but there was no indication of adiabatic shear bands, cracks, or twinning."

"Deformation mechanisms at quasistatic and dynamic strain rates were primarily characterized by extensive dislocation slip and low density of stacking faults."

"Deformation mechanisms at ballistic strain rates were characterized by grain rotation, disordering of the $L1_2$ phase, and high density of stacking faults."

Introduction

"Low strain rate (in the range of 10^{-4} s^{-1}–10^{-2} s^{-1}) and dynamic strain rate (in the range of 10^1 s^{-1}–10^3 s^{-1}) response of HEAs has been extensively studied for the fundamental understanding of their plastic deformation mechanisms [15, 51–53]."

"Previous studies focused on single phase HEAs have demonstrated that the deformation mechanisms, strain hardening, flow stress, and strain rate sensitivity are strongly influenced by the strain rate [54–56], owing to the unique combination of solid solution strengthening, twinning, dislocation mediated plasticity, and stress-induced phase transitions [57]."

"There have been several studies focused on the deformation mechanisms in Al_xCoCrFeNi alloys, but there is a lack of systematic analysis of the deformation mechanism as a function of strain rate in compositionally and microstructurally complex alloys."

"The deformation mechanism in several HEAs has been reported to transition from dislocation-mediated to twinning-induced plasticity with increase in strain rate."

"The response of thermomechanically processed $Al_{0.3}$CoCrFeNi HEA was studied as a function of wide range of strain rates."

Experimental

"Quasi-static (strain rate of 10^{-3} s^{-1}) mini-tensile experiments were performed using a computer-controlled mini-tensile testing frame containing a 500-pound load cell with specimen dimensions of gauge length ~ 5 mm and width ~ 1 mm."

"Compression tests at strain rates of 10^{-3}, 10^{-2}, and 10^{-1} s^{-1} were conducted using a MTS 322 load frame equipped with a 500 kN load cell."

"High strain rate compression experiments were performed using a Split Hopkinson Pressure Bar (SHPB) from REL Inc SureTest Systems at strain rates of 700, 1200, and $1750s^{-1}$."

"All the compression tests were performed at room temperature on cubic specimens having a side length of 4 mm."

"The compression experiments at low strain rates (10^{-3}, 10^{-2}, and 10^{-1} s^{-1}) and high strain rate (700, 1200, and 1750 s^{-1}) were recorded at a sampling rate of 10 Hz and 1 MHz, respectively."

Results and Discussion

"The resistance to shear localization for single phase $Al_{0.3}CoCrFeNi$ alloy, at high strain rates was attributed to the excellent strain hardenability from forest dislocation hardening, mechanical twinning, and solid solution strengthening [57]."

"Even without any sign of micro- or nano-scale twinning, the excellent work hardening shown by this alloy during dynamic deformation may be attributed to solid solution strengthening, grain boundary hardening due to the presence of fine recrystallized FCC grains in the microstructure, and precipitation strengthening from the grain boundary B2 precipitates and nano-scale $L1_2$ precipitates [58–60]."

"There is a competition between the formation of microbands and twins in low SFE materials at high strain rates."

"The overall deformation mechanism during ballistic impact of our alloy was characterized by the formation of high density of stacking faults, grain rotation, and disordering of $L1_2$ phase."

Summary

"The microstructural features and mechanical response of $Al_{0.3}CoCrFeNi$ high entropy alloy was characterized in a wide range of strain rates including quasi-static deformation, Split-Hopkinson pressure bar testing, and ballistic impact."

"The microstructure after quasi-static and SHPB strain rates showed highly deformed grains with deformation bands and small density of stacking faults at dynamic strain rate, without any evidence of shear localization and twins."

"The target material failed by ductile hole growth in ballistic impact and the deformed microstructure was characterized by ASBs, ASB induced cracks, microbands, and dynamically recrystallized ultrafine grains near ASBs."

"Microstructure of the alloy after ballistic impact showed high density of dislocations, profuse stacking faults, grain boundary rotation, and disordering of $L1_2$ phase."

Aluminum Alloying Effects on Lattice Types, Microstructures, and Mechanical Behavior of High-Entropy Alloys Systems [210]

This is a machine-generated summary of:

Tang, Zhi; Gao, Michael C.; Diao, Haoyan; Yang, Tengfei; Liu, Junpeng; Zuo, Tingting; Zhang, Yong; Lu, Zhaoping; Cheng, Yongqiang; Zhang, Yanwen; Dahmen, Karin A.; Liaw, Peter K.; Egami, Takeshi: Aluminum Alloying Effects on Lattice

Types, Microstructures, and Mechanical Behavior of High-Entropy Alloys Systems [210]

Published in: JOM (2013)
Link to original: https://doi.org/10.1007/s11837-013-0776-z

If you want to cite the papers, please refer to the original.

For technical reasons we could not place the page where the original quote is coming from.

Abstract-Summary

"The crystal lattice type is one of the dominant factors for controlling the mechanical behavior of high-entropy alloys (HEAs)."

"As in conventional alloys with one or two principal elements, the addition of minor alloying elements to HEAs can further alter their mechanical properties, such as strength, plasticity, hardness, etc. Excessive alloying may even result in the change of lattice types of HEAs."

"We first review alloying effects on lattice types and properties of HEAs in five Al-containing HEA systems: $Al_xCoCrCuFeNi$, $Al_xCoCrFeNi$, $Al_xCrFe_{1.5}MnNi_{0.5}$, $Al_xCoCrFeNiTi$, and $Al_xCrCuFeNi_2$."

"It is found that Al acts as a strong bcc stabilizer, and its addition enhances the strength of the alloy at the cost of reduced ductility."

Introduction

"[16, 17, 61–65] HEA formation is widely conceived due to the contribution from the significantly higher entropy, compared to conventional alloys based on one or two principal elements, according to the definition of Gibbs free energy of mixing (ΔG_m) for a solution phase: where ΔH_m is the enthalpy of mixing, ΔS_m is the entropy of mixing, and T is the absolute temperature."

"The reality is that there are only a few compositions that form high-entropy disordered solid solutions [66, 67], while the majority of the alloys claimed to be HEAs actually contains more than one phase, and elemental partitioning among phases is apparent."

"This fact suggests that (I) the configurational entropy of mixing of multicomponent alloys with an equal composition for each component is simply not high enough to compensate the enthalpy effect of competing compound phases and (II) the effect of other entropy sources cannot be ignored."

Review on Effects of Al Addition

"[16, 68, 69] Those alloys with low Al contents ($0 \leq x \leq 0.5$) showed a simple fcc structure; as x exceeded ~ 0.8, a bcc structure appeared in addition to the fcc structure."

"Whether a mixed microstructure of both disordered bcc and B2 structures at high Al contents can still be classified as an HEA is somewhat questionable because partitioning of alloying elements between these two phases reduces the configurational entropy in each phase."

"[70] As the Al content increases, the hardening ratio (HR) after 50% deformation varies from 2.0 to 0.4, which suggests that the hardening ability of the fcc structure is about twice that of the bcc structure in this HEA system."

"In the $Al_xCoCrFeNiTi$ HEA system, due to the increased strengthening effect of lattice strains caused by the lattice-sites occupation of Al, the compressive strength and elastic modulus at room temperature reach the highest values of 2.28 GPa and 147.6 GPa, respectively, for the equimolar alloy ($x = 1.0$) in the as-cast condition."

Discussion

"Al acts as a strong bcc stabilizer in the 3d TM-based HEAs."

"[16, 71] The structural change of the HEA with the Al addition can be partially explained by the local topologies of the fcc and bcc structures."

"To relax the lattice distortion energy, the metastable fcc phase prefers to transform to a relatively stabilized bcc structure as the ratio of Al in the alloy is increased."

"For the seven binary systems shown, the interaction between Al and Cr is the weakest and repulsive in both fcc and bcc structures."

"This critical Al content (or the Al solubility in the fcc phase) and the compositional range over which duplex fcc/bcc structure forms, in turn, depend on the constituent elements in the alloy."

"The apparent contrast in Al-Ni and Al-Cr interactions strongly suggests that Cr will be excluded in the NiAl-type B2 structure (if it forms) in Al-containing HEAs, based on 3d TMs."

Suggested Future Work

"A high Al molar ratio (e.g., 1.0 in the $Al_xCoCrFeNi$ HEA system) will cause the formation of the brittle B2 phase."

"An optimal Al ratio should be carefully studied for each HEA system to achieve optimized mechanical properties."

"Like conventional alloys, heat-treated HEAs certainly have different microstructures and mechanical properties than as-cast ones."

"While the compression, hardness, and even corrosion behavior in an aqueous solution have been reported for these Al-containing HEAs, other mechanical properties, such as creep and fatigue [17], are needed to evaluate whether these HEAs meet the requirements for gas turbines that require structure stability and mechanical sustainability for substantially long hours at high temperatures."

Conclusions

"We reviewed the effects of Al alloying on the lattice types, microstructures, and mechanical behavior of five HEA systems, namely $Al_xCoCrCuFeNi$, $Al_xCoCrFeNi$, $Al_xCrFe_{1.5}MnNi_{0.5}$, $Al_xCoCrFeNiTi$, and $Al_xCrCuFeNi_2$."

"To quantify the interatomic interaction, the accurate enthalpy data of seven representative Al-TM binary systems for fcc, bcc, and intermetallic compounds were calculated using the CALPHAD method."

"Al behaves as a potent bcc stabilizer in 3d TM-based HEAs."

"Among the seven Al-TM binary systems, Al is a bcc stabilizer in the descending order: Ni > Co > Mn > Fe > Cr."

"The interaction between Al and Cr is the weakest and repulsive in both fcc and bcc structures, whereas it is the strongest for Al-Ni, followed by Al-Co. 3."

"The bcc phase formed in these five HEA systems at higher Al contents is actually a mixture of A2 and B2 phases for which element partitioning is apparent."

Atomistic Clustering-Ordering and High-Strain Deformation of an $Al_{0.1}CrCoFeNi$ High-Entropy Alloy [211]

This is a machine-generated summary of:

Sharma, Aayush; Singh, Prashant; Johnson, Duane D.; Liaw, Peter K.; Balasubramanian, Ganesh: Atomistic clustering-ordering and high-strain deformation of an $Al_{0.1}CrCoFeNi$ high-entropy alloy [211]

Published in: Scientific Reports (2016)

Link to original: https://doi.org/10.1038/srep31028

If you want to cite the papers, please refer to the original.

For technical reasons we could not place the page where the original quote is coming from.

Abstract-Summary

"Computational investigations of structural, chemical, and deformation behavior in high-entropy alloys (HEAs), which possess notable mechanical strength, have been limited due to the absence of applicable force fields."

"The simulation results are validated by a comparison to first-principles Korringa-Kohn-Rostoker (KKR)—Coherent Potential Approximation (CPA) [KKR-CPA] calculations for the HEA structural properties (lattice constants and bulk moduli), relative stability, pair probabilities, and high-temperature short-range ordering."

"We study $Al_xCrCoFeNi$ for Al ranging from $0 \leq x \leq 2$ mol fractions, and find that the HEA shows large chemical clustering over a wide temperature range for $x < 0.5$."

"At various temperatures high-strain compression promotes atomistic rearrangements in $Al_{0.1}$CrCoFeNi, resulting in a clustering-to-ordering transition that is absent for tensile loading."

Introduction

"Knowledge of high-strain deformation is particularly important for applications where strain rates higher than ~ 10^2 s^{-1} are encountered [72–78], such as high-temperature mechanical strength of lightweight armor materials, blast impact of debris on aircraft composite panels, satellites and spacecraft, as well as high-speed machining, all potentially relevant opportunities for employing HEAs."

"Because atomic arrangements in the alloy crystal significantly contribute to the material phase and deformation mechanics, we employ a combination of the computationally-demanding first-principles calculations and classical molecular-dynamics (MD) simulations to explore the high-strain deformation in a model HEA."

"Besides structural properties, the KKR-CPA directly predicts the global and local relative stability of phases in HEAs, the Warren-Cowley short-range order (SRO) in the most stable, high-temperature solid-solution phase, as well as the underlying electronic origin for the chemical clustering or ordering behavior—pair by pair."

Results and Discussion

"In MD calculations, $Al_{0.1}$CrCoFeNi is quenched from the melt phase, starting at 2200 K. This arrangement ensures that the solid-solution alloy formed at 300 K is through the high-temperature melt phase at 2200 K, as the experimental [79] melting temperature reported for CrCoFeNi and $Al_{0.3}$CrCoFeNi are around 1690 K and 1655 K, respectively."

"These energies are manifested in the observable $\alpha_{\alpha\beta}$ (k) through the Al-Ni pair with a peak at k = [000] (i.e., Al-Al or Ni-Ni pairs clusters in real space), and a weak Al-Cr nearest-neighbor ordering at k = [001] (X-point)."

"The diverging Al-Ni pair of $\alpha_{\alpha\beta}$ (k = [000]) at the spinodal temperature (T_{sp}) indicates the absolute instability of the alloy that forces clustering in $Al_{0.1}$CrCoFeNi, manifest in the Al-Al and Ni-Ni pairs, for example, and that can be compared to MD simulations from new force fields."

Methods

"The MD simulations for investigating the structure of the $Al_{0.1}$CrCoFeNi HEA [80, 81] have previously employed LJ potentials that cannot offer reliable predictions for such multi-component systems."

"Only the Al-Cr and Co-Cr cross interactions are modeled, using the Lennard-Jones (LJ) potential [80–82], due to the lack of the available EAM/alloy parameters for these interactions."

"The optimized structure is initialized at 2200 K under an isothermal-isobaric (NPT) ensemble at a pressure of 0 MPa for 90 picoseconds (ps) to melt the alloy using equilibrium MD simulations."

"This step is followed by rapid quenching of the alloy under the NPT ensemble at 0 MPa with two different cooling rates of 21.11 K/ps and 5.42 K/ps, respectively, to reach the desired temperatures between 77 K and 1500 K. We employ the Nosé-Hoover thermostat and barostat, each with a coupling time of 0.001 ps."

Deviation from High-Entropy Configurations in the Atomic Distributions of a Multi-principal-element Alloy [212]

This is a machine-generated summary of:

Santodonato, Louis J.; Zhang, Yang; Feygenson, Mikhail; Parish, Chad M.; Gao, Michael C.; Weber, Richard J.K.; Neuefeind, Joerg C; Tang, Zhi; Liaw, Peter K: Deviation from high-entropy configurations in the atomic distributions of a multi-principal-element alloy [212]

Published in: Nature Communications (2015)

Link to original: https://doi.org/10.1038/ncomms6964

If you want to cite the papers, please refer to the original.

For technical reasons we could not place the page where the original quote is coming from.

Abstract-Summary

"The alloy-design strategy of combining multiple elements in near-equimolar ratios has shown great potential for producing exceptional engineering materials, often known as 'high-entropy alloys'."

"The results suggest that the high-entropy alloy-design strategy may be applied to a wide range of complex materials, and should not be limited to the goal of creating single-phase solid solutions."

Introduction

"At high temperatures, disordered solid solutions are especially favoured in such alloys because of the high mixing entropy effect."

"As temperature decreases, the high-entropy effect becomes less important in relation to the enthalpy effect, and phases with a higher degree of order are favoured, including ordered solid solutions [83]."

"Depending on the detailed entropy–enthalpy balance, a number of phases may evolve as the alloy cools from high temperatures, including a random solid solution, partially ordered solid solution, ordered solid solution or even a stoichiometric compound."

"The term 'high-entropy' has been used for alloys with multiple principal elements (that is, five or more elements, with concentrations between 5 and 35 atomic percent) because of their inherent high configurational entropy of mixing in the random solution state and because of the emphasis on enhancing the formation of solid solution phases [16, 84]."

Results

"Taking all of the above Cu-rich FCC phases into account, the adjusted volume phase fractions of the alloy are 50% B2, 34% BCC and 16% FCC structures."

"Despite the distinct types of Cu-rich precipitates observed, the neutron diffraction patterns contain only a single family of FCC peaks (in addition to the B2/BCC peaks), with a room temperature lattice parameter a = 3.624 ± 0.003 Å. This structure is similar to the pure Cu (Ref. [85]), which is FCC, with a = 3.615 Å. Furthermore, the intensities of the FCC neutron diffraction peaks are consistent with an FCC phase fraction formed by all (or most) of the Cu in the alloy."

"In this simplified model (more detailed characterizations of the room temperature phases in the Al-Co-Cr-Cu-Fe-Ni system are found in the literature) [86, 87], the present B2 phase composition, in the temperature range of 600–1080 °C, is given by $Al_{1.3}CoCrCu_{0.1}FeNi$."

Discussion

"Multiphase microstructures should not be dismissed, not only because they provide valuable test cases for understanding phase separation in high-entropy alloys, but also because many are technologically significant in their own right."

"The success of using multiple principal elements in the development of these spinodal alloys is relevant to the high-entropy alloy research community and encouraging because it demonstrates that the strategy of using multiple principal elements is effective for developing a broad range of disordered and partially ordered alloys."

"Strong evidence of the local structural similarity between the liquid and solid phases has been provided by the present combined AIMD, experimental PDF and complementary temperature-dependent neutron and synchrotron X-ray diffraction studies."

"The results suggest that the high-entropy alloy-design strategy (that is, using multiple principal elements) should be considered, not only for the goal of producing single-phase solid solutions, but also for producing more complex multiphase materials with the benefit of the entropy-enhanced stability."

Methods

"Consider a B2 alloy composed of n different elements, where the overall molar fractions are given by x_i, for i = 1 to n. Equations (7)–(10), may be replaced by and where $x_{i\alpha}$ and $x_{i\beta}$ are the molar fractions of the element, i, on the α and β sublattices, respectively, and η_i is the order parameter of the element, i. Because the molar fractions, $x_{i\alpha}$ and $x_{i\beta}$, cannot be negative, all η_i must be within the range of − 1 to 1."

"In the present $Al_{1.3}CoCrCuFeNi$ alloy, one scenario, which gives good agreement with the experimental data at high temperatures, has the five transition metals (Co, Cr, Cu, Fe and Ni) sharing the same ordering parameter, η_{TM}, which is related to the Al-ordering parameter, η_{Al}, by where x_{Al} is the molar fraction of Al in the B2 phase."

4.5 Eutectic High Entropy Alloys

Machine generated keywords: ehea, eutectic, alcocrfeni ehea, alcocrfeni, alloy ehea, eutectic alloy, lamellar, castability, bcc phase, elongation, dualphase, mpa, particle, entropy alloy, solution strengthen

Effects of Chromium on the Microstructures and Mechanical Properties of $AlCoCr_xFeNi_{2.1}$ Eutectic High Entropy Alloys [213]

This is a machine-generated summary of:

Jiang, Hui; Li, Li; Wang, Rui; Han, Kaiming; Wang, Quanwei: Effects of Chromium on the Microstructures and Mechanical Properties of $AlCoCr_xFeNi_{2.1}$ Eutectic High Entropy Alloys [213]

Published in: Acta Metallurgica Sinica (English Letters) (2021)

Link to original: https://doi.org/10.1007/s40195-021-01303-4

If you want to cite the papers, please refer to the original.

For technical reasons we could not place the page where the original quote is coming from.

Abstract-Summary

"A series of $AlCoCr_xFeNi_{2.1}$ (x = 0, 0.25, 0.5, 0.75, 1.0) eutectic high entropy alloys (EHEAs) have been designed and prepared."

"The effect of Cr content on the microstructures and mechanical properties of the $AlCoCr_xFeNi_{2.1}$ alloys was systematically investigated."

"The results indicate that the $AlCoCr_xFeNi_{2.1}$ (x > 0) alloys exhibit almost complete lamellar eutectic microstructures with a mixture structure of FCC and B2 phases."

"The room temperature compressive tests results show that with an increase in Cr content (from x = 0 to x = 1.0), the yield strength will first decrease, and thereafter increase."

"The $AlCoCr_{0.5}FeNi_{2.1}$ (Cr0.5) alloy shows the best comprehensive mechanical properties."

"The Cr0.5 alloy also displays a high strength with a yield strength of 362 MPa at 700 °C. In summary, by changing the Cr content, $AlCoCr_xFeNi_{2.1}$ eutectic high entropy alloys with excellent comprehensive mechanical properties were obtained and prepared."

Introduction

"Lu and others designed the $AlCoCrFeNi_{2.1}$ EHEA with attractive castability and mechanical properties over a wide temperature range [88]."

"How to further improve the comprehensive mechanical properties of $AlCoCrFeNi_{2.1}$ EHEA and realize the rational allocation of strength, plasticity, and toughness has become a key problem to promote the industrial application of EHEAs."

"The variation of composition in alloys is expected to show a significant effect on its microstructure and mechanical properties."

"Jin and others [89] designed a Cr-free $Fe_{20}Co_{20}Ni_{41}Al_{19}$ EHEA based on the $AlCoCrFeNi_{2.1}$ EHEA, and the alloy shows a complete and regular nano-lamellar eutectic organization."

"The effects of alloying of Cr element on the microstructure and properties of the $AlCoCrFeNi_{2.1}$ EHEA are rarely reported [90]."

"The effect of alloying of Cr element on the formation of eutectic phase is still a problem."

"The $AlCoCr_xFeNi_{2.1}$ EHEA was selected and studied the effects of Cr on the microstructure and properties."

Experimental

"The $AlCoCr_xFeNi_{2.1}$ alloys with $x = 0, 0.25, 0.5, 0.75$, and 1.0 are denoted as Cr0, Cr0.25, Cr0.5, Cr0.75, and Cr1.0, respectively."

"The characterization of the crystal structures of the $AlCoCr_xFeNi_{2.1}$ alloys was carried out by using an X-ray diffractometer (XRD, Bruker Corporation, Ettlingen, Germany), with Cu Kα radiation at voltage of 40 kV, and by using a transmission electron microscope (TEM, JEOL-2100, JEOL, Japan)."

"The thermal properties of the $AlCoCr_{0.5}FeNi_{2.1}$ alloy were measured by using a differential scanning calorimeter (DSC) at a heating/cooling rate of 10 K/min."

"The strain rate was 10^{-3} s^{-1}, and cylindrical samples of $\varphi5$ mm $\times$ 10 mm and $\varphi8$ mm $\times$ 12 mm, respectively, were used in these studies."

Results

"When further decreasing the Cr content to $x = 0$, the alloy will be composed of a primary phase and a lamellar eutectic structure."

"Solid solution of Cr in the FCC phase have also been reported for some traditional FCC/B2 eutectic alloys [91, 92]."

"These results indicate that the Cr0.5 alloy fracture mode change from a brittle fracture to ductile fracture when increasing the temperature up to 800 °C. The FCC + B2 composition phase of the Cr0.5 EHEA is most probably responsible for these mechanical properties."

"The Cr0 alloy exhibits a hypoeutectic microstructure with a hard B2 primary phase, which leads to a higher yield strength."

"For the Cr0.25, Cr0.5, Cr0.75, and Cr1.0 alloys, having a eutectic microstructures without hard B2 primary phase exhibited lower yield strength and better compressive ductility will be obtained."

Conclusions

"The Cr-induced alloying effects on the crystal phase, microstructure, and mechanical properties (at both room temperature and high temperatures) were characterized and discussed."

"From the room temperature compressive tests, the yield strength decreases first and then increases with the increase of Cr content (from $x = 0$ to $x = 1.0$)."

"Wherein, the Cr0.25 and Cr0.5 alloys exhibited excellent compressive mechanical properties, the yield strengths, fracture strengths, and plastic strains are as high as 617 MPa vs. 617 MPa, 3161 MPa vs. 3167 MPa, and 47.5 vs. 47.6%, respectively."

"The Cr0.5 EHEA exhibited excellent tensile mechanical properties, with a yield strength of 536.5 MPa, a fracture strength of 1062 MPa, and an elongation of 13.8%."

"From the high temperature compressive test results, the compressive yield strength, and maximum strength, of the $AlCoCr_{0.5}FeNi_{2.1}$ EHEA were observed to be decreased at high temperature (700 °C, 800 °C, and 900 °C)."

Enhancing the Mechanical Properties of Casting Eutectic High Entropy Alloys with Mo Addition [214]

This is a machine-generated summary of:

Yang, Zhongsheng; Wang, Zhijun; Wu, Qingfeng; Zheng, Tao; Zhao, Purui; Zhao, Jinkai; Chen, Jiayin: Enhancing the mechanical properties of casting eutectic high entropy alloys with Mo addition [214]

Published in: Applied Physics A (2019)

Link to original: https://doi.org/10.1007/s00339-019-2506-z

If you want to cite the papers, please refer to the original.

For technical reasons we could not place the page where the original quote is coming from.

Abstract-Summary

"Mo element was chosen to further strengthen the EHEAs by solid solution strengthening."

"The microstructures, phase compositions and mechanical properties of $Co_{30}Cr_{10}Fe_{10}Al_{18}Ni_{32-x}Mo_x$ ($x = 0, 1, 2, 3$) alloys were investigated."

"That Mo acts as a solid solution strengthening element in the face-centered cubic (FCC) and body-centered cubic (BCC) phases."

Introduction

"The eutectic alloy concept to design HEAs is proposed, called EHEAs, aiming to solve the above problems with dual-phase structures [88]."

"Eutectic HEAs are expected to have the advantages of excellent mechanical properties and castability, which can be directly used in the as-cast condition."

"It has been shown that a series of EHEAs with a lamellar combination of FCC and BCC phases is also a kind of promising structural material [88, 93, 94]."

"Previous investigations have explored the variations of Al [95–97], Ni [88, 97], Fe [98] and Cr [92] on the microstructural and mechanical properties of HEAs or EHEAs with FCC and BCC phases."

"Mo was added to single FCC HEAs as the solid solution strengthening element and can improve corrosion resistance combined with Cr [99–101]."

"The effects of Mo element on the microstructures and mechanical properties of EHEAs $Co_{30}Cr_{10}Fe_{10}Al_{18}Ni_{32-x}Mo_x$ were investigated."

Experiments

"Five multi-principal component alloys with a small content of Mo, $Co_{30}Cr_{10}Fe_{10}Al_{18}Ni_{32-x}Mo_x$ (x = 0, 1, 2, 3, in molar ratio, denoted as Mo0, Mo1, Mo2, Mo3) were prepared by arc melting under a high-purity argon atmosphere simultaneously."

"Alloy ingots were re-melted four times to improve their homogeneities."

"The ingots were finally casted in a copper mold with a size of 80 mm in length, 10 mm in width and 2 mm in thickness."

"Specimens with a gauge length of 48 mm and a thickness of 3 mm were subjected to tensile tests at room temperature, using the Instron 3382 tensile testing machine at a constant strain rate of 1×10^{-3} s^{-1}."

Results and Discussion

"Both peaks of FCC and BCC phases shift towards the smaller 2θ angle with the increase of Mo content, indicating an increase in the lattice parameters."

"The trend of lattice parameters of FCC and BCC phases with respect to Mo content can certify the increasing solid solution effect of Mo. The lattice parameters of FCC phases get larger with the increase of Mo concentration, while there is only a small difference in Mo2 and Mo3, indicating that the Mo element has reached its solid solubility limit in FCC phase."

"The improvement is attributed to the solid solution strengthening of Mo in FCC phase and the increased volume fraction of primary BCC phase."

"Combining the microstructure evolution and mechanical properties from Mo0-Mo3, the effects of Mo in strengthening $Co_{30}Cr_{10}Fe_{10}Al_{18}Ni_{32-x}Mo_x$ alloys contain the solid solution strengthening and formation of BCC primary phase."

"With the content of Mo increasing, the volume fraction of primary BCC phase increased."

Conclusions

"According to the study, the following conclusions can be obtained: 1. Mo0 exhibits a full eutectic microstructure, while Mo1, Mo2, Mo3 alloys consist of primary BCC phase and eutectic phase. 2. Mo acts as a solid solution strengthening element in the FCC and BCC phases."

"It promotes the formation of BCC phase and the solid solubility of Mo in FCC phase is about 3 at.%. 3. Mo2 has the highest tensile strength of 1250.8 MPa in the as-cast condition due to the solid solution strengthening and the increase of BCC volume fraction."

Wear Properties of Spark Plasma-Sintered AlCoCrFeNi2.1 Eutectic High Entropy Alloy with NbC Additions [215]

This is a machine-generated summary of:

Jiang, Hui; Li, Li; Wang, Jianming; Wei, Chengbin; Zhang, Qiang; Su, Chunjian; Sui, Huaiming: Wear Properties of Spark Plasma-Sintered AlCoCrFeNi2.1 Eutectic High Entropy Alloy with NbC Additions [215]

Published in: Acta Metallurgica Sinica (English Letters) (2023)

Link to original: https://doi.org/10.1007/s40195-023-01529-4

If you want to cite the papers, please refer to the original.

For technical reasons we could not place the page where the original quote is coming from.

Abstract-Summary

"$AlCoCrFeNi_{2.1}$-xNbC (x = 0, 2.5, 7.5, and 10 wt%, denoted as NbC0, NbC2.5, NbC7.5, and NbC10) high-entropy alloy (HEA) matrix composites were fabricated by mechanical alloying (MA) and spark plasma sintering (SPS) methods."

"The effect of NbC content on the microstructures, Vickers hardness, and wear properties of $AlCoCrFeNi_{2.1}$ eutectic high entropy alloy (EHEA) was systematically investigated."

"With the addition of NbC, the average friction coefficient and wear rate of the $AlCoCrFeNi_{2.1}$-xNbC alloys decrease from 0.59 to 0.42 and from 1.5×10^{-5} mm^3 N^{-1} m^{-1} to 2.4×10^{-6} mm^3 N^{-1} m^{-1}, respectively."

"Wherein, NbC10 alloy shows the smallest average friction coefficient of 0.42 and lowest wear rate of 2.4×10^{-6} mm^3 N^{-1} m^{-1}, indicating that the NbC10 alloy displays the best wear resistance."

Introduction

"Many studies have shown that it is feasible to the addition of ceramic particles increase the wear resistance of alloys by refining the grains and improving hardness [102–108]."

"Ye and others [105] fabricated $AlCoCrFeNi_{2.1}$ EHEA with different TiC contents by micro-plasma cladding technique, and the results showed that the addition of TiC greatly improved the hardness and wear resistance of the cladding layer."

"Wherein, the NbC particles having much better wettability can significantly refine the microstructure of alloy and effectively improve mechanical properties [109–113]."

"Li and others [112] added NbC to $Fe_{50}Mn_{30}Co_{10}Cr_{10}$ HEAs coatings and found that nano-scale NbC particles can restrict the dendrite growth, making the microhardness and wear resistance of the coatings are significantly improved."

"In this study, the NbC was selected added into $AlCoCrFeNi_{2.1}$ EHEA to improve the wear resistance."

"The microstructures and mechanical properties including hardness and wear performance of $AlCoCrFeNi_{2.1}$-xNbC HEA composites were system investigated."

Experimental

"The milled powders were subjected to spark plasma sintering furnace at 900 °C for 15 min with an applied pressure of 50 MPa to obtain cylindrical blocks of 15 mm diameter and 11 mm height."

"The Vickers hardness was measured with a Vickers hardness tester (MH-50) at a loading of 500 g and a dwell time of 15 s. The dry wear tests were carried out on multifunctional friction and wear testing machine (Rtec MFT-5000, USA) with ball-on-disk dry sliding reciprocating friction at room temperature."

"The round trip distance, sliding speed, loads, and sliding time of tests were 5 mm, 1 Hz, 10 N, and 30 min, respectively."

"The corresponding wear rates (mm^3 N^{-1} m^{-1}) were calculated by the equation: W = V/FL, where V is the worn volume (mm^3), F is the normal load (N) and L is the sliding distance (m)."

Results and Discussion

"It can be seen that the hardness of the $AlCoCrFeNi_{2.1}$-xNbC alloys first decreases slightly from 679 HV (x = 0) to 664 HV (x = 2.5) and then, increases to 695 HV (x = 10) with the increase in NbC content."

"The NbC particles gradually emerge from the surface of the $AlCoCrFeNi_{2.1}$ EHEA during the wear process, reducing the contact area between the alloy surface and the friction pair to prevent further wear of the surface."

"It is clear that the NbC10 alloy has the best wear resistance and the lightest surface wear."

"This indicates that the oxidation reaction occurred on the wear surface, which is formed by the frictional heat generated by the reciprocal squeezing of the alloy surface by the counterpart ball."

"For the dry wear experiment, oxidation wear resulted from the oxygen to react with the metal elements in the alloy surface to produce metal oxides due to elevated temperature at the friction interface."

Conclusions

"The effect of NbC particles addition on the microstructure, Vickers hardness, and wear resistance of the $AlCoCrFeNi_{2.1}$-xNbC HEAs was investigated."

"The crystal structure of $AlCoCrFeNi_{2.1}$-xNbC alloys changed from the B2 + FCC main phases and Al_2O_3 and Cr_7C_3 minor phases to the NbC + FCC main phases and Al_2O_3 and Cr_7C_3 minor phases. In addition, with the increase in NbC content, the volume fraction of NbC phase increased, and the NbC particles in the NbC2.5, NbC7.5, and NbC10 alloys present the biscale distribution with micro-/nano-scale."

"With the addition of NbC particles, the average friction coefficient of the $AlCoCrFeNi_{2.1}$-xNbC alloys gradually decreases from 0.59 to 0.42, and the wear rate increases first from 1.5×10^{-5} mm^3 N^{-1} m^{-1} to 1.7×10^{-5} mm^3 N^{-1} m^{-1} and then, decrease to 2.4×10^{-6} mm^3 N^{-1} m^{-1}."

Phase Evolution and Mechanical Behaviour of Co–Fe–Mn–Ni–Ti Eutectic High Entropy Alloys [216]

This is a machine-generated summary of:

Jain, Reliance; Rahul, M. R.; Jain, Sandeep; Samal, Sumanta; Kumar, Vinod: Phase Evolution and Mechanical Behaviour of Co–Fe–Mn–Ni–Ti Eutectic High Entropy Alloys [216]

Published in: Transactions of the Indian Institute of Metals (2018)

Link to original: https://doi.org/10.1007/s12666-018-1437-2

If you want to cite the papers, please refer to the original.

For technical reasons we could not place the page where the original quote is coming from.

Abstract-Summary

"For Co–Fe–Mn–Ni–Ti HEAs, at first BCC (β) phase primary dendritic phase is created from the liquid, followed by peritectic reaction to form FCC (α) phase (i.e. BCC (β) + L → FCC (α))."

"The left over liquid undergoes eutectic reaction to form FCC (α) and Ti_2(Ni, Co) phases (i.e. L → FCC (α) + Ti_2(Ni, Co))."

"The pseudo quasi-peritectic reactions i.e. L + BCC (β) → FCC (α) + Ti_2(Ni, Co) has been proposed based on the microstructure evolution study in case of Co–Fe–Mn–Ni–Ti HEAs."

"Co–Fe–Mn–Ni–Ti HEAs show eutectics between FCC (α) and Ti_2(Ni, Co) phases."

Introduction

"The concentrated multicomponent high entropy alloys (HEAs) are nowadays considered as new research frontier in the materials science society."

"Results have proposed that a high value of configuration entropy in such multi-component alloys leads to the development of simple BCC, FCC, HCP structures with solid solution phases and thus reduces the chances of formation of complex phases."

"A few eutectic high entropy alloys (EHEAs) have so far been made because of the lack of multi-element phase diagrams."

"The EHEAs combining the advantages of both high entropy alloys and eutectic alloys, and at same time eliminating the drawbacks of these two alloys, leads to the promising new class of high-temperature alloys [114]."

"CALPHAD based thermodynamic calculation can also be utilized to design the eutectic high entropy alloys [115, 116]."

"The high-temperature mechanical properties of the newly designed multicomponent Co–Fe–Mn–Ni–Ti HEAs have also been investigated."

Experimental Materials and Procedures

"The multicomponents $Co_{20}Fe_{20}Mn_{20}Ni_{20}Ti_{20}$ and $Co_{25}Fe_{25}Mn_5Ni_{25}Ti_{20}$ HEAs were prepared by arc melting cum casting machine from highly pure ($\geq$ 99.7%) Co, Fe, Mn, Ni and Ti elements in a Ti-gettered ultra high-purity argon atmosphere to get arc melted alloy button."

"Further, both alloy buttons were suction casted in the form of cylindrical rod having 6 mm diameter."

"The phase evolution of studied HEA during solidification was explored using both X-ray diffraction (XRD) and scanning electron microscopy (SEM) coupled with energy dispersive spectroscopy (EDS)."

"The constant temperature hot compressive deformation experiments of suction casted rod (ϕ = 6 mm and height: diameter ratio of 1.5:1) were performed using Gleeble 3800® thermo-mechanical simulator at different deformation temperatures of 800 °C, 900 °C and 1000 °C with constant strain rate of 10^{-3} s^{-1}."

Results and Discussion

"It is observed from microstructural characterization of $Co_{20}Fe_{20}Mn_{20}Ni_{20}Ti_{20}$ HEA that there are one peritectic reaction i.e. BCC (β) + L $\rightarrow$ FCC (α) and one eutectic reaction i.e. L $\rightarrow$ FCC (α) + Ti_2(Ni, Co) features in the microstructure of studied HEA."

"The solidification pathways for the investigated multicomponent EHEA has been described as follows; (i) at first BCC (β) primary dendritic phase is evolved from the liquid during solidification, then (ii) the formation of secondary dendritic phase such as FCC (α) solid solution phase due to peritectic reaction i.e. BCC (β) + L $\rightarrow$ FCC (α) and finally (iii) the remaining liquid forms FCC (α) and Laves phase by eutectic reaction i.e. L $\rightarrow$ FCC (α) + Ti_2(Ni, Co) Based upon the microstructural features observed in the studied multicomponent EHEA, the four phase reaction i.e. L + BCC (β) $\rightarrow$ FCC (α) + Ti_2(Ni, Co) has been proposed for the investigated HEA."

Conclusion

"Multicomponent HEAs have been made effectively by vacuum arc melting cum suction casting technique and the detailed phase evolution in the newly designed multicomponent HEAs have been understood during solidification."

"Multicomponent HEAs reveal the unique microstructural features consisting of peritectic and eutectic morphologies."

"The non-equiatomic multicomponent eutectic $Co_{25}Fe_{25}Mn_5Ni_{25}Ti_{20}$ HEA shows the retention of better mechanical strength at elevated temperature."

Enhanced Strength–Ductility Synergy in Ultrafine-Grained Eutectic High-Entropy Alloys by Inheriting Microstructural Lamellae [217]

This is a machine-generated summary of:

Shi, Peijian; Ren, Weili; Zheng, Tianxiang; Ren, Zhongming; Hou, Xueling; Peng, Jianchao; Hu, Pengfei; Gao, Yanfei; Zhong, Yunbo; Liaw, Peter K.: Enhanced strength–ductility synergy in ultrafine-grained eutectic high-entropy alloys by inheriting microstructural lamellae [217]

Published in: Nature Communications (2019)

Link to original: https://doi.org/10.1038/s41467-019-08460-2

If you want to cite the papers, please refer to the original.

For technical reasons we could not place the page where the original quote is coming from.

Abstract-Summary

"Realizing improved strength–ductility synergy in eutectic alloys acting as in situ composite materials remains a challenge in conventional eutectic systems, which is why eutectic high-entropy alloys (EHEAs), a newly-emerging multi-principal-element eutectic category, may offer wider in situ composite possibilities."

"This work provides a pathway for strengthening eutectic alloys and widens the design toolbox for high-performance materials based upon EHEAs."

Introduction

"Eutectic high-entropy alloys (EHEAs) [88, 89, 117], first proposed by Lu and others in 2014 [118], were designed based on the above eutectic-alloy concept."

"Previous studies showed that engineering an ultrafine-grained duplex microstructure by severe cold-rolling and annealing could dramatically strengthen eutectic alloys although degrading ductility [93]."

"The full deformation constraint imposed by the hard lamella matrix enables the soft lamellae to be almost as strong as the hard matrix, thus making materials with high strength [119, 120]."

"A remarkable strength–ductility enhancement is expected if we could inherit the heterogeneous lamellar nature via an appropriate thermo-mechanical treatment, instead of just tailoring a dual-phase ultrafine-grained structure in eutectic alloys."

"Compared to the aforementioned heterogeneous lamella structure, our current structure features the strength heterogeneity with soft/hard phases instead of bimodal grains [120], and simultaneously exhibits a higher lamella density arising from the full lamella nature of eutectic alloys."

Results

"To emphasize the markedly improved properties after engineering the DPHL structure, the curves of the ultrafine-grained EHEA [93] and the as-cast EHEA are also shown."

"Our DPHL700 and DPHL740 with the inherited lamellar geometry exhibit a simultaneous strength–ductility enhancement, and even higher strengths over that of the ultrafine-grained EHEA."

"The dimensions of the tensile specimens in Ref. [94] show that the sample is small, and the gage length, width, and thickness are 2 mm, 1 mm, and 300 μm, respectively, while only one annealing condition, 800 °C for 1 h, is applied."

"We use three annealing conditions of 660, 700, and 740 °C for 1 h, and the dimensions of tensile specimens are 15 mm in gage length, 3.2 mm in width, and 600 μm in thickness, which are typically consistent with the frequently reported dimensions in the literature [119, 121]."

Discussion

"This process produces a long-range back-stress, making dislocations difficult to move in FCC grains until B2 grains start to yield deformation [122]."

"Although these regions allow significant strain partitioning to induce back-stress strengthening, the soft FCC phase also allows the material to yield at low stress, owning to the absence of the lamellar constraint effect [120]."

"Yang and others reported work in a single-phase medium-entropy alloy [123], where they purposely architectured a three-level heterogeneous grain structure with grain sizes spanning the nanometer-to-micrometer range via partial recrystallization annealing following conventional cold-rolling, achieving large uniform tensile strain (~22%) after yielding even at the gigapascal stress."

"The heterogeneous structure becomes even more heterogeneous during tensile straining as more twins, faults, and nano-grains with high-angle grain boundaries are generated dynamically owing to the low stacking fault energy, as well as the dislocations, leading to increased inhomogeneous plastic deformation."

Methods

"Small pieces [dimensions: 50 mm (length) × 25 mm (width) × 4 mm (thickness)] were extracted from the as-cast material and subjected to multi-pass cold-rolling to 84–86% reduction in thickness (the final thickness of ~ 600 μm) using a laboratory-scale two-high rolling machine."

"That by the extraction of small pieces from the as-cast material, we can obtain high-quality samples with no or few surface defects (such as oxide films), smooth surfaces, and uniform thickness, for the subsequent cold-rolling and annealing treatment, which is able to prevent other factors to influence and even deteriorate mechanical properties."

"All tensile specimens were dog-bone-shaped, with a gauge length of 15 mm, a width of 3.2 mm, and a thickness of 600 μm."

"Tensile tests were carried out at room temperature using an MTS Criterion Model 44 with an initial strain rate of 2.5×10^{-4} s^{-1}."

"All tensile tests were conducted, using a 12-mm extensometer to monitor the strain."

4.6 Metastable/transformative High Entropy Alloys

Machine generated keywords: trip, chemistry, transformation, martensite, hcp phase, work harden, hcp, friction stir, stir, metastable, dimension, deformation, friction, processing, dualphase

Microstructural Evolution and Deformation Behavior of Ni-Si- and Co-Si-Containing Metastable High Entropy Alloys [218]

This is a machine-generated summary of:

Sinha, S.; Nene, S. S.; Frank, M.; Liu, K.; Mishra, R. S.; McWilliams, B. A.; Cho, K. C.: Microstructural Evolution and Deformation Behavior of Ni-Si- and Co-Si-Containing Metastable High Entropy Alloys [218]

Published in: Metallurgical and Materials Transactions A (2018)

Link to original: https://doi.org/10.1007/s11661-018-4968-6

If you want to cite the papers, please refer to the original.

For technical reasons we could not place the page where the original quote is coming from.

Abstract-Summary

"Two high entropy alloy (HEA) compositions were compared to explore the individual effects of Co and Ni on phase stability and resultant deformation response in Fe-Mn-Cr-Si-containing HEAs."

"It was observed that Co-Si-containing HEA depicted responsive phase evolution upon friction stir processing owing to decreased γ (f.c.c.) matrix stability as against Ni-Si-containing HEA."

"The Co-Si HEA showed the presence of dual-phase microstructure with the dominance of ε (h.c.p.) phase (52 pct), whereas the Ni-Si HEA showed single-phase γ microstructure under similar processing condition."

Introduction

"The broad objective of the present research effort is to design nonequiatomic HEAs, the microstructure and deformation response of which can be tailored through their chemistry and processing, to incorporate the effect of various mechanisms like phase transformation, twinning, dislocation, and precipitation strengthening, while maintaining the core effects of HEAs [84, 124, 125]. Friction stir processing (FSP) of these metastable HEAs enables microstructural refinement with some control over temperature, strain, and strain rate [126]. Previously, Nene and others [127, 128] discussed FSP-engineered dual-phase HEAs where enhanced strength and ductility were the result of pronounced TRIP in the material."

"Those previous studies highlighted the advantages that can be obtained by martensitic transformation in terms of strength–ductility tradeoff, generating interest in studying various aspects of the transformation like its dependence on alloy chemistry, fundamental understanding of the ε (h.c.p.) martensite phase in HEAs, and the kinetics of martensite formation."

"This study outlines the dependence of martensitic transformation on alloy chemistry by specifically comparing two friction stir processed HEAs to illustrate the effect of Co or Ni on the γ (f.c.c.) metastability and concurrent mechanical response."

Experimental Methodology

"FSP was carried out on the 6-mm as-cast HEA sheets using tungsten-rhenium (W-Re) tool."

"Flat (rectangular) dog-bone-shaped mini-tensile specimens with gage length of 5 mm, width of 1.25 mm, and thickness of 1 mm were machined out using a mini computer numerical control machine from the top surface of the as-cast ingot and from 1 mm below the surface from the stirred region of the friction stir processed material."

"Microstructures of the as-cast, friction stir processed, and tensile-deformed specimens were characterized using electron backscattered diffraction (EBSD) on a FEI Nova Nano SEM 230 equipped with Hikari Super EBSD detector using an operating voltage of 20 kV. The EBSD scan area for the as-cast material was roughly 350 μm $\times$ 250 μm, and the step size of acquisition was 0.5 μm."

"X-ray diffraction (XRD) measurements of the as-cast and tensile-deformed specimens of Ni-Si HEA and Co-Si HEA were carried out using Rigaku Ultima III diffractometer and Cu Kα radiation operating at 40 kV and 44 mA."

Results

"While Co-Si HEA shows ε (h.c.p.)-dominated dual-phase microstructure, highlighting the metastability of the γ (f.c.c.) phase in the latter alloy that triggered transformation during deformation."

"The inherent dependence of martensitic transformation on the phase stability is also proven by the fact that the as-cast Co-Si HEA material showed higher amount of transformation after tensile deformation (9 pct h.c.p.)".

"Ni-Si HEA with the increased γ (f.c.c.) stability showed no transformation in the as-cast material (100 pct f.c.c.)".

"Ni-Si HEA [single-phase γ (f.c.c.)] shows lower overall GND density (~ 1.5×10^{15} m^{-2} in the D-pass-deformed condition) than Co-Si HEA."

"Absent any transformation, the available strain accommodating mechanism in Ni-Si HEA is through f.c.c."

Discussion

"The as-cast Ni-Si HEA only displayed peaks for the γ (f.c.c.) phase indicating the presence of a single-phase microstructure, whereas the as-cast Co-Si HEA specimen shows peaks for both γ (f.c.c.) and ε (h.c.p.) phases."

"Higher metastability of γ (f.c.c.) phase in Co-Si HEA than Ni-Si HEA is expected to result in a more favorable microstructure for invoking transformation-driven plasticity in the Co-Si HEA and twinning assisted deformation in the Ni-Si HEA [129]. Microstructural evaluation of Ni-Si HEA after tensile test showed no change in the fraction of γ (f.c.c.) phase."

"The sustained work hardening of Co-Si HEA suggests that, unlike single-phase Ni-Si HEA specimen, dual-phase Co-Si HEA specimen should have undergone sustained work hardening owing to the active participation of both ε (h.c.p.) and γ (f.c.c.) phases during deformation."

Conclusions

"(1) The addition of Ni or Co substantially altered the phase evolution of Fe-Mn-Cr-Si-containing HEAs, and led to a single-phase microstructure in Ni-containing HEA (Ni-Si HEA), but dual-phase microstructure in Co-containing HEA (Co-Si HEA) in the as-cast condition. (2) FSP with two overlapping passes on Ni-Si HEA resulted in grain-refined γ (f.c.c.) microstructure as against the ε (h.c.p.)-dominant dual-phase microstructure in Co-Si HEA."

"This variation in microstructural evolution is a strong indication of higher γ phase metastability in Co-Si HEA than Ni-Si HEA. (3) Co-Si HEA possesses higher work-hardening capability due to controlled γ (f.c.c.) to ε (h.c.p.) transformation and uniform strain partitioning in the γ (f.c.c.) and ε (h.c.p.) phases compared with Ni-Si HEA, which is coarse-grained and has limited twinning-mediated strain-hardening ability."

Extremely High Strength and Work Hardening Ability in a Metastable High Entropy Alloy [219]

This is a machine-generated summary of:

Nene, S. S.; Frank, M.; Liu, K.; Mishra, R. S.; McWilliams, B. A.; Cho, K. C.: Extremely high strength and work hardening ability in a metastable high entropy alloy [219]

Published in: Scientific Reports (2018)

Link to original: https://doi.org/10.1038/s41598-018-28383-0

If you want to cite the papers, please refer to the original.

For technical reasons we could not place the page where the original quote is coming from.

Abstract-Summary

"Inspired by this, here we present Si containing dual phase $Fe_{42}Mn_{28}Co_{10}Cr_{15}Si_5$ high entropy alloy (DP-5Si-HEA) exhibiting very high strength (1.15 GPa) and work hardening (WH) ability."

"Higher yet sustained WH ability in fine grained DP-5Si-HEA is associated with the uniform strain partitioning among the metastable γ phase and resultant h.c.p. (ε) phase thereby resulting in total elongation of 12%."

"Design of dual phase HEAs for improved strength and work hardenability can be attained by tuning the metastability of γ matrix through proper choice of alloy chemistry from the abundant compositional space of HEAs."

Introduction

"These results suggested that, either the change in alloy chemistry or the processing path could result in enhanced YS but may lose the metastability driven TRIP or twinning induced plasticity (TWIP) effects in DP-HEAs."

"TRIP literature [130–134] suggested that (h.c.p.) ε phase is harder than the γ matrix and hence evolution of martensite dominant microstructure would result in increased YS of the alloy while retaining the TRIP or TWIP effect."

"Recent results on TRIP assisted HEAs [127, 130, 131, 135] showed that presence of both Fe and Mn is needed for altering the γ phase stability during deformation and hence should be an integral part of the alloy chemistry."

"Design of dual phase HEA with increased metastability of the γ matrix: This altered phase stability in the HEA can be achieved by changing the alloy chemistry."

Alloy Design Approach

"Alloys with metastable γ matrix would result in lower equilibrium γ fraction at room temperature (25 °C) owing to higher driving force for transformation."

"According to their work, C and Al increased the $\Delta G^{\gamma \to \varepsilon}$ values (more γ phase fraction) whereas Si decreased it (less γ phase fraction)."

"It appears that 5 at.% Si would result in effective increase in the metastability of the γ matrix up-to high temperature and also corresponds to the lowering of SFE of the HEA system under consideration."

"In order to compare the effect of Si addition on the γ phase metastability as against the equivalent addition of the Ni or Al in the HEA matrix, T_o values and equilibrium γ fractions were obtained for $Fe_{42}Mn_{28}Co_{10}Cr_{15}Y_5$ (where Y_5 stands for 5 at.% of elements like Si or Al or Ni) alloys from simulations."

Microstructure and Phase Evolution in As-Cast and FSP Conditions

"The as-cast microstructure was refined by multi-pass friction stir processing (M-pass, FSP carried out with 3 overlapping passes starting with 650 rotations per minute (RPM) down to 350 RPM in the second pass and the last pass being carried out at 250 RPM)."

"Stored deformation during each pass promotes the conversion of metastable γ phase to ε phase as a result of TRIP."

"The starting of lower strain rate [126] processing at 250 RPM during third overlapping pass of FSP limits further nucleation of the ε phase and stabilizes the metastable γ phase in the microstructure at the onset of cooling."

"This kinetic stability of phases triggers the formation of γ phase (88%) dominant microstructure with limited fraction of ε phase (12%) upon M-pass (multi-pass friction stir processing)."

Improved Strength in As-Cast and FSP DP-5Si-HEA

"When compared with the values of YS reported by Li and others [130, 135] for DP-HEA, DP-5Si-HEA showed almost 100 and ~ 600 MPa increase in YS for as-cast and M-pass specimens respectively."

"The significant increase in the YS of M-pass specimen is mainly attributed to the severe grain refinement from the as-cast grain size of 100 μm to 1.3 μm attained upon FSP."

"M-pass specimen showed very fine grain size and almost two times more ε phase fraction in comparison with the as-cast material."

"The onset of yielding was increased in M-pass specimen thereby reaching to a YS of 950 MPa from as-cast YS of 400 MPa."

Rapid Work Hardening and Resultant Mechanical Properties

"Rate of work hardening during this sub-stage II (i.e. TRIP) depends on the γ phase stability during deformation for a given grain size and ε phase fraction [127, 130, 131, 135]."

"This is because, higher grain boundary area exerts more back stress and results in more controlled $\gamma \rightarrow \varepsilon$ transformation and thus sustained work hardening in the very fine grained condition [127, 130, 131, 135, 136]."

"Controlled interplay between the TRIP in metastable γ phase and the twinning in ε phase for grain refined HEA resulted in almost double elongation (12%) than coarse grained as-cast material."

"This is because the change in the γ proportion upon complete deformation for a given grain size and prior ε fraction indicates higher driving force ($\Delta G^{\gamma \rightarrow \varepsilon}$) and hence the preferential tendency for the $\gamma \rightarrow \varepsilon$ transformation (i.e. γ phase metastability) [132, 137]."

Conclusions

"Friction stir processing of Si containing HEA resulted in extremely refined metastable γ phase dominant microstructure thereby exhibiting a very high YS and UTS of 950 MPa and 1.15 GPa, respectively, with an elongation of 12%."

"These improved mechanical properties in comparison with TRIP assisted HEAs are attributed to the extremely high work hardening ability of the material."

"This rapid work hardening ability is associated with the increased metastability of γ phase which facilitated the transformation of γ phase and twinning in resultant ε phase during deformation."

"Design of metastable DP-5Si-HEA and its microstructural engineering via FSP shows a new pathway for obtaining strong and highly work hardenable HEAs."

Materials and Methods

"These sheets of 5 mm were subjected to multi-pass friction stir processing (M-pass)."

"The processing tool had a shoulder diameter of 12 mm with tapered pin."

"The root diameter, pin diameter, and length for the tool were 7.5 mm, 6 mm, and 3.5 mm, respectively."

"Rectangular 1 mm-thick, dog-bone-shaped mini-tensile specimens were machined using a mini computer numerical control (CNC) machine from 1 mm below from the surface within the nugget region of the FSP specimen whereas from the top surface of the as-cast ingot."

On the Evolving Nature of c/a Ratio in a Hexagonal Close-Packed Epsilon Martensite Phase in Transformative High Entropy Alloys [220]

This is a machine-generated summary of:

Sinha, Subhasis; Nene, Saurabh S.; Frank, Michael; Liu, Kaimiao; Agrawal, Priyanka; Mishra, Rajiv S.: On the evolving nature of c/a ratio in a hexagonal close-packed epsilon martensite phase in transformative high entropy alloys [220]

Published in: Scientific Reports (2019)
Link to original: https://doi.org/10.1038/s41598-019-49904-5

If you want to cite the papers, please refer to the original.

For technical reasons we could not place the page where the original quote is coming from.

Abstract-Summary

"Alloys where there is no diffusion-less phase transformation and associated transformation volume change with deformation, the c/a ratio remains constant during deformation."

"C/a ratio and transformation volume change of h.c.p."

"The study revealed that while intrinsic c/a is dependent on alloying elements, c/a of epsilon in transformative HEAs changes with processing and deformation."

"This is attributed to transformation volume change induced dependence of h.c.p."

"Lower than ideal c/a ratio promotes non-basal pyramidal ⟨c + a⟩ slip and deformation twinning in epsilon phase of transformative HEAs."

Introduction

"The ε-martensitic transformation and the presence of the hexagonal close packed (h.c.p.) ε phase in the microstructure of transformative HEAs gained importance rapidly [134, 135, 138, 139]."

"Transformative HEAs provide opportunities to investigate the difference in the response of this h.c.p."

"Structured phase compared with well-known h.c.p."

"The present study investigates the c/a of ε (h.c.p.) phase in five transformative HEAs as a function of alloy composition (in the as-cast condition)."

"The effect of subsequent friction stir processing and deformation on the c/a ratio of the ε (h.c.p.) phase is investigated for all the alloys."

"This study aims to elucidate the ε-phase controlled deformation mechanisms in transformative Fe-Mn-Co-Cr-Si containing HEAs through quantification of the c/a ratio of the ε (h.c.p.) phase and transformation volume change in these alloys."

Results and Discussion

"SF analysis suggests that FSP followed by tensile deformation induces a textural evolution (of the ε-phase) that is not entirely random but is rather a dual-textured microstructure where adjacent grains are preferentially oriented for either basal or prismatic slip; and higher strain is accommodated in both types of grains via pyramidal $\langle c + a \rangle$ slip."

"In the present study, both CS-HEA and Al-HEA show increase in $\rho_{\langle c+a \rangle}$ from as-cast to as-FSP condition with decrease in c/a. Also, Cu-HEA shows increased $\rho_{\langle c+a \rangle}$ with decreasing c/a from as-cast to deformed condition."

"After tensile deformation, all three alloys transform to ε-dominant microstructures, wherein $\rho_{\langle c+a \rangle}$ increases similarly for Cu-HEA and Al-HEA; while $\rho_{\langle c+a \rangle}$ decreases slightly in CS-HEA due to decreased c/a ratio, as explained before."

"Characteristic deformation twins were observed in the ε (h.c.p.) phase in tensile deformed specimens of as-cast Cu-HEA that illustrate the contribution of deformation twinning to strain accommodation."

Summary

"The present study showed that the ε (h.c.p.) phase in transformative HEAs exhibits c/a ratio similar to Mg in the as-cast condition."

"The c/a ratio and transformation volume are responsive to alloy chemistry, microstructure and stress state."

"Microstructural evolution of the ε-phase during FSP followed by deformation is such that a dual texture is developed with progressive strain, wherein alternate grains orient themselves for basal and prismatic slip to accommodate initial strain; subsequently, higher strain is accommodated by pyramidal $\langle c + a \rangle$ slip and deformation twinning."

Methods

"X-ray diffraction (XRD) measurements of as-cast, as-FSP and tensile-deformed specimens were performed with a Rigaku Ultima III diffractometer and Cu Kα radiation operating at 40 kV and 44 mA. Lattice parameters of γ (f.c.c.) and ε (h.c.p.) phases and c/a ratio of ε (h.c.p.) phase were calculated from XRD peak positions using the following procedure."

"The c/a ratio of ε (h.c.p.) phase in as-cast CS-HEA was also verified from TEM diffraction pattern (DP)."

"Microstructural characterization by electron backscatter diffraction (EBSD) of as-cast, as-FSP and tensile-deformed specimens was performed using a FEI Nova Nano SEM 230 equipped with Hikari Super EBSD detector at an operating voltage of 20 kV. TEAM™ software enabled data acquisition and TSL OIM Version 8 software was used for data analysis."

Correlating Work Hardening with Co-activation of Stacking Fault Strengthening and Transformation in a High Entropy Alloy Using In-situ Neutron Diffraction [221]

This is a machine-generated summary of:

Frank, M.; Nene, S. S.; Chen, Y.; Gwalani, B.; Kautz, E. J.; Devaraj, A.; An, K.; Mishra, R. S.: Correlating work hardening with co-activation of stacking fault strengthening and transformation in a high entropy alloy using in-situ neutron diffraction [221]

Published in: Scientific Reports (2020)

Link to original: https://doi.org/10.1038/s41598-020-79492-8

If you want to cite the papers, please refer to the original.

For technical reasons we could not place the page where the original quote is coming from.

Abstract-Summary

"Transformation induced plasticity (TRIP) leads to enhancements in ductility in low stacking fault energy (SFE) alloys, however to achieve an unconventional increase in strength simultaneously, there must be barriers to dislocation motion."

"While stacking faults (SFs) contribute to strengthening by impeding dislocation motion, the contribution of SF strengthening to work hardening during deformation is not well understood; as compared to dislocation slip, twinning induced plasticity (TWIP) and TRIP."

"We used in-situ neutron diffraction to correlate SF strengthening to work hardening behavior in a low SFE $Fe_{40}Mn_{20}Cr_{15}Co_{20}Si_5$ (at%) high entropy alloy, SFE ~ 6.31 mJ m^{-2}."

Introduction

"Stacking faults (SFs) formed in metals and alloys are important to their deformation behavior and mechanical properties [140–142]."

"This understanding gap likely stems from challenges in measuring SF dimensions, i.e. width or interspacing, in a specimen of a size that work hardening behavior can be simultaneously measured and correlated with deformation mechanisms."

"The number of studies which correlate the SF strengthening deformation mechanism to work hardening behavior using in-situ approaches of this type are limited."

"The activation of SF strengthening and transformation induced plasticity (TRIP) clearly provided real-time evidence of their cooperative behavior and contributions to simultaneous enhancement of strength and ductility."

"While the relationship between SF strengthening and SF interspacing was previously established [143–145], we use in-situ neutron diffraction to capture this phenomenon in real time for the first time and demonstrate the combined importance of activating multiple deformation mechanisms to overcoming the strength-ductility tradeoff."

Results and Discussion

"The changes in intensity of the ε-h.c.p. {10.1} and {00.2} were observed, where the intensity of the {00.2} peak decreased significantly after FSP."

"The {10.1} peak was not observed in the AC condition whereas it was clearly captured after FSP as a result of severe plastic deformation at high temperatures."

"We observed that less transformation corresponded to less ductility (by comparison of fractured AC and AC + FSP conditions)."

"It would be expected that SFs would have contributed to strengthening in TRIP steels, while transformed fraction would have been decreasing with the increase in strength."

"The observations from previous studies and the comparison of ductility in the AC and AC + FSP conditions suggest that the contributions of both SF strengthening and TRIP are important to enabling the synergistic increase in strength and ductility in metastable HEAs."

Conclusions

"In-situ neutron diffraction was used to quantify stacking fault strengthening and reveal the onset of the γ-f.c.c."

"The present study demonstrates the application of in situ diffraction techniques for real-time quantification of SF strengthening in low stacking fault energy metals and alloys."

"The changes in L_{SF} and σ_{SF} indicate that stacking faults also contribute to strengthening during plastic deformation alongside other mechanisms."

"Phase transformation being accompanied by the stacking fault strengthening deformation mode is also important to attaining simultaneous increases in both strength and ductility."

Methods

"The cast plate 300 × 100 × 6 mm was subjected to friction stir processing (FSP) for microstructural refinement, henceforth denoted as AC + FSP."

"To characterize the extent of microstructural refinement following processing, optical microscopy was carried out on both conditions."

"For secondary electron (SE) images a FEI Quanta 3D FEG microscope (FEI Company, Hillsboro, Oregon) was operated at 20 kV to investigate the microstructures of both AC and AC + FSP conditions."

"Reconstruction of APT data was carried out in CAMECA's Integrated Visualization and Analysis Software (IVAS) version 3.8.2 for determination of composition, visualization of ion maps and subsequent analysis of the spatial distribution of elements."

"Single peak fitting was used to determine both lattice spacing and integrated intensities for various family of grains."

"In time-of-flight neutron diffraction, deformation induced changes in the lattice spacing, d^{hkl}, of families of grains, for example {111} and {222}, in polycrystalline materials are measured."

4.7 Refractory High Entropy Alloys

Machine generated keywords: refractory, rhea, bcc, refractory high, mpa, yield strength, pct, yield, microstructure mechanical, anneal, ductile, roll, high yield, matrix, compressive

Surface Oxidation Mechanism of a Refractory High-Entropy Alloy [222]

This is a machine-generated summary of:

Osei-Agyemang, Eric; Balasubramanian, Ganesh: Surface oxidation mechanism of a refractory high-entropy alloy [222]

Published in: npj Materials Degradation (2019)

Link to original: https://doi.org/10.1038/s41529-019-0082-5

If you want to cite the papers, please refer to the original.

For technical reasons we could not place the page where the original quote is coming from.

Abstract-Summary

"We examine and explain the reaction process of oxygen on a representative refractory HEA surface using first principles and atomistic thermodynamic models."

"The HEA surface is highly reactive to oxygen yielding a full monolayer coverage at temperatures between 300 and 1500 K. The preferential adsorption of oxygen to specific sites of the HEA surface is attributed to the electronic configuration of the bonding shells of the constituent surface atoms."

"Manipulation of temperature and oxygen pressure reveals that it is difficult to rid the alloy surface of oxygen even at extremely low pressures of 10^{-9} bar at 2000 K."

Introduction

"High-entropy alloys (HEAs), a subset of multi-principal element alloys (MPEAs), contain N principal elements typically in near equiatomic concentrations with $N \geq 5$."

"[39, 61] Ordered solid-solutions have, nevertheless, been observed to exist in some HEAs though they are typically less stable, which affects their properties and microstructures, e.g., through increase in density, higher melting temperatures and increased brittleness of the resulting alloys."

"The z value of 0.5 indicates each of Mo and W is 0.425 mol fraction in that HEA, while $z = 0.05$ indicates Mo is 0.8075 and W is 0.0425 by mole fraction in the corresponding alloy."

"As these refractory alloys are potentially useful for operations in high temperature environments, knowledge of their oxidation behavior is required to develop oxidation-resistant HEAs."

"The data on the oxidation mechanisms of HEAs is sparse, and more so for the Mo-W-Ta-Ti-Zr refractory HEA."

Results and Discussion

"The calculated adsorption energy (E_{ads}) is 5.45 eV. An earlier investigation on oxidation of ZrC found O to adsorb strongly at fcc three-fold hollow sites mostly between surface Zr atoms with $E_{ads} = 5.83$ eV on the (111) surface that primarily had only Zr atoms."

"We further examine the diffusion of oxygen into the bulk structure of the HEA1 alloy by adding a 25th O atom to the fully oxidized surface layer."

Methods

"The errors introduced by representing random solid solutions with small periodically repeated unit cells can be mitigated by constructing SQS structures to reproduce the correlation functions between the first few nearest neighbors."

"As solid-solution HEAs have structures with random distribution of constituent atoms, there are additional challenges in selecting the most stable surface for surface adsorption studies."

"Different surface layers containing different atomic compositions are generated and their surface energies calculated for the (100), (110), and (111) facets."

"To study the effect of oxygen on all five constituent elements of the alloy, a surface model containing Mo, W, Ta, Ti, and Zr atoms with the lowest surface energy is subsequently generated and employed for the oxidation studies."

"The distribution of the calculated Gibbs free energies ($\Delta_r G$) as a function of pressure p at different temperatures with the different coverages of O_2 molecules exemplify the thermodynamic stability of these surfaces."

Grain Growth and Hall–Petch Relationship in $Ti_{37}V_{15}Nb_{22}Hf_{23}W_3$ Refractory High-Entropy Alloys [223]

This is a machine-generated summary of:

Huang, Wenjun; Yin, Shaochun; Wang, Xuejiao; Guo, Ruipeng; Wu, Yucheng; Qiao, Junwei: Grain growth and Hall–Petch relationship in $Ti_{37}V_{15}Nb_{22}Hf_{23}W_3$ refractory high-entropy alloys [223]

Published in: Journal of Materials Research (2023)

Link to original: https://doi.org/10.1557/s43578-023-00924-y

If you want to cite the papers, please refer to the original.

For technical reasons we could not place the page where the original quote is coming from.

Abstract-Summary

"The effects of annealing on the mechanical characteristics of non-equiatomic $Ti_{37}V_{15}Nb_{22}Hf_{23}W_3$ RHEAs that are 80% cold rolled are investigated in this study."

"The friction stress of 786 MPa in $Ti_{37}V_{15}Nb_{22}Hf_{23}W_3$ RHEAs is higher than that of most metals and alloys reported previously, indicating that the current BCC RHEA has a high intrinsic stress."

"The coefficient, 712 MPa $\mu m^{0.5}$, is higher than that in the HfNbTaTiZr RHEA and lower than in the HfNbTiZr RHEA, indicating greater grain-boundary strengthening."

Introduction

"The smaller the grain size, the greater the stress needed for dislocation slip due to frequent dislocation packing [146, 147]."

"The smaller the grain size, the shorter the distance that dislocations move, which require more dislocation slipping to satisfy the corresponding strain."

"By adjusting the grain size, the free control of the yield strength of CoCrFeMnNi HEAs from 165 to 361 MPa can be easily achieved [7]."

"For RHEAs, the yield strengths of the HfNbTiZr alloy with different grain sizes would range from 681 to 763 MPa [148]."

"So far there is no relevant study on the dependence of mechanical properties on the grain size for non-equiatomic RHEAs."

Results

"The predicted result is that the grain size grows as the annealing time rises."

"A significant number of defects, such as dislocations, grain boundaries, and other deformation heterogeneities, like deformation bands, are generated inside the deformed material as a result of plastic deformation during the cold-rolling process as the thickness lowers [149]."

"This might be due to the fact that during boundary migration, numerous solutes of various sizes would alter their relative locations in situ, leading to the low-energy structure of grain borders [150]."

"Annealing the RHEA specimens at 1000 °C for different times produces an average grain sizes of 22.5, 40.5, 66.5, and 71.7 μm, respectively."

"The RHEA with the minimum grain size, 22.5 μm, has the highest yield strength, 942 MPa, next by specimens with grain sizes of 40.2 μm, 930 MPa, and 66.5 μm, 874 MPa, and 71.7 μm, 841 MPa."

Discussion

"In pure systems or solid solutions with substantial solubility of alloying elements, the grain-boundary mobility is inversely related to the grain size, and the parameter $n = 2$ is typically associated with normal (self-similar) grain formation [151, 152]."

"According to previous studies [151, 153], $n = 3$ is another more frequently observed value for the situation of the segregation of low-solubility alloying elements or impurities to grain borders, for which the grain-boundary mobility is inversely related to the grain diameter squared."

"In the cold-rolled and annealed $Ti_{37}V_{15}Nb_{22}Hf_{23}W_3$ RHEAs investigated in the present study, $n = 3.59$ is a little bit bigger than the typical value of 3.5 for the drag of secondary-phase precipitates or the segregation of low-solubility alloying elements/impurities to grain boundaries, suggesting that slow diffusion effect in HEAs might play some role in slowing grain growth [125, 154]."

Conclusion

"After cold-rolling and annealing at 1000 °C for 10 min to 24 h, the grain growth of the cast $Ti_{37}V_{15}Nb_{22}Hf_{23}W_3$ RHEA was investigated."

"The grain size of this alloy increases with increasing annealing time."

"The yield strength of the alloy gradually decreases with increasing annealing time."

Materials and Methods

"Energy-dispersive spectroscopy (EDS) and scanning electron microscopy (SEM) using a Phenom XL were used to determine the alloy compositions, which were calculated as follows (in at.%): 37.13 Ti, 15.54 V, 22.19 Nb, 22.56 Hf, and 2.58 W. After drop casting into the mold, the cast specimens with the dimensions of 6 × 12 × 36 mm were cold rolled to a plate with a deformation of 80% from 6 to 1.2 mm."

"The cold-rolled plates were annealed at 1000 °C for 10 min, 30 min, 1 h, 5 h, and 24 h, respectively, in a covered quartz tube with Ar atmosphere, following water quenching (WQ) to get entirely recrystallized grains."

"The microstructures were examined using energy-dispersive spectroscopy (EDS) and scanning electron microscopy (SEM) with a Phenom XL."

"For the SEM observation, the final polished specimens were etched with a $HF:HNO_3:H_2O = 1:2:7$ solution (in a volume percent, vol%)."

Phase-Transition Assisted Mechanical Behavior of $TiZrHfTa_x$ High-Entropy Alloys [224]

This is a machine-generated summary of:

Huang, Shuo; Li, Wei; Holmström, Erik; Vitos, Levente: Phase-transition assisted mechanical behavior of $TiZrHfTa_x$ high-entropy alloys [224]

Published in: Scientific Reports (2018)

Link to original: https://doi.org/10.1038/s41598-018-30892-x

If you want to cite the papers, please refer to the original.

For technical reasons we could not place the page where the original quote is coming from.

Abstract-Summary

"Recent developments of high-entropy alloys with high strength and high ductility draw attention to the metastability-engineering strategy."

"Using first-principle theory, here we demonstrate that reducing the Ta level in the refractory $TiZrHfTa_x$ system destabilizes the body-centered cubic (bcc) phase and leads to the appearance of the hexagonal close-packed (hcp) phase embedded in the bcc matrix."

"The present atomic-level insight provides strong evidence to the experimental observation, and emphasizes the significance of quantum-design for advanced multi-phase high-entropy alloys with excellent strength-ductility combinations."

Introduction

"HEAs are composed of multi-principal elements with equal or near-equal molar ratios, and most of them show preference to stabilize in simple solid-solution phases with face-centered cubic (fcc), body-centered cubic (bcc), or hexagonal close-packed (hcp) underlying lattices [162]."

"The bcc HEAs based on refractory elements usually possess good phase stability and mechanical strength [163–165], which makes them potential high-temperature structural materials."

"These alloys consist of multiple phases (i.e., fcc and hcp phases), and show improved strength and ductility as compared to many single-phase HEAs."

"Understanding the intrinsic elastic properties of individual structures is critical for optimizing the mechanical performance of such dual-phase HEAs."

Results and Discussion

"The present theoretical results indicate that the hcp phase is energetically stable in the low-Ta region, and the bcc phase becomes favorable in the upper part of the actual concentration range."

"The small decrease of the bcc DOS near E_F is likely to be associated with the decrease of the hcp-stabilizing elements and the addition of Ta which in pure bcc phase has a pseudo-gap slightly above E_F."

"This explains the relative stability of the bcc phase over hcp with increasing Ta content."

"Taking into account the mechanical stability results and those found for the thermodynamic stability, below we discuss the elastic behavior of $TiZrHfTa_x$ in the bcc and hcp phases only."

"The total-energy and electronic structure calculations predict that the hcp phase remains stable in the low-Ta region, whereas Ta addition makes the bcc phase thermodynamically favorable."

Methods

"The present ab initio calculations were based on the exact muffin-tin orbitals (EMTO) method [166], in combination with the coherent potential approximation (CPA) [167–169]."

"The equilibrium volume and bulk modulus were extracted from the equation of state fitted to the ab initio total energies for a series of different volumes."

"The single-crystal elastic constants were obtained by straining the lattice and evaluating the energy changes due to the strain as a function of its magnitude at equilibrium volume [166, 170, 171]."

"All calculations were performed for static lattice and for completely disordered solid solutions."

"The chemical disorder for both lattices was treated by CPA."

Evaluation of the Microstructural Aspects, Mechanical Properties and Dry Sliding Wear Response of MoTaNbVTi Refractory High Entropy Alloy [225]

This is a machine-generated summary of:

Poulia, Anthoula; Georgatis, Emmanuel; Karantzalis, Alexander: Evaluation of the Microstructural Aspects, Mechanical Properties and Dry Sliding Wear Response of MoTaNbVTi Refractory High Entropy Alloy [225]

Published in: Metals and Materials International (2019)

Link to original: https://doi.org/10.1007/s12540-019-00283-6

If you want to cite the papers, please refer to the original.

For technical reasons we could not place the page where the original quote is coming from.

Abstract-Summary

"MoTaNbVTi refractory high entropy alloy was synthesized by the vacuum arc melting technique in its equi-atomic composition."

"In terms of sliding wear response, the use of a metallic counterbody partner combined with the increase in sliding distance, led to possible thermal fatigue effects, while the presence of a ceramic counterbody ball was associated with the lubrication action of the formed oxides and the improvement of the alloy's tribological response, as the sliding distance increased."

Introduction

"The development of new metallic alloys has been rapidly evolved, due to the need of new materials that combine effective structures and functional properties."

"Named as High Entropy Alloys (HEAs), these materials are of great interest to both the scientific and the industrial community [16, 175] since they denoted the commonly alloy concept and transformed the traditional view of metallic materials into an original scientific trend."

"Mechanisms of all properties, features and characteristics of these systems should be thoroughly understood, so that this category of metallic materials could be explored to a wider extent [84]."

"The basic findings suggest a significantly higher wear resistance of both the HEAs and RHEAs compared to the state of the art materials (e.g. steels or superalloys), mainly attributed either to the addition of various constituent elements or to the nature of the oxides formed under different experimental conditions tested."

Materials and Methods

"Compression testing was performed using a Galdabini Sun 10 (100 kN) machine at room temperature on rectangular specimens (4 mm $\times$ 4 mm $\times$ 6 mm)."

"Three compression samples were tested at room temperature, while further analysis of the specimens deformed was carried out to examine the fracture surfaces with the aid of SEM analysis."

"Concerning the tribological investigation of the alloy, dry-sliding unlubricated wear tests were performed using a ball-on disc configuration (CSM Instruments) at room temperature, with relative humidity of about 70%."

"Test runs were interrupted every 100 m (in the case of a 400 m total sliding distance) or every 200 m (in the cases of the longer distances e.g. 1000 m and 2000 m) and specimens were reweighted."

"Three repeating trials were performed for each specimen in every testing condition."

Results and Discussion

"As observed, MoTaNbVW and MoTaNbZrTi HEAs demonstrate a relatively similar trend in their derived compression curves, while MoTaNbVTi HEA exhibits a significant increase in its maximum compression strength, as well as in its strain values up to the fracture point."

"The observation of the tracks surfaces and the first indications for the material's response are further supported by the calculation of the mass loss data, as a function of the three sliding distances, as well as by the derived wear rate estimations."

"The observed increase in the wear rate values associated with increasing the sliding distance to 2000 m, is probably related to thermal fatigue phenomena on the contact surfaces being examined."

"With a further increase of the sliding distance (as in the case of 2000 m), the temperature further increases, resulting in a possible fragmentation of these particles aggregation and their final removal from the surface of the sample, due to subsurface softening phenomena, thermal mismatch effects and a limitation of the plastic deformation mechanism."

Conclusions

"As derived by the alloy's tribological investigation when using a metallic counterbody partner, different wear track landscapes were associated with the corresponding mass loss and wear rate values."

"The increase of the sliding distance up to 2000 m was followed by a wear degradation of the system, possibly associated with some thermal fatigue effects along the contact points."

"The system improved its mass loss data and consequently the wear rate estimations, as the testing developed over longer distances, due to the formation of oxidized phases with good lubrication effects."

"The improvement of the tribological response of the alloy by increasing the testing distance was associated with the existence of a relatively smooth landscape of the wear tracks (especially at longer distances) and a gradual increase of the Al formed oxides in the debris."

Microstructure and Mechanical Properties of the Ductile Al–Ti–Mo–Nb–V Refractory High Entropy Alloys [226]

This is a machine-generated summary of:

Cieślak, Grzegorz; Dąbrowa, Juliusz; Jawańska, Monika; Parzuchowska, Agnieszka; Oleszak, Dariusz: Microstructure and Mechanical Properties of the Ductile Al–Ti–Mo–Nb–V Refractory High Entropy Alloys [226]

Published in: Metallurgical and Materials Transactions A (2021)

Link to original: https://doi.org/10.1007/s11661-021-06543-8

If you want to cite the papers, please refer to the original.

For technical reasons we could not place the page where the original quote is coming from.

Abstract-Summary

"A number of non-equimolar refractory high entropy alloys (RF HEAs) from the Al–Ti–Mo–Nb–V system are synthesized, with the selected compositions aimed to balance the conflicting requirements of the low-temperature ductility and high-temperature corrosion protection."

"For the selected $Al_{20}Ti_5Mo_{25}Nb_{25}V_{25}$ and $Al_{10}Ti_{30}Mo_{20}Nb_{20}V_{20}$ alloys, the measured stress–strain curves indicate the highly coveted, ductile room temperature behavior, with the values of ultimate strain measured under compression mode being 9.17 and 9.00 pct, respectively, and compressive fracture strain of 13.38 and 13.25 pct, respectively."

"The obtained results suggest that it is possible to include Al as a vital component of refractory HEAs without compromising their low-temperature ductility."

Introduction

"In the case of high entropy alloys, by far the most prospective application are the high-temperature structural materials."

"Thanks to these exceptional properties, high entropy alloys are nowadays considered potential materials for energy, automotive, aerospace, and nuclear industries."

"[32, 176] While the development of HEAs started with the transition metal-based alloys, during the last decade another group of HEAs has gathered ever-increasing attention, namely the refractory metal-based high entropy alloys (RF HEAs)."

"Developed by Senkov and others [66] in 2010, over the years they have proved to possess an outstanding potential as the next-generation, high-temperature materials, in many cases displaying properties outclassing the conventional superalloys."

"Based on the typical empirical criteria of HEAs solid solution formation, thermodynamic calculations, as well as the above-described factors affecting the performance of these materials, the non-equimolar alloys of satisfactory low-temperature ductility behavior are obtained, with a chemical composition at least theoretically optimized toward the requirements of the high-temperature corrosion."

Materials and Methods

"The selection of these three elements was based on their excellent mutual solubility as well as moderate values of oxides' enthalpies of formation, when compared with, e.g., IV group elements."

"The use of aluminum was justified by the requirements of the high-temperature corrosion resistance, as it is the only viable element with the capability to form a protective scale (not considering the possibility of the formation of complex, spinel, protective scales as in Ref. [177]), since the stability of Cr_2O_3 appears to be too low in relation to the oxide phases of refractory elements."

"All ingots were obtained in a form of round droplets, weighing around 15 g. The microstructure of the as-cast alloys was examined by, both, optical and scanning electron microscopy (SEM)."

"Five measurements were performed for each alloy."

"The compression tests were performed with a standard strain rate 1×10^{-3} s^{-1}, using the Zwick compression machine."

Results and Discussion

"Considering the as-cast state of the alloys presented in this study, based on the results of thermodynamic modeling, it can be expected that the alloys should exhibit a single-phase structure, with the ordered body-centered cubic B2 phase being present in all cases."

"Two alloys were selected for further studies: $Al_{20}Ti_5Mo_{25}Nb_{25}V_{25}$, in which Al is the main element decreasing the VEC value of the alloy, and $Al_{10}Ti_{30}Mo_{20}Nb_{20}V_{20}$, in which Ti acted as the main plasticizer ($VEC_{Ti} = 4$), more in-line with the original postulates by Sheikh and others [178]. As a result, the effectiveness of both elements in this role was investigated, which is of high significance, considering the crucial importance of aluminum presence in terms of the high-temperature resistance."

"For the first alloy, $Al_{20}Ti_5Mo_{25}Nb_{25}V_{25}$, despite the relatively high content of Al, and low content of the assumed plasticizer (Ti), the average fracture strain ε_f reaches the value of 13.38 pct, with the ultimate strain ε_u value of 9.17 pct."

Conclusions

"In the presented study, two series of the refractory high entropy alloys (RF HEA) from Al–Mo–Nb–Ti–V system: $Al_{20-x}Ti_xMo_{25}Nb_{25}V_{25}$ and $Al_{20-x}Ti_xMo_{20}Nb_{20}V_{20}$, were successfully produced by arc-melting technique."

"The thermodynamic modeling indicates the presence of a single-phase, B2 structure in all considered as-cast alloys, with the $Mo_{25}Nb_{25}V_{25}$-based series being characterized by the tendency toward the formation of Mo-rich A15 phase, while the precipitation of the secondary, Ti-rich B2 phase is predicted for $Al_{20}Ti_{20}Mo_{20}Nb_{20}V_{20}$."

"Basing on the measured stress-strain curves for the selected $Al_{20}Ti_5Mo_{25}Nb_{25}V_{25}$ and $Al_{10}Ti_{30}Mo_{20}Nb_{20}V_{20}$, the studied alloys possess ductile behavior, a highly coveted feature for refractory HEAs."

"Based on the literature data, the corrosion behavior of RF HEAs appears to be highly complex, therefore despite the content of Al at the level typically accepted as sufficient to support the selective oxidation process, further studies must proceed in order to verify this capability in the presented materials."

Effect of Cold Deformation and Annealing on the Microstructure and Tensile Properties of a HfNbTaTiZr Refractory High Entropy Alloy [227]

This is a machine-generated summary of:

Senkov, O. N.; Pilchak, A. L.; Semiatin, S. L.: Effect of Cold Deformation and Annealing on the Microstructure and Tensile Properties of a HfNbTaTiZr Refractory High Entropy Alloy [227]

Published in: Metallurgical and Materials Transactions A (2018)

Link to original: https://doi.org/10.1007/s11661-018-4646-8

If you want to cite the papers, please refer to the original.

For technical reasons we could not place the page where the original quote is coming from.

Abstract-Summary

"The microstructure and tensile properties of HfNbTaTiZr after cold working and annealing were investigated."

"The cold-worked material retained a single-phase microstructure and had a room temperature tensile yield stress $\sigma_{0.2} = 1438$ MPa, peak true stress $\sigma_p = 1495$ MPa, and true fracture strain $\varepsilon_f = 5$ pct."

"Annealing at 800 °C for up to 256 h resulted in the precipitation of Nb and Ta rich particles with a BCC crystal structure inside a Hf-and-Zr-enriched BCC matrix."

"The second phase particles nucleated heterogeneously inside deformation bands and slip lines and coarsened during annealing."

"After heat treatment at this temperature, the alloy was characterized by a single-phase BCC structure with $\sigma_{0.2} = 1110$ to 1115 MPa, $\sigma_p = 1160$ to 1195 MPa, and $\varepsilon_f = 12$ to 19 pct with the maximum values attained after annealing for 1 h."

Introduction

"Notable exceptions include RHEAs such as HfNbTaTiZr [64, 179–181], HfNbTiZr [182], and HfMoxNbTaTiZr [183]. It should be emphasized, however, that the mechanical properties of most RHEAs reported in the literature have been for material in the as-cast or cast-and-annealed condition."

"Noticeable improvements in the tensile properties of a HfNbTaTiZr RHEA after cold working and annealing have been reported recently [180]. For example, this alloy had a yield stress of 820 MPa and tensile fracture strain between 0 and 9.5 pct (depending on the sample position in the casting) [181] in the as-cast condition with a dendritic microstructure."

"Following cold rolling to 86 pct thickness reduction and annealing at 1000 °C for 2 h, the same alloy exhibited a homogeneous recrystallized microstructure with a yield stress of 1145 MPa and a fracture strain of 10 pct [180]. The HfNbTaTiZr alloy has a single-phase BCC crystal structure in the as-cast condition as well as after cold rolling and annealing at temperatures above 1000 °C [179–181, 184]. By contrast, second-phase precipitates with a BCC crystal structure, enriched in Nb and Ta, were identified in the material that was rolled to a reduction of 86 pct and annealed at 800 °C for 2 h [180] or processed by high-pressure torsion (HPT) and annealed at 800 °C or 900 °C for 1 h [185]. It is interesting to note that without preliminary cold deformation, the alloy retained a single phase BCC structure during annealing at 800 °C for up to 10 h while a hexagonal phase enriched in Hf and Ta precipitated at sub-grain boundaries during longer annealing times [186]. A hexagonal phase enriched in Hf and Zr was also reported in the HPT-processed alloy, but at much lower (500 °C) temperature [185]. These results indicate that in an equilibrium condition the HfNbTaTiZr alloy is a single-phase BCC structure at temperatures $\geq$ 1000 °C, but has multiple phases below ~ 1000 °C."

Material and Experimental Procedures

"Two rectangular blocks, each measuring 20.3 mm × 34.7 mm × 9.55 mm, were extracted from the annealed button and initially cold forged in a 1000-ton press with 355.6 mm diameter H-13 steel dies; Fel-Pro C-300 lubrication was used to reduce friction."

"After forging, the sides of the samples were re-machined to obtain preforms measuring 31.3 mm × 48.0 mm × 4.3 mm which were then rolled in a two-high mill without lubrication along the 31.3-mm direction."

"The cross-section of samples extracted for metallographic analysis contained rolling and normal directions of the rolled sheet. The tension and metallography samples were sealed in Ar-filled glass capsules and annealed at 800 °C for 2, 4, 16, 64, or 256 h, or at 1100 °C for 0.5, 1, 2 or 4 h."

Results

"An increase in annealing time at 800 °C to 256 h caused a further slight decrease in the peak stress."

"The tensile true fracture strain exhibited a maximum of 7.6 pct after annealing for 16 h; after annealing for 64 and 256 h, the ductility decreased to 6.0 pct and 4.6 pct, respectively."

"All of the tension specimens extracted from the as-rolled or rolled-and-800 °C-annealed sheets showed strong strain localization in the through-thickness direction (localized through-thickness necking)."

"The fracture behavior of tension specimens extracted from rolled sheet that was annealed at 1100 °C was different from that for the two previous conditions."

"The magnitude of the strain localization was between 1.4 and 2.5, being smallest for the specimen annealed for 1 h which also showed the highest tensile ductility (18.5 pct)."

Discussion

"pct Nb are 219, 219 and 235 kJ/mol, respectively [187]. One can therefore surmise that the activation energy of diffusion of the alloying elements in equiatomic HfNbTaTiZr might be comparable to an average value of the self-diffusion activation energies (~ 246 kJ/mol), with the activation energies of Nb and Ta only slightly higher than those of other elements."

"Because the dislocation density was not measured experimentally, $\Delta\sigma_{df}$ for the as-cold-worked specimen and specimens annealed at 800 °C, which have an unrecrystallized structure, were calculated by subtracting all other strengthening contributions from the measured respective yield-stress values."

"This estimated ρ_{df} is in good agreement with reported data for cold-worked BCC materials [188, 189]. A similar analysis for samples annealed at 800 °C provides $\Delta\sigma_{df}$ in the range of 49 to 72 MPa, which corresponds to a dislocation density of ~ 10^{13} m^{-2}."

Summary and Conclusions

"A single-phase BCC NbTaHfTiZr refractory high entropy alloy was cold worked to a total reduction of 89 pct (effective true strain = 2.55) using upset forging (54.5 pct thickness reduction, true strain = −0.91) followed by rolling (75.8 pct thickness reduction, true strain = −1.64)."

"The cold-worked alloy retained its single-phase structure and had room temperature yield stress $\sigma_{0.2} = 1438$ MPa, peak stress $\sigma_p = 1495$ MPa, and fracture strain $\varepsilon_f = 5$ pct."

"Ductile fracture occurred due to through-thickness strain localization followed by shear along a plane which was ~ 55 deg to the tensile axis and whose trace in the rolling plane was almost normal to the rolling/tensile direction."

"$\sigma_{0.2}$ and σ_p of the single-phase BCC structure developed during annealing at 1100 °C were 1110 to 1115 MPa and 1160 to 1195 MPa, respectively; ε_f lay in the range 12 to 19 pct with a maximum value for the 1-h anneal."

Microstructural Design for Improving Ductility of An Initially Brittle Refractory High Entropy Alloy [228]

This is a machine-generated summary of:

Soni, V.; Senkov, O. N.; Gwalani, B.; Miracle, D. B.; Banerjee, R.: Microstructural Design for Improving Ductility of An Initially Brittle Refractory High Entropy Alloy [228]

Published in: Scientific Reports (2018)

Link to original: https://doi.org/10.1038/s41598-018-27144-3

If you want to cite the papers, please refer to the original.

For technical reasons we could not place the page where the original quote is coming from.

Abstract-Summary

"Refractory high-entropy alloys (RHEAs), comprising a two-phase ordered B2 + BCC microstructure, exhibit extraordinarily high yield strengths, but poor ductility at room temperature, limiting their engineering application."

Introduction

"Extending this theory to RHEAs, Sheikh, and others [178] found that single-phase BCC RHEAs consisting of subgroup IV, V and VI metals are intrinsically ductile if the valence electron concentration (VEC) is less than 4.5 and intrinsically brittle if VEC $\geq$ 4.6."

"Huang, and others [190] and Lilensten, and others [191] used a "metastability-engineering" approach, which is similar to well-known transformation-induced plasticity (TRIP), to improve tensile ductility of BCC RHEAs by tailoring the stability of constituent phases."

"Although they showed exceptionally good strength at both room and elevated temperatures, substantially exceeding those of single-phase BCC RHEAs, these novel two-phase RHEAs have very limited room temperature compressive ductility, which can be explained by the inherent brittleness of the ordered B2 matrix phase [192]."

"The present work is the first demonstration of enhancing the ductility of high-strength BCC + B2 two-phase RHEAs by controlling their microstructure."

Results

"The highest hardness was found in Condition (1) due to its fine-scale B2 (hard) + BCC decomposed microstructure, as compared to the single-phase weakly ordered BCC microstructure in Condition (2)."

"Intermediate hardness was observed in Condition (3) due to the presence of discrete B2 precipitates within the continuous BCC matrix, since the scale of the two-phase BCC + B2 microstructure in Condition (3) is substantially coarser than Condition (1)."

"In Condition (1) the alloy had high room temperature (RT) strength (yield stress $\sigma_{0.2} = 2032$ MPa, true peak stress $\sigma_p = 2035$ MPa) but showed a true compressive fracture strain ε_f of only 4.7%."

"The brittle nature of Condition (1) is due to the presence of continuous B2 channels, whereas Condition (2) is largely a single-phase BCC alloy with a small degree compositional fluctuations, and thus has more ductility."

"Further, the scale of the two-phase BCC + B2 microstructure in Condition (3) is substantially coarser than Condition (1)."

Discussion

"Condition (1), the initial "as-processed" condition, exhibited a "superalloy-like" microstructure consisting of cuboidal pockets of BCC within a continuous matrix of the ordered B2 phase."

"Isothermal annealing of Condition (1) at 1400 °C for 20 min resulted in disordering of the microstructure to form a single BCC phase, followed by early stages of B2 ordering during water-quenching, resulting in Condition (2)."

"This "superalloy-like" microstructure in Condition (1) changes substantially during a subsequent isothermal anneal at 600 °C for 120 h, resulting in discrete B2 precipitates within a continuous BCC matrix (Condition (3))."

"It appears that the as-processed Condition (1), exhibiting a "superalloy-like" microstructure, with cuboidal pockets of BCC within a continuous matrix of the ordered B2 phase, was not an equilibrium microstructure for this alloy at 600 °C."

"The results reveal that the homogenized and slow cooled (~10 °C/min) microstructure (Condition (1)) has an ordered B2 phase as the continuous matrix, interspersed with discrete cuboidal-like pockets of a disordered BCC phase."

Methods

"It was homogenized by holding at 1200 °C for 24 h in a tube furnace with a continuous flow of high purity argon and then slow cooled at 10 °C/min."

"A piece of the sample in Condition (1) was sealed in a quartz tube filled with high purity argon and additionally annealed at 1400 °C for 20 min and water quenched."

"Another piece of the sample in Condition (2) was also sealed in a quartz tube filled with high purity argon and additionally annealed at 600 °C for 120 h and water quenched."

"The sample was then heated to the test temperature at 50 °C/min, soaked at the temperature for 15 min under 20 N force control, and then compressed at a constant ram speed that provided an initial strain rate of 0.001 s^{-1}."

Microstructures and Mechanical Properties of Ta–Nb–Zr–Ti–Al Refractory High Entropy Alloys with Varying Ta/Ti Ratios [229]

This is a machine-generated summary of:

Dong, Ya-Guang; Chen, Shang; Jia, Nan-Nan; Zhang, Qiu-Hong; Wang, Liang; Xue, Yun-Fei; Jin, Ke: Microstructures and mechanical properties of Ta–Nb–Zr–Ti–Al refractory high entropy alloys with varying Ta/Ti ratios [229]

Published in: Tungsten (2021)

Link to original: https://doi.org/10.1007/s42864-021-00111-8

If you want to cite the papers, please refer to the original.

For technical reasons we could not place the page where the original quote is coming from.

Abstract-Summary

"The refractory high entropy alloys (RHEAs) containing disordered body-centered cubic (BCC) and ordered BCC (B2) structures often exhibit high strength but low ductility at room temperature, even in the compressive tests."

"All the three alloys exhibit a single BCC structure at the as-cast and solution-treated states, while the basket weave-like microstructures consisting of cuboidal and strip-like phases are formed after aging at 600 °C."

"The microstructure of the aged alloys is sensitive to the compositions: only disordered BCC1 + BCC2 dual phases are observed in the $(Ta_{25}Nb_{25}Zr_{25}Ti_{25})_{95}Al_5$ alloy; additional B2 phases are formed by replacing 5 at% Ta with Ti; a complex microstructure containing BCC1 + BCC2 + B2 + Omega phases are identified in the $(Ta_{15}Nb_{25}Zr_{25}Ti_{35})_{95}Al_5$ alloy."

Introduction

"With the addition of Mo and Ta elements, the $AlMo_{0.5}NbTa_{0.5}TiZr$ alloy is composed of continuous channels of B2 matrix and periodic cuboidal and plate-like BCC precipitates, achieving an ultrahigh yield strength of 2197 MPa but a low fracture strain of 4.7% [192]."

"Reducing the Al concentration in this alloy may result in the spherical BCC-structured precipitates uniformly distributed in the B2 matrix, exhibiting a higher yield strength of 2350 MPa but lower fracture strain of 3.2% [192]."

"A special phase inversion, from the B2 matrix and cuboidal BCC precipitates to the BCC matrix with discrete B2 precipitates, can be observed in the $Al_{0.5}NbTa_{0.8}Ti_{1.5}V_{0.2}Zr$ alloy after annealing at 600 °C for 120 h, which improves the fraction strain from 4.7% to 16.9% while keeping a high yield strength of over 1300 MPa [193, 194]."

"A novel $Al_{0.4}Nb_{0.5}Ta_{0.5}TiZr_{0.8}$ alloy containing cuboidal B2 precipitates in BCC matrix is designed with a compressive yield strength of 1232 MPa and a fracture strain of ~ 10% [195]."

Experimental

"The alloys with three designed compositions, $(Ta_{25\text{-}x}Nb_{25}Zr_{25}Ti_{25+x})_{95}Al_5$ (x = 0, 5, 10), were fabricated, and are denoted in short as $Ta_{25}Ti_{25}$, $Ta_{20}Ti_{30}$, and $Ta_{15}Ti_{35}$ in the following text, respectively."

"The Ta, Nb, Zr, Ti, and Al elemental metals with the purity of > 99.9% were used as raw materials, which were mixed by arc melting in the Ti-gettered high purity argon atmosphere."

"The as-cast alloys were solution-treated (ST) at 1300 °C for 24 h, followed by water quenching, and denoted as ST-1300/24 in the following text."

"The alloys were then aged at 600 °C for 6, 12, and 24 h, which were depicted as 600–6, 600–12, and 600–24, respectively."

"Transmission electron microscopy (TEM) observations of the aged samples were performed with FEI Talos F200X microscopy operated at 200 kV with an energy dispersive spectrometer (EDS), and the TEM samples were prepared using ion milling after mechanical grounding down to the thickness of 30 μm."

Results and Discussion

"After 6 h aging, the $Ta_{25}Ti_{25}$ and $Ta_{20}Ti_{30}$ alloys still exhibit a single BCC structure, while two sets of BCC peaks can be observed in the $Ta_{15}Ti_{35}$ alloy."

"With increasing the aging time to 12 h, the peaks of secondary BCC phases can be observed in both the $Ta_{25}Ti_{25}$ and $Ta_{20}Ti_{30}$ alloys, while a B2 phase can be identified in the $Ta_{15}Ti_{35}$ alloy."

"Although high strengths are commonly seen in the RHEAs containing basket-weave-like BCC/B2 microstructures, most previous alloys exhibited very low compressive ductility (< 10%) [192, 196]."

"Although the content of each phase can hardly been quantified, the content of B2 phase in $Ta_{15}Ti_{35}$ could be less than that in $Ta_{20}Ti_{30}$, since the Omega phases observed in $Ta_{15}Ti_{35}$ alloy may be transformed from the B2 phases, which has been commonly observed in the metastable Ti alloys [197, 198]."

Conclusion

"The microstructures and mechanical properties of three RHEAs, $(Ta_{25\text{-}x}Nb_{25}Zr_{25}Ti_{25+x})_{95}Al_5$ (x = 0, 5, 10), have been investigated at their as-cast, solution-treated, and aged states."

"At the solution-treated state, single-phase BCC structures with equiaxed grain microstructures are observed in all the three alloys, but the average grain size increases from 130 to 542 μm with decreasing Ta/Ti ratio."

"Extension of the aging time to 24 h results in considerable differences in the microstructures and mechanical properties between the three alloys: the average size of cuboidal phases increases from 47 to 103 nm with decreasing Ta/Ti ratio, while, more importantly, B2 and B2/Omega phases are formed when 5 and 10 at% of Ta is replaced with Ti, respectively."

"Although the mechanical properties are similar of these alloys at their solution-treated states, significant differences are observed after aging."

Effect of Nb Content on Microstructure and Mechanical Properties of $Mo_{0.25}V_{0.25}Ti_{1.5}Zr_{0.5}Nb_x$ High-Entropy Alloys [230]

This is a machine-generated summary of:

Zhang, Fengqi; Xiang, Chao; Han, En-Hou; Zhang, Zijian: Effect of Nb Content on Microstructure and Mechanical Properties of $Mo_{0.25}V_{0.25}Ti_{1.5}Zr_{0.5}Nb_x$ High-Entropy Alloys [230]

Published in: Acta Metallurgica Sinica (English Letters) (2022)

Link to original: https://doi.org/10.1007/s40195-022-01399-2

If you want to cite the papers, please refer to the original.

For technical reasons we could not place the page where the original quote is coming from.

Abstract-Summary

"High-entropy alloys (HEAs) have significant application prospects as promising candidate materials for nuclear industry due to their excellent mechanical properties, corrosion resistance and irradiation resistance."

"The alloys were prepared by vacuum arc melting, and all the ingots were annealed at 1200 °C for 24 h. The microstructures, crystal structures, hardness and mechanical properties of $Mo_{0.25}V_{0.25}Ti_{1.5}Zr_{0.5}Nb_x$ HEAs were investigated."

"The single BCC phase is retained after annealing at 1200 °C for 24 h. The compressive and microhardness tests show that the strength and hardness of the alloys decrease gradually with the increase of Nb content, while the plasticity increases and the fracture mode of the alloy changes from brittle fracture to ductile fracture, which is mainly owing to the decrease of grain size."

"Nb1.0 alloy can reach 28.32% strain without fracture, which exhibits promising potential in industrial application."

Introduction

"The contents of Mo and V were reduced, while the contents of Ti and Zr were increased, and the effect of Nb on the microstructure and properties of the alloy was studied."

"Non-equimolar $Mo_{0.25}V_{0.25}Ti_{1.5}Zr_{0.5}Nb_x$ (x = 0, 0.25, 0.5, 0.75, and 1.0) HEAs were designed and Nb element was considered as a variable based on the following reasons: (1) the thermal neutron absorption cross-section of Nb element is relatively low (1.15 barn), the content of Nb element in the alloy will greatly affect the neutron absorption cross-section of the alloy; (2) the addition of Nb element with high melting point (2468 °C) can significantly improve the melting point and thermal stability of the alloy; (3) it is necessary to explore the effect of Nb on the corrosion resistance of the alloy."

Materials and Methods

"Ingots of $Mo_{0.25}V_{0.25}Ti_{1.5}Zr_{0.5}Nb_x$ (x = 0, 0.25, 0.5, 0.75 and 1.0, denoted as Nb0, Nb0.25, Nb0.5, Nb0.75 and Nb1.0, respectively) were prepared by vacuum arc melting under high-purity argon atmosphere in a water-cooled Cu crucible."

"The prepared button-shaped ingots were approximately 45 mm in diameter and 10 mm in height."

"The TEM sample was prepared by a focused ion beam (FIB, FEI Helios Nanolab 600i)."

"The cylindrical samples of Φ 3 mm × 6 mm were prepared and the compressive tests were carried out on the INSTRON 5582 machine with a stain rate of 1×10^{-3} s^{-1} at room temperature."

Results and Discussion

"For Nb0 alloy, the primary BCC phase decomposes to BCC#1 phase and HCP phase at the decomposition temperature (T_{dec}) of 557 °C."

"With the increase of Nb content, the decomposition temperature gradually increases from 557 °C of Nb0 alloy to 700 °C of Nb1.0 alloy, indicating that the addition of Nb will reduce the stability of BCC phase."

"It can be seen that the microstructure of as-cast Nb0, Nb0.5, and Nb1.0 alloys are dendritic structure, including white dendritic phase and gray interdendritic phase."

"With the increase of Nb content, the hardness of the alloy is significantly reduced when $0 \leq x \leq 0.5$."

"The main reason is that the second phase appears in the alloy and the volume fraction increases gradually with the increase of Nb content in the reported system."

Conclusions

"The dendritic phase of the Nb0 alloy is rich in Zr, while the interdendritic phase is rich in Mo and Ti, and the distribution of V in the alloy is relatively uniform."

"The dendritic phases of Nb0.5 and Nb1.0 alloys are rich in Mo and Nb, and the interdendritic phases are rich in V, Ti and Zr."

"With the increase of Nb content, the hardness of the alloy is significantly reduced when $0 \leq x \leq 0.5$."

"The strength of the alloys increases with the increase of Nb content when $0 \leq x \leq 0.25$, while the strength decreases gradually when $0.25 \leq x \leq 1.0$."

"With the increase of Nb content, the plasticity of alloys increases gradually."

"The fracture mode of Nb0.5, Nb0.75 and Nb1.0 alloys is plastic fracture."

4.8 Light-weight High Entropy Alloys

Machine generated keywords: al, isothermal, etc., aluminum alloy, magnesium, heat treatment, shape, hcp, density, hcp phase, aluminum, hold, entropy alloy, copper, microscope

Microstructural Evolution and Mechanical Properties of Semi-Solid $Al_{15}Mg_{45}Li_{39}Ca_{0.5}Si_{0.5}$ Light-Weight High Entropy Alloys Fabricated by Isothermal Heat Treatment [231]

This is a machine-generated summary of:

Hu, Yong; Liu, Fei; Zhao, Longzhi; Tang, Yanchuan; Jiao, Haitao: Microstructural Evolution and Mechanical Properties of Semi-Solid $Al_{15}Mg_{45}Li_{39}Ca_{0.5}Si_{0.5}$ Light-Weight High Entropy Alloys Fabricated by Isothermal Heat Treatment [231]

Published in: Metals and Materials International (2022)

Link to original: https://doi.org/10.1007/s12540-022-01303-8

If you want to cite the papers, please refer to the original.

For technical reasons we could not place the page where the original quote is coming from.

Abstract-Summary

"The non-dendritic globular microstructures of $Al_{15}Mg_{45}Li_{39}Ca_{0.5}Si_{0.5}$ LW-HEAs were prepared by isothermal heat treatment, and the effects of isothermal temperatures and holding times on the microstructures of semi-solid $Al_{15}Mg_{45}Li_{39}Ca_{0.5}Si_{0.5}$ LW-HEAs were investigated by optical microscope and scanning electron microscope."

"The results indicate that the phases of $Al_{15}Mg_{45}Li_{39}Ca_{0.5}Si_{0.5}$ LW-HEAs consist of HCP phases and intermetallic compounds."

"The semi-solid $Al_{15}Mg_{45}Li_{39}Ca_{0.5}Si_{0.5}$ LW-HEAs have a low coarsening coefficient."

"Compared with the as-cast $Al_{15}Mg_{45}Li_{39}Ca_{0.5}Si_{0.5}$ LW-HEAs, the ultimate strength, plasticity and hardness of semi-solid $Al_{15}Mg_{45}Li_{39}Ca_{0.5}Si_{0.5}$ LW-HEAs are improved by 25.7%, 18.5% and 89.7%, respectively."

Introduction

"Scholars have begun to study the semi-solid forming of high entropy alloys."

"Rogal [199] is the first to propose the semi-solid forming of high entropy alloys in 2017."

"He obtained semi-solid CoCrCuFeNi high entropy alloy by RAP, and the compression yield strength, compression strength and hardness of thixo-cast CoCr-CuFeNi high entropy alloy were 30% higher in comparison with those of as-cast CoCrCuFeNi high entropy alloy."

"Zhang and others [200] and Jiang and others [201] prepared the semi-solid CoCr-CuFeNi high entropy alloy and semi-solid $CoCrCu_{1.2}FeNi$ high entropy alloy by RAP, respectively."

"They found that the semi-solid high entropy alloys have lower coarsening coefficient than traditional alloys."

"Semi-solid high entropy alloys are mainly prepared by RAP."

"Other methods for preparing semi-solid high entropy alloys have not been reported in the literatures."

Experimental Procedures

"The differential thermal analysis (DTA) curve was measured using NETZSCH STA449 F3 at the heating rate of 1 °C/min in an argon atmosphere."

"The solidus–liquidus temperature and liquid volume fraction were determined from the DTA curve."

"According to the DTA curve, the isothermal temperatures were selected as 420 °C, 425 °C, 430 °C and 440 °C, and the holding times were selected as 19 ~ 29 min."

"The samples were heated in a box-type resistance furnace with temperature fluctuation range of ± 1 °C, followed by water quenching at room temperature."

"The density of the alloy was measured by a DeAnt ES-D electronic density balance."

Results and Discussion

"Solid solution phases can be formed when $\delta \leq 6.6\%$ for the traditional high entropy alloys which are based on 3d and/or 4d transition metal elements [32]."

"For LW-HEAs, a modified value of atomic size difference ($\delta \leq 4.5\%$) was proposed to predict the formation of solid solution phases."

"The semi-solid CoCrCuFeNi HEA [200] and semi-solid $CoCrCu_{1.2}FeNi$ HEA [201] have the similar high value of the shape factor compared with this work."

"It seems that the high entropy alloys are more likely to form semi-solid microstructures with high spheroidization."

"The higher ultimate strength and hardness of semi-solid alloy, in comparison with as-cast alloy, can be a result of Hall Petch dependence, which led to grain boundary strengthening of the refined microstructure [202]."

"The shape factor of the semi-solid alloy is better obviously than that of as-cast alloy, which is helpful for obtaining higher mechanical properties."

Formation Mechanism of Semi-Solid Microstructure

"During the isothermal heat treatment, the atoms will diffuse from high concentration to low concentration, resulting in the dendrite root with smaller curvature radius melt into grains with different shapes and sizes."

"The grain coarsening in semi-solid state is a solid–liquid coarsening process, which is driven by the reduction of interfacial energy."

"The grain coarsening model of semi-solid alloy can be described by a classical LSW theory [201]: where D is the average grain size at time t, D_0 the initial average grain size at time t = 0, k is the coarsening coefficient, n is the coarsening exponent."

"The linear correlation coefficient R^2 is 0.88, which is close to 1, indicating LSW theory is suitable to describe the grain coarsening for semi-solid $Al_{15}Mg_{45}Li_{39}Ca_{0.5}Si_{0.5}$ LW-HEAs."

"The solid–liquid contacts in semi-solid state have a great influence on the grain coarsening."

Conclusions

"The phases of the $Al_{15}Mg_{45}Li_{39}Ca_{0.5}Si_{0.5}$ LW-HEAs consist of α-Mg phase of HCP structure and intermetallic compounds AlLi and $MgAlLi_2$. (2) The average grain size increases with the increase of isothermal temperature, but first increases, then decreases and then increases with the increase of holding time."

"The shape factor increases first and then decreases with the increase of isothermal temperature and holding time."

"The microstructures with an average grain size of 22 μm and a shape factor of 0.92 are obtained at 430 °C for 25 min."

"The coarsening coefficient of $Al_{15}Mg_{45}Li_{39}Ca_{0.5}Si_{0.5}$ LW-HEAs in semi-solid state is lower due to the sluggish diffusion effect. (3) The ultimate strength, Plasticity and hardness of the $Al_{15}Mg_{45}Li_{39}Ca_{0.5}Si_{0.5}$ LW-HEAs increase from 105 MPa, 2.7% and 78 HV in as-cast state to 132 MPa, 3.2% and 148 HV in semi-solid state, which are improved by 25.7%, 18.5% and 89.7%, respectively."

References

1. Z. Wu, H. Bei, G.M. Pharr, and E.P. George, Temperature Dependence of the Mechanical Properties of Equiatomic Solid Solution Alloys with Face-Centered Cubic Crystal Structures, Acta Mater., 2014, 81, p 428–441.
2. Zeng, Y., Cai, X. & Koslowski, M. Effects of the stacking fault energy fluctuations on the strengthening of alloys. Acta Mater. 164, 1–11 (2019).
3. Smith, T. M. et al. Atomic-scale characterization and modeling of 60 dislocations in a high-entropy alloy. Acta Mater. 110, 352–363 (2016).
4. Schneider, T. & Stoll, E. Molecular-dynamics study of a three-dimensional one-component model for distortive phase transitions. Phys. Rev. B 17, 1302–1322 (1978).
5. Vítek, V. Intrinsic stacking faults in body-centred cubic crystals. Philos. Mag. 18, 773–786 (1968).
6. E. Pickering, N. Jones, High-entropy alloys: a critical assessment of their founding principles and future prospects. Int. Mater. Rev. 61(3), 183–202 (2016)

7. F. Otto, A. Dlouhý, C. Somsen, et al., "The influences of temperature and microstructure on the tensile properties of a CoCrFeMnNi high-entropy alloy," Acta Mater., 61, No. 15, 5743–5755 (2013).
8. A. Gali and E.P. George, Tensile Properties of High- and Medium-Entropy Alloys, Intermetallics, 2013, 39, p 74-78
9. Laplanche, G., Kostka, A., Horst, O., Eggeler, G. & George, E. Microstructure evolution and critical stress for twinning in the CrMnFeCoNi high-entropy alloy. Acta Mater. 118, 152–163 (2016).
10. G. Laplanche, A. Kostka, C. Reinhart, J. Hunfeld, G. Eggeler, and E.P. George, Reasons for the Superior Mechanical Properties of Medium-Entropy CrCoNi Compared to High-Entropy CrMnFeCoNi, Acta Mater., 2017, 128, p 292–303. https://doi.org/10.1016/j.actamat.2017.02.036
11. W. Huo, H. Zhou, F. Fang, X. Hu, Z. Xie, and J. Jiang, Strain-Rate Effect upon the Tensile Behavior of CoCrFeNi High-Entropy Alloys, Mater. Sci. Eng. A, 2017, 689, p 366–369
12. Gludovatz, B. et al. A fracture-resistant high-entropy alloy for cryogenic applications. Science 345, 1153–1158 (2014).
13. J.J. Licavoli, M.C. Gao, J.S. Sears, P.D. Jablonski, and J.A. Hawk, Microstructure and Mechanical Behavior of High-Entropy Alloys, J. Mater. Eng. Perform., 2015, 24(10), p 3685–3698. https://doi.org/10.1007/s11665-015-1679-7
14. G.R. Holcomb, J. Tylczak, and C. Carney, Oxidation of CoCrFeMnNi High Entropy Alloys, JOM, 2015, 67, p 2326–2339
15. N. Kumar, Q. Ying, X. Nie, R.S. Mishra, Z. Tang, P.K. Liaw, R.E. Brennan, K.J. Doherty, and K.C. Cho, High Strain-Rate Compressive Deformation Behavior of the Al0.1CrFeCoNi High Entropy Alloy, Mater. Des., 2015, 86, p 598–602
16. J.W. Yeh, S.K. Chen, S.J. Lin, J.Y. Gan, T.S. Chin, T.T. Shun, C.H. Tsau, and S.Y. Chang, Adv. Eng. Mater. 6, 299 (2004).
17. M.A. Hemphill, T. Yuan, G.Y. Wang, J.W. Yeh, C.W. Tsai, A. Chuang, P.K. Liaw, Acta Mater., 2012, vol. 60, pp. 5723–5734.
18. H.Y. Diao, R. Feng, K.A. Dahmen, and P.K. Liaw, Fundamental Deformation Behavior in High-Entropy Alloys: An Overview, Curr. Opin. Solid State Mater. Sci., 2017, 21(5), p 252–266. https://doi.org/10.1016/j.cossms.2017.08.003
19. B. Cantor, I. T. H. Chang, P. Knight, et al., "Microstructural development in equiatomic multicomponent alloys," Mater. Sci. Eng. A, 375, 213–218 (2004).
20. P.D. Jablonski, J.J. Licavoli, M.C. Gao, and J.A. Hawk, Manufacturing of High Entropy Alloys, JOM, 2015, 67(10), p 2278–2287. https://doi.org/10.1007/s11837-015-1540-3
21. B. Wang, A. Fu, X. Huang, B. Liu, Y. Liu, Z. Li, and X. Zan, Mechanical Properties and Microstructure of the CoCrFeMnNi High Entropy Alloy Under High Strain Rate Compression, J. Mater. Eng. Perform., 2016, 25(7), p 2985–2992
22. J.M. Park, J. Moon, J.W. Bae, M.J. Jang, J. Park, S. Lee, and H.S. Kim, Strain Rate Effects of Dynamic Compressive Deformation on Mechanical Properties and Microstructure of CoCrFeMnNi High-Entropy Alloy, Mater. Sci. Eng. A, 2018, 719, p 155–163
23. M. Komarasamy, N. Kumar, R.S. Mishra, and P.K. Liaw, Anomalies in the Deformation Mechanism and Kinetics of Coarse-Grained High Entropy Alloy, Mater. Sci. Eng. A, 2016, 654, p 256–263
24. J. Moon, S.I. Hong, J.W. Bae, M.J. Jang, D. Yim, and H.S. Kim, On the Strain Rate-Dependent Deformation Mechanism of CoCrFeMnNi High-Entropy Alloy at Liquid Nitrogen Temperature, Mater. Res. Lett., 2017, 5, p 427–477
25. M.A. Meyers, Dynamic Behavior of Materials, Wiley, New York, 1994
26. Schuh, B. et al. Mechanical properties, microstructure and thermal stability of a nanocrystalline CoCrFeMnNi high-entropy alloy after severe plastic deformation. Acta Mater. 96, 258–268 (2015).
27. Stepanov, N. D. et al. Mechanical properties of a new high entropy alloy with a duplex ultra-fine grained structure. Mater. Sci. Eng. A 728, 54–62 (2018).

28. Yasuda, H. Y., Miyamoto, H., Cho, K. & Nagase, T. Formation of ultrafine-grained microstructure in Al0.3CoCrFeNi high entropy alloys with grain boundary precipitates. Mater Lett 199, 120–123 (2017).
29. Bae, J. W. et al. Trade-off between tensile property and formability by partial recrystallization of CrMnFeCoNi high-entropy alloy. Mater. Sci. Eng. A 703, 324–330 (2017).
30. Klimova, M. V., Shaysultanov, D. G., Zherebtsov, S. V. & Stepanov, N. D. Effect of second phase particles on mechanical properties and grain growth in a CoCrFeMnNi high entropy alloy. Mater. Sci. Eng., A 748, 228–235 (2019).
31. Patrick. Matlab_Linecut. Preprint at (2022).
32. Zhang, Y. et al. Microstructures and properties of high-entropy alloys. Prog. Mater. Sci. 61, 1–93 (2014).
33. X.-W. Qiu and C.-G. Liu, J. Alloy. Compd. 553, 216 (2013).
34. Z. Chen, W. Chen, B. Wu, X. Cao, L. Liu, and Z. Fu, Effects of Co and Ti on Microstructure and Mechanical Behavior of Al0.75FeNiCrCo High Entropy Alloy Prepared by Mechanical Alloying and Spark Plasma Sintering, Mater. Sci. Eng. A, 2015, 648, p 217–224
35. K.B. Zhang, Z.Y. Fu, J.Y. Zhang, J. Shi, W.M. Wang, H. Wang, Y.C. Wang, and Q.J. Zhang, Nanocrystalline CoCrFeNiCuAl High-Entropy Solid Solution Synthesized by Mechanical Alloying, J. Alloys Compd, 2009, 485, p 31–34
36. G.A. Salishchev, M.A. Tikhonovsky, D.G. Shaysultanov, N.D. Stepanov, A.V. Kuznetsov, I.V. Kolodiy, A.S. Tortika, and O.N. Senkov, Effect of Mn and v on Structure and Mechanical Properties of High-Entropy Alloys Based on CoCrFeNi System, J. Alloys Compd., 2014, 591, p 11–21
37. G.D. Sathiaraj and P.P. Bhattacharjee, Effect of Starting Grain Size on the Evolution of Microstructure and Texture During Thermo-Mechanical Processing of CoCrFeMnNi High Entropy Alloy, J. Alloys Compd, 2015, 647, p 82–96
38. F.J. Zerilli and R.W. Armstrong, Dislocation-Mechanics-Based Constitutive Relations for Material Dynamics Calculations, Appl. Phys., 1987, 61, p 1816–1825
39. Y. Zhang, Y.J. Zhou, J.P. Lin, G.L. Chen, and P.K. Liaw, Adv. Eng. Mater. 10, 534 (2008).
40. Guo, S. & Liu, C. T. Phase stability in high entropy alloys: Formation of solid-solution phase or amorphous phase. Prog. Nat. Sci.: Mater. Int. 21, 433–446 (2011).
41. S. Mandal, K.G. Pradeep, S. Zaefferer, and D. Raabe, Scripta Mater. 81, 16 (2014).
42. W. Ji, Z.Y. Fu, W.M. Wang, H. Wang, J.Y. Zhang, Y.C. Wang, F. Zhang, J. Alloys Compd. 589, 61 (2014)
43. S. Mohanty, T.N. Maity, S. Mukhopadhyay, S. Sarkar, N.P. Gurao, S. Bhowmick, K. Biswas, Mater. Sci. Eng. A 679, 299 (2017)
44. M. Vaidya, A. Prasad, A. Parakh, B.S. Murty, Mater. Design 126, 37 (2017)
45. P.P. Ding, A.Q. Mao, X. Zhang, X. Jin, B. Wang, M. Liu, X.L. Gu, J. Alloys Compd. 721, 609 (2017)
46. I. Kunce, M. Polanski, K. Karczewski, T. Plocinski, K.J. Kurzydlowski, J. Alloys Compd. 648, 751 (2015)
47. H. Shiratori, T. Fujieda, K. Yamanaka, Y. Koizumi, K. Kuwabara, T. Kato, A. Chiba, Mater. Sci. Eng. A 656, 39 (2016)
48. A. Munitz, S. Salhov, S. Hayun, N. Frage, J. Alloys Compd. 683, 221 (2016)
49. Z. Tang, O.N. Senkov, C.M. Parish, C. Zhang, F. Zhang, L.J. Santodonato, G.Y. Wang, G.F. Zhao, F.Q. Yang, P.K. Liaw, Mater. Sci. Eng. A 647, 229 (2015)
50. W.R. Wang, W.L. Wang, J.W. Yeh, J. Alloys Compd. 589, 143 (2014)
51. Choudhuri, D. et al. Ballistic response of a FCC-B2 eutectic AlCoCrFeNi2.1 high entropy alloy. J. Dyn. Behav. Mater. 5(4), 495–503. https://doi.org/10.1007/s40870-019-00220-z (2019).
52. Sadeghilaridjani, M., Muskeri, S., Hasannaeimi, V., Pole, M. & Mukherjee, S. Strain rate sensitivity of a novel refractory high entropy alloy: Intrinsic versus extrinsic effects. Mater. Sci. Eng. A 766, 138326. https://doi.org/10.1016/j.msea.2019.138326 (2019).
53. Haridas, R. S. et al. Strain rate sensitive microstructural evolution in a TRIP assisted high entropy alloy: Experiments, microstructure and modeling. Mech. Mater. 156, 103798. https://doi.org/10.1016/j.mechmat.2021.103798 (2021).

54. Muskeri, S. et al. Ballistic impact response of Al0.1CoCrFeNi high-entropy alloy. Adv. Eng. Mater. 22(6), 1–7. https://doi.org/10.1002/adem.202000124 (2020).
55. Wang, L. et al. Mechanical response and deformation behavior of Al0.6CoCrFeNi high-entropy alloys upon dynamic loading. Mater. Sci. Eng. A 727, 208–213. https://doi.org/10.1016/j.msea.2018.05.001 (2018).
56. Zhang, T. W., Jiao, Z. M., Wang, Z. H. & Qiao, J. W. Dynamic deformation behaviors and constitutive relations of an AlCoCr1.5Fe1.5NiTi0.5 high-entropy alloy. Scr. Mater. 136, 15–19. https://doi.org/10.1016/j.scriptamat.2017.03.039 (2017).
57. Li, Z., Zhao, S., Diao, H., Liaw, P. K. & Meyers, M. A. High-velocity deformation of Al0.3CoCrFeNi high-entropy alloy: Remarkable resistance to shear failure. Sci. Rep. 7(1), 42742. https://doi.org/10.1038/srep42742 (2017).
58. He, J. Y. et al. A precipitation-hardened high-entropy alloy with outstanding tensile properties. Acta Mater. 102, 187–196 (2016).
59. Tian, Y. Z. et al. Significant contribution of stacking faults to the strain hardening behavior of Cu-15%Al alloy with different grain sizes. Sci. Rep. 5(1), 16707. https://doi.org/10.1038/srep16707 (2015).
60. Wang, Z. G. et al. Effect of coherent L12 nanoprecipitates on the tensile behavior of a fcc-based high-entropy alloy. Mater. Sci. Eng. A 696, 503–510. https://doi.org/10.1016/j.msea.2017.04.111 (2017).
61. Zhang, Y., Yang, X. & Liaw, P. K. Alloy design and properties optimization of high-entropy alloys. JOM 64, 830–838 (2012).
62. Senkov, O. N., Senkova, S. V., Woodward, C. & Miracle, D. B. Low-density, refractory multi-principal element alloys of the Cr-Nb-Ti-V-Zr system: microstructure and phase analysis. Acta Mater. 61, 1545–1557 (2013).
63. Y. Zhang, T. Zuo, Y. Cheng, and P.K. Liaw, Sci. Rep. 3, 1455 (2013).
64. O.N. Senkov, J.M. Scott, S.V. Senkova, D.B. Miracle, and C.F. Woodward, J. Alloys Compd. 509, 6043 (2011).
65. P.K. Huang, J.W. Yeh, T.T. Shun, and S.K. Chen, Adv. Eng. Mater. 6, 74 (2004).
66. O.N. Senkov, G.B. Wilks, D.B. Miracle, C.P. Chuang, and P.K. Liaw, Intermetallics 18, 1758 (2010).
67. M.S. Lucas, G.B. Wilks, L. Mauger, J.A. Munoz, O.N. Senkov, E. Michel, J. Horwath, S.L. Semiatin, M.B. Stone, D.L. Abernathy, and E. Karapetrova, Appl. Phys. Lett. 100, 251907 (2012).
68. C.J. Tong, M.R. Chen, S.K. Chen, J.W. Yeh, T.T. Shun, S.J. Lin, and S.Y. Chang, Metall. Mater. Trans. A 36A, 1263 (2005).
69. R. Sriharitha, B.S. Murty, and R.S. Kottada, Intermetallics 32, 119 (2013).
70. Y.-F. Kao, T.-J. Chen, S.-K. Chen, and J.-W. Yeh, Microstructure and Mechanical Property of As-Cast,-Homogenized, and-Deformed AlxCoCrFeNi ($0 \le x \le 2$) High-Entropy Alloys, J. Alloy. Compd., 2009, 488(1), p 57–64
71. T. Egami, Y. Waseda, and J. Non-Cryst, Solids 64, 113 (1984).
72. Odeshi, A. G., Al-ameeri, S. & Bassim, M. N. Effect of high strain rate on plastic deformation of a low alloy steel subjected to ballistic impact. Journal of Materials Processing Technology 162–163, 385–391 (2005).
73. Owolabi, G., Odoh, D., Peterson, A., Odeshi, A. & Whitworth, H. Measurement of the deformation of aluminum alloys under high strain rates using high speed digital cameras. World Journal of Mechanics 3, 112–121 (2013).
74. Xu, Y. & Meyers, M. A. A transition of the crystalline lattice to the amorphous in the shear bands of Fe-Cr-Ni single crystal induced under high-strain rate. Scientific Journal of Materials Science 2, 9–12 (2012).
75. Wright, S. C., Fleck, N. A. & W. J. Stronge. Ballistic impact of polycarbonate-An experimental investigation. International Journal of Impact Engineering 13, 1–20 (1993).
76. Meyer Jr, H. W. & Kleponis, D. S. Modeling the high strain rate behavior of titanium undergoing ballistic impact and penetration. International Journal of Impact Engineering 26, 509–521 (2001)

77. Gray, G. T. High-Strain-Rate deformation: mechanical behavior and deformation substructures induced. Annual Review of Materials Research 42, 285–303 (2012).
78. Lee, J.-H. et al. High strain rate deformation of layered nanocomposites. Nature Communications 3, 1–9 (2012).
79. H.-P. Chou, Y.-S. Chang, S.-K. Chen, and J.-W. Yeh, Microstructure, Thermophysical and Electrical Properties in AlxCoCrFeNi ($0 \le x \le 2$) High-Entropy Alloys, Mater. Sci. Eng., B, 2009, 163(3), p 184–189
80. Huang, J.-C. Evaluation of tribological behavior of Al-Co-Cr-Fe-Ni high entropy alloy using molecular dynamics simulation. Scanning 34, 325–331 (2012).
81. Xie, L., Brault, P., Thomann, A. L. & Bauchire, J. M. AlCoCrCuFeNi high entropy alloy cluster growth and annealing on silicon: A classical molecular dynamics simulation study. Applied Surface Science 285P, 810–816 (2013).
82. Zhou, X. W., Johnson, R. A. & Wadley, H. N. G. Misfit-energy-increasing dislocations in vapor-deposited CoFe/NiFe multilayers. Physical Review B 69, 1–10 (2004).
83. Pike, L. M., Anderson, I. M., Liu, C. T. & Chang, Y. A. Site occupancies, point defect concentrations, and solid solution hardening in B2 (Ni,Fe)Al. Acta Mater. 50, 3859–3879 (2002).
84. J.W. Yeh, Ann. Chim. Sci. Mater. 31, 633 (2006).
85. Straumanis, M. E. & Yu, L. S. Lattice parameters, densities, expansion coefficients and perfection of structure of Cu and of Cu-In α phase. Acta Crystallogr. A25, 676–682 (1969).
86. Singh, S., Wanderka, N., Murty, B. S., Glatzel, U. & Banhart, J. Decomposition in multi-component AlCoCrCuFeNi high-entropy alloy. Acta Materialia 59, 182–190 (2011).
87. B.A. Welk, R.E.A. Williams, G.B. Viswanathan, M.A. Gibson, P.K. Liaw, and H.L. Fraser, Ultramicroscopy 134, 193 (2013).
88. Y. Lu, X. Gao, L. Jiang, Z. Chen, T. Wang, J. Jie, H. Kang, Y. Zhang, S. Guo, and H. Ruan, Directly Cast Bulk Eutectic and Near-Eutectic High Entropy Alloys with Balanced Strength and Ductility in a Wide Temperature Range, Acta Mater., 2017, 124, p 143–150
89. X. Jin, Y. Zhou, L. Zhang, X.Y. Du, B.S. Li, Mater. Lett. 216, 144 (2018)
90. Y. Dong, Z. Yao, X. Huang, F. Du, C. Li, A. Chen, F. Wu, Y. Cheng, and Z. Zhang: J. Alloys Compd., 2020, vol. 823, 153886.
91. F. Meng, S.F. Bauer, Y. Liao, I. Baker, Intermet. 56, 28 (2015)
92. F. Meng, J. Qiu, I. Baker, Mater. Sci. Eng. A 586, 45 (2013)
93. I.S. Wani, T. Bhattacharjee, S. Sheikh, Y.P. Lud, S. Chatterjee, P.P. Bhattacharjeea, S. Guo, and N. Tsujib: Mater. Res. Lett., 2016, vol. 4, pp. 174–79.
94. T. Bhattacharjee, I.S. Wani, S. Sheikh, I.T. Clark, T. Okawa, S. Guo, P.P. Bhattacharjee, N. Tsuji, Simultaneous strength-ductility enhancement of a nano-lamellar AlCoCrFeNi2.1 eutectic high entropy alloy by cryo-rolling and annealing. Sci Rep. 8, 3276 (2018)
95. F. Meng, J. Qiu, I. Baker, Effect of Al content on the microstructure and mechanical behavior of two-phase FeNiMnAl alloys. J. Mater. Sci. 49, 1973–1983 (2013)
96. W.-R. Wang, W.-L. Wang, S.-C. Wang, Y.-C. Tsai, C.-H. Lai, and J.-W. Yeh, Effects of Al Addition on the Microstructure and Mechanical Property of AlxCoCrFeNi High-Entropy Alloys, Intermetallics, 2012, 26, p 44–51
97. X. Jin, J. Bi, L. Zhang, Y. Zhou, X. Du, Y. Liang, B. Li, A new CrFeNi2Al eutectic high entropy alloy system with excellent mechanical properties. J. Alloys Compd. 770, 655–661 (2019)
98. Liu, N. et al. Rapid solidification and liquid-phase separation of undercooled CoCrCuFexNi high-entropy alloys. Intermetallics 72, 44–52 (2016).
99. W.H. Liu, Z.P. Lu, J.Y. He, J.H. Luan, Z.J. Wang, B. Liu, Y. Liu, M.W. Chen, and C.T. Liu, Ductile CoCrFeNiMox High Entropy Alloys Strengthened by Hard Intermetallic Phases, Acta Mater., 2016, 116, p 332–342
100. F. He, Z. Wang, M. Zhu, J. Li, Y. Dang, J. Wang, The phase stability of Ni2CrFeMox multi-principal-component alloys with medium configurational entropy. Mater. Design. 85, 1–6 (2015)

101. X.-L. Shang, Z.-J. Wang, Q.-F. Wu, J.-C. Wang, J.-J. Li, J.-K. Yu, Effect of Mo addition on corrosion behavior of high-entropy alloys CoCrFeNiMox in aqueous environments. Acta Metall. Sin. 32(1), 41–51 (2019)
102. Y. Guo, C.G. Li, M. Zeng, J.Q. Wang, P. Deng, Y. Wang, Mater. Chem. Phys. 242, 122522 (2020)
103. H. Wu, S.R. Huang, C.Y. Zhu, J.F. Zhang, H.G. Zhu, Z.H. Xie, Acta Metall. Sin. -Engl. Lett. 33, 1091 (2020)
104. M. Ghanbariha, M. Farvizi, T. Ebadzadeh, A.A. Samiyan, Wear 484, 43 (2021)
105. F.X. Ye, Y. Yang, Z. Lou, L.Z. Feng, L. Guo, J.X. Yu, Mater. Lett. 284, 128859 (2021)
106. Y.B. Peng, W. Zhang, T.C. Li, M.Y. Zhang, B. Liu, Y. Liu, L. Wang, S.H. Hu, Surf. Coat. Technol. 385, 125326 (2020)
107. Y.M. Zou, Z.G. Qiu, C.J. Huang, D.C. Zeng, R. Lupoi, N. Zhang, S. Yin, Surf. Coat. Technol. 434, 128205 (2022)
108. A. Hegazy, M. Bhlol, O. Dawood, O. Elkady, J. Mech. Eng. 7, 1127 (2022)
109. X.F. Li, Y.H. Feng, B. Liu, D.H. Yi, X.H. Yang, W.D. Zhang, G. Chen, Y. Liu, P.K. Bai, J. Alloys Compd. 788, 485 (2019)
110. H. Wu, S.R. Huang, C.M. Zhao, H.G. Zhu, Z.H. Xie, C.L. Tu, X.D. Li, Intermetallics 127, 106983 (2020)
111. Q.S. Gou, J. Xiong, Z.X. Guo, J.B. Liu, L. Yang, X.R. Li, Int. J. Refract. Met. Hard Mater. 94, 105375 (2021)
112. X.F. Li, X.H. Yang, D.H. Yi, B.S. Liu, J.C Zhu, J.X Li, C. Gao, L.Z. Wang, Intermetallics 138, 107309 (2021)
113. Y. Shao, Z.H. Guo, Y.B. Wang, H. Ma, Int. J. Refract. Met. Hard Mater. 94, 105388 (2021)
114. Yiping L, Yong D, Sheng G, Jiang L, Kang H, Wang T, Wen B, Wang Z, Jie J, Cao Z, Ruan H, and Li T. Met Alloys Struct Prop Sci Rep 4 (2014) 6200, 10.1038.
115. F. Zhang, C. Zhang, S. L. Chen, J. Zhu, W. S. Cao, and U. R. Kattner, "An understanding of high entropy alloys from phase diagram calculations, "CALPHAD 45, 1–10 (2014).
116. C. Zhang, F. Zhang, S. Chen, and W. Cao, Computational Thermodynamics Aided High-Entropy Alloy Design, JOM, 2012, 64(7), p 839–845.
117. Guo S, Ng C, and Liu C T, Mater Res Lett 1 (2013) 228.
118. Y.P. Lu, Y. Dong, S. Guo, L. Jiang, H.J. Kang, T.M. Wang, B. Wen, Z.J. Wang, J.C. Jie, Z.Q. Cao, H.H. Ruan, T.J. Li, Sci. Rep. 4, 6200 (2014)
119. Wu, X. L. et al. Heterogeneous lamella structure unites ultrafine-grain strength with coarse-grain ductility. Proc. Natl Acad. Sci. USA 112, 14501–14505 (2015).
120. Wu, X. L. & Zhu, Y. T. Heterogeneous materials: a new class of materials with unprecedented mechanical properties. Mater. Res. Lett. 5, 527–532 (2017).
121. Yang, M. X. et al. Strain hardening in Fe–16Mn–10Al–0.86C–5Ni high specific strength steel. Acta Mater. 109, 213–222 (2016).
122. Ma, E. & Zhu, T. Towards strength–ductility synergy through the design of heterogeneous nanostructures in metals. Mater. Today 20, 323–331 (2017).
123. Yang, M. X. et al. Dynamically reinforced heterogeneous grain structure prolongs ductility in a medium-entropy alloy with gigapascal yield strength. Proc. Natl Acad. Sci. USA 115, 7224–7229 (2018).
124. M.C. Gao, J.W. Yeh, P.K. Liaw, Y. Zhang, High-entropy alloys (Springer, New York, 2016)
125. Miracle, D. B. & Senkov, O. N. A critical review of high entropy alloys and related concepts. Acta Mater. 122, 448–511 (2017).
126. [10] S. Palanivel, A. Arora, K.J. Doherty, and R.S. Mishra: Mater. Sci. Eng. A, 2016, vol. 678, pp. 308–314.
127. S. S. Nene, K. Liu, M. Frank, R.S. Mishra, R.E. Brennan, K. Cho, Z. Li, D. Raabe, Sci. Rep., 2017, vol. 7, pp. 16167.
128. S.S. Nene, M. Frank, K. Liu, R.S. Mishra, B.A. McWilliams, and K.C. Cho: Sci. Reports, 2018, vol. 8, pp. 9920–1–9920–8.
129. [23] E.I. Galindo-Nava and P.E.J. Rivera-Díaz-del-Castillo: Acta Mater., 2017, vol. 128, pp. 120–134.

130. [5] Z. Li, C.C. Tasan, K.G. Pradeep, and D. Raabe: Acta Mater., 2017, vol. 131, pp. 323–335.
131. Z. Li, C.C. Tasan, H. Springer, B. Gault, D. Raabe, Sci. Rep., 2017, vol. 7, pp. 40704.
132. [24] R. Xiong, H. Peng, H. Si, W. Zhang, and Y. Wen: Mater. Sci. Eng. A, 2014, vol. 598, pp. 376–386.
133. [20] D.T. Pierce, J.A. Jiménez, J. Bentley, D. Raabe, C. Oskay, and J.E. Wittig: Acta Mater., 2014, vol. 68, pp. 238–253.
134. [3] Z. Li and D. Raabe: JOM, 2017, vol. 69, pp. 2099–2106.
135. Li, Z., Pradeep, K. G., Deng, Y., Raabe, D. & Tasan, C. C. Metastable high-entropy dual-phase alloys overcome the strength–ductility trade-off. Nature 534, 227–230 (2016).
136. Calcagnotto, M. et al. Orientation gradients and geometrically necessary dislocations in ultra-fine grained dual-phase steels studied by 2D and 3D EBSD. Mater. Sci. Eng. A 527, 2738–2746 (2010).
137. Papula, S. et al. Strain hardening of cold-rolled lean-alloyed metastable ferritic-austenitic stainless steels. Mater. Sci. Eng. A 677, 11–19 (2016).
138. S.S. Nene, M. Frank, K. Liu, R.S. Mishra, B. McWilliams, K.C. Cho, Sci. Rep., 2018, vol. 8, pp. 9920.
139. Sinha, S. et al. Microstructural evolution and deformation behavior of Ni-Si- and Co-Si-containing metastable high entropy alloys. Metall. Mater. Transac. A 50, 179–190 (2019).
140. Hirth, J. P., Lothe, J. & Mura, T. Theory of Dislocations (2nd ed.). J. Appl. Mech. (1983). https://doi.org/10.1115/1.3167075
141. Warren, B. E. X-Ray Diffraction. (1990).
142. Smallman, R. E. & Bishop, R. J. Modern Physical Metallurgy and Materials Engineering Science, process, applications (Butterworth Heinemann, Oxford, 1999).
143. Yamanaka, K., Mori, M., Sato, S. & Chiba, A. Stacking-fault strengthening of biomedical Co-Cr-Mo alloy via multipass thermomechanical processing. Sci. Rep. 7 (2017).
144. Jian, W. W. et al. Ultrastrong Mg alloy via nano-spaced stacking faults. Mater. Res. Lett. 1, 61–66 (2013).
145. Jian, W. W., Cheng, G. M., Xu, W. Z., Koch, C. C., Wang, Q. D., Zhu, Y. T. & Mathaudhu, S. N. Physics and model of strengthening by parallel stacking faults. Appl. Phys. Lett. 103 (2013).
146. L. Jiang, H. Fu, H. Zhang, J. Xie, Physical mechanism interpretation of polycrystalline metals' yield strength via a data-driven method: A novel Hall–Petch relationship, Acta Materialia 231 (2022).
147. A. Navarro, E.R. de los Rios, An alternative model of the blocking of dislocations at grain boundaries. Philos. Mag. A 57(1), 37–42 (1988)
148. C.-H. Tu, Y.-C. Lai, S.-K. Wu, Y.-H. Lin, The effects of annealing on severely cold-rolled equiatomic HfNbTiZr high entropy alloy. Mater. Lett. 303, 130526 (2021)
149. G.D. Sathiaraj, P.P. Bhattacharjee, Effect of cold-rolling strain on the evolution of annealing texture of equiatomic CoCrFeMnNi high entropy alloy. Mater. Charact. 109, 189–197 (2015)
150. C.-C. Juan, M.-H. Tsai, C.-W. Tsai, W.-L. Hsu, C.-M. Lin, S.-K. Chen, S.-J. Lin, J.-W. Yeh, Simultaneously increasing the strength and ductility of a refractory high-entropy alloy via grain refining. Mater. Lett. 184, 200–203 (2016)
151. H.V. Atkinson, Overview no. 65: Theories of normal grain growth in pure single phase systems, Acta Metall. 36(3), 469–491 (1988)
152. O.N. Senkov, A.L. Pilchak, S.L. Semiatin, Effect of cold deformation and annealing on the microstructure and tensile properties of a HfNbTaTiZr refractory high entropy alloy. Metall. Mater. Trans. A 49(7), 2876–2892 (2018)
153. S. Chen, K.-K. Tseng, Y. Tong, W. Li, C.-W. Tsai, J.-W. Yeh, P. Liaw, Grain growth and Hall-Petch relationship in a refractory HfNbTaZrTi high-entropy alloy. J. Alloy Compd. 795, 19–26 (2019)
154. George, E. P., Raabe, D. & Ritchie, R. O. High-entropy alloys. Nat. Rev. Mater. 4, 515–534 (2019).
155. Huang, S. et al. Thermal expansion in FeCrCoNiGa high-entropy alloy from theory and experiment. Appl. Phys. Lett. 110, 241902 (2017).

156. Huang, S., Vida, Á., Heczel, A., Holmström, E. & Vitos, L. Thermal expansion, elastic and magnetic properties of FeCoNiCu-based high-entropy alloys using first-principle theory. JOM 69, 2107–2112 (2017).
157. Huang, S. et al. Mechanism of magnetic transition in FeCrCoNi-based high entropy alloys. Mater. Des. 103, 71–74 (2016).
158. Huang, S. et al. Phase stability and magnetic behavior of FeCrCoNiGe high-entropy alloy. Appl. Phys. Lett. 107, 251906 (2015).
159. [27] S. Huang, W. Lei, S. Lu, F. Tian, J. Shen, E. Holmstrom, and L. Vitos: Scripta Mater., 2015, vol. 108, pp. 44-47.
160. Huang, S., Holmström, E., Eriksson, O. & Vitos, L. Mapping the magnetic transition temperatures for medium- and high-entropy alloys. Intermetallics 95, 80–84 (2018).
161. Huang, S. et al. Twinning in metastable high-entropy alloys. Nat. Commun. 9, 2381 (2018).
162. Gao, M. C., Yeh, J. W., Liaw, P. K. & Zhang, Y. High-entropy alloys: fundamentals and applications (Springer, Switzerland, 2016).
163. O.N. Senkov, G.B. Wilks, J.M. Scott, and D.B. Miracle, Intermetallics 19, 698 (2011).
164. Zou, Y., Maiti, S., Steurer, W. & Spolenak, R. Size-dependent plasticity in an Nb25Mo25Ta25W25 refractory high-entropy alloy. Acta Mater. 65, 85–97 (2014).
165. Han, Z. D. et al. Effect of Ti additions on mechanical properties of NbMoTaW and VNbMoTaW refractory high entropy alloys. Intermetallics 84, 153–157 (2017).
166. Vitos, L. Computational quantum mechanics for materials engineers (Springer, London, 2007).
167. Soven, P. Coherent-potential model of substitutional disordered alloys. Phys. Rev. 156, 809–813 (1967).
168. Győrffy, B. L. Coherent-potential approximation for a nonoverlapping-muffin-tin-potential model of random substitutional alloys. Phys. Rev. B 5, 2382–2384 (1972).
169. Vitos, L., Abrikosov, I. A. & Johansson, B. Anisotropic lattice distortions in random alloys from first-principles theory. Phys. Rev. Lett. 87, 156401 (2001).
170. Steinle-Neumann, G., Stixrude, L. & Cohen, R. E. First-principles elastic constants for the hcp transition metals Fe, Co, and Re at high pressure. Phys. Rev. B 60, 791–799 (1999).
171. Fast, L., Wills, J. M., Johansson, B. & Eriksson, O. Elastic constants of hexagonal transition metals: theory. Phys. Rev. B 51, 17431–17438 (1995).
172. R.S. Mishra, N. Kumar, M. Komarasamy, J. Mater. Sci. Technol. 31, 1259 (2015)
173. H.W. Yao, J.W. Qiao, J.A. Hawk, H.F. Zhou, M.W. Chen, M.C. Gao, J. Alloys Compd. 696, 1139 (2017)
174. T. Yang, Y. Zhao, W. Liu, J. Kai, C. Liu, J. Mater. Res. 33, 2983 (2018)
175. B. Cantor, I.T.H. Chang, P. Knight, A.J.B. Vincent, Mater. Sci. Eng. A 375, 213 (2004)
176. [14] D.B. Miracle, J.D. Miller, O.N. Senkov, C. Woodward, M.D. Uchic, and J. Tiley: Entropy 2014, vol. 16, pp. 494-525.
177. B. Gorr, F. Müller, S. Schellert, H.J. Christ, H. Chen, A. Kauffmann, and M. Heilmaier: Corr. Sci., 2020, vol. 166, p. 108475.
178. Sheikh S, Shafeie S, Hu Q et al (2016) Alloy design for intrinsically ductile refractory high-entropy alloys. J Appl Phys 120:164902. https://doi.org/10.1063/1.4966659
179. 4. J. Ph Couzinie, L. Lilensten, Y. Champion, G. Dirras, L. Perriere and I. Guillot: Mater. Sci. Eng. A 2015, vol. 645, pp. 255-263.
180. O.N. Senkov, S.L. Semiatin, Microstructure and properties of a refractory high-entropy alloy after cold working. J. Alloy Compd. 649, 1110–1123 (2015)
181. 6. G. Dirras, L. Lilensten, P. Djemia, M. Laurent-Brocq, D. Tingaud, J. P. Couzinie, L. Perriere, T. Chauveau and I. Guillot: Mater. Sci. Eng. A 2016, vol. 654, pp. 30–38.
182. Wu YD, Cai YH, Wang T et al (2014) A refractory Hf25Nb25Ti25Zr25 high-entropy alloy with excellent structural stability and tensile properties. Mater Lett 130:277–280. https://doi.org/10.1016/j.matlet.2014.05.134
183. 8. C.-C. Juan, K.-K. Tseng, W.-L. Hsu, M.-H. Tsai, C.-W. Tsai, C.-M. Lin, S.-K. Chen, S.-J. Lin and J.-W. Yeh: Materials Letters 2016, vol. 175, pp. 284–287.

184. 9. C.-M. Lin, C.-C. Juan, C.-H. Chang, C.-W. Tsai and J.-W. Yeh: Journal of Alloys and Compounds 2015, vol. 624, pp. 100–107.
185. Schuh B, Völker B, Todt J et al (2018) Thermodynamic instability of a nanocrystalline, single-phase TiZrNbHfTa alloy and its impact on the mechanical properties. Acta Mater 142:201–212. https://doi.org/10.1016/J.ACTAMAT.2017.09.035
186. 11. N. D. Stepanov, N. Yu Yurchenko, S. V. Zherebtsov, M. A. Tikhonovsky and G. A. Salishchev: Materials Letters 2018, vol. 211, pp. 87–90.
187. 19. P. Knorr and C. Herzig: J. Phys: Condens. Matter 1995, vol. 7, pp. 9185–9200.
188. 39. J.G. Sevillano, P. van Houtte and E. Aernoudt: Progr. Mater. Sci. 1980, vol. 25, pp. 69–412.
189. 40. A. Belyakov, Y. Kimura, Y. Adachi and K. Tsuzaki: Mater. Trans. 2004, vol. 45, pp. 2812–2821.
190. Huang, H. L. et al. Phase-transformation ductilization of brittle high-entropy alloys via metastability engineering. Adv. Mater. 29, 1701678 (2017).
191. Lilensten, L. et al. Design and tensile properties of a bcc Ti-rich high-entropy alloy with transformation-induced plasticity. Mater. Res. Letters 5, 110–116 (2017).
192. Senkov, O. N., Jensen, J. K., Pilchak, A. L., Miracle, D. B. & Fraser, H. L. Compositional variation effects on the microstructure and properties of a refractory high-entropy superalloy AlMo0.5NbTa0.5TiZr. Materials & Design 139, 498–511 (2018).
193. Soni V, Senkov ON, Gwalani B, Miracle DB, Banerjee R. Microstructural design for improving ductility of an initially brittle refractory high entropy alloy. Sci Rep. 2018;8(1):8816.
194. Soni V, Gwalani B, Alam T, Dasari S, Zheng Y, Senkov ON, Miracle D, Banerjee R. Phase inversion in a two-phase, BCC+B2, refractory high entropy alloy. Acta Mater. 2020;185:89.
195. Wang Q, Han J, Liu Y, Zhang Z, Dong C, Liaw PK. Coherent precipitation and stability of cuboidal nanoparticles in body-centered-cubic $Al_{0.4}Nb_{0.5}Ta_{0.5}TiZr_{0.8}$ refractory high entropy alloy. Scripta Mater. 2021;190:40ss.
196. O. N. Senkov, C. Woodward, and D. B. Miracle, "Microstructure of aluminum-containing refractory high-entropy alloys," JOM 66, 2030–2042 (2014).
197. Bendersky L, Boettinger W, Burton B, Biancaniello F, Shoemaker C. The formation of ordered ω-related phases in alloys of composition Ti4Al3Nb. Acta Metall Mater. 1990;38(6):931.
198. Wang X, Yang J, Song L, Kou H, Li J, Fu H. Evolution of B2 (ω) region in high-Nb containing TiAl alloy in intermediate temperature range. Intermetallics. 2017;82:32.
199. Ł Rogal, Mater. Design 119, 406 (2017)
200. L.J. Zhang, J.T. Fan, D.J. Liu, M.D. Zhang, P.F. Yu, Q. Jing, M.Z. Ma, P.K. Liaw, G. Li, R.P. Liu, J. Alloy. Compd. 745, 75 (2018)
201. J. Jiang, M. Huang, Y. Wang, Y. Liu, Y. Zhang, J. Mater. Sci. Technol. 120, 172 (2022)
202. T. Chen, Z. Xie, Z. Luo, Q. Yang, S. Tan, Y. Wag, Y. Luo, T. Nonferr. Metal. Soc. 24, 3421 (2014)
203. Utt, Daniel; Lee, Subin; Xing, Yaolong; Jeong, Hyejin; Stukowski, Alexander; Oh, Sang Ho; Dehm, Gerhard; Albe, Karsten The origin of jerky dislocation motion in high-entropy alloys. Nature Communications (2022). https://doi.org/10.1038/s41467-022-32134-1
204. Shabani, Mitra; Indeck, Joseph; Hazeli, Kavan; Jablonski, Paul D.; Pataky, Garrett J. Effect of Strain Rate on the Tensile Behavior of CoCrFeNi and CoCrFeMnNi High-Entropy Alloys. Journal of Materials Engineering and Performance (2019). https://doi.org/10.1007/s11665-019-04176-y
205. Huang, Chang-Wei; Su, Pei-Ying; Yu, Chi-Hua; Wang, Chia-Ling; Lo, Yu-Chieh; Jang, Jason Shian-Ching; Hu, Hsuan-Teh A micromechanical study on the effects of precipitation on the mechanical properties of CoCrFeMnNi high entropy alloys with various annealing temperatures. Scientific Reports (2023). https://doi.org/10.1038/s41598-023-30508-z
206. Wang, Bingfeng; Fu, Ao; Huang, Xiaoxia; Liu, Bin; Liu, Yong; Li, Zezhou; Zan, Xiang Mechanical Properties and Microstructure of the CoCrFeMnNi High Entropy Alloy Under High Strain Rate Compression. Journal of Materials Engineering and Performance (2016). https://doi.org/10.1007/s11665-016-2105-5

207. Tasan, C. C.; Deng, Y.; Pradeep, K. G.; Yao, M. J.; Springer, H.; Raabe, D. Composition Dependence of Phase Stability, Deformation Mechanisms, and Mechanical Properties of the CoCrFeMnNi High-Entropy Alloy System. JOM (2014). https://doi.org/10.1007/s11837-014-1133-6
208. Zhou, P. F.; Xiao, D. H.; Yuan, T. C. Microstructure, Mechanical and Corrosion Properties of AlCoCrFeNi High-Entropy Alloy Prepared by Spark Plasma Sintering. Acta Metallurgica Sinica (English Letters) (2019). https://doi.org/10.1007/s40195-019-00962-8
209. Muskeri, Saideep; Gwalani, Bharat; Jha, Shristy; Yu, Anqi; Jannotti, Philip A.; Haridas, Ravi Sankar; Schuster, Brian E.; Lloyd, Jeffrey T.; Mishra, Rajiv S.; Mukherjee, Sundeep Excellent ballistic impact resistance of $Al_{0.3}CoCrFeNi$ multi-principal element alloy with unique bimodal microstructure. Scientific Reports (2021). https://doi.org/10.1038/s41598-021-02209-y
210. Tang, Zhi; Gao, Michael C.; Diao, Haoyan; Yang, Tengfei; Liu, Junpeng; Zuo, Tingting; Zhang, Yong; Lu, Zhaoping; Cheng, Yongqiang; Zhang, Yanwen; Dahmen, Karin A.; Liaw, Peter K.; Egami, Takeshi Aluminum Alloying Effects on Lattice Types, Microstructures, and Mechanical Behavior of High-Entropy Alloys Systems. JOM (2013). https://doi.org/10.1007/s11837-013-0776-z
211. Sharma, Aayush; Singh, Prashant; Johnson, Duane D.; Liaw, Peter K.; Balasubramanian, Ganesh Atomistic clustering-ordering and high-strain deformation of an $Al_{0.1}CrCoFeNi$ high-entropy alloy. Scientific Reports (2016). https://doi.org/10.1038/srep31028
212. Santodonato, Louis J.; Zhang, Yang; Feygenson, Mikhail; Parish, Chad M.; Gao, Michael C.; Weber, Richard J.K.; Neuefeind, Joerg C; Tang, Zhi; Liaw, Peter K Deviation from high-entropy configurations in the atomic distributions of a multi-principal-element alloy. Nature Communications (2015). https://doi.org/10.1038/ncomms6964
213. Jiang, Hui; Li, Li; Wang, Rui; Han, Kaiming; Wang, Quanwei Effects of Chromium on the Microstructures and Mechanical Properties of $AlCoCr_xFeNi_{2.1}$ Eutectic High Entropy Alloys. Acta Metallurgica Sinica (English Letters) (2021). https://doi.org/10.1007/s40195-021-013 03-4
214. Yang, Zhongsheng; Wang, Zhijun; Wu, Qingfeng; Zheng, Tao; Zhao, Purui; Zhao, Jinkai; Chen, Jiayin Enhancing the mechanical properties of casting eutectic high entropy alloys with Mo addition. Applied Physics A (2019). https://doi.org/10.1007/s00339-019-2506-z
215. Jiang, Hui; Li, Li; Wang, Jianming; Wei, Chengbin; Zhang, Qiang; Su, Chunjian; Sui, Huaiming Wear Properties of Spark Plasma-Sintered AlCoCrFeNi2.1 Eutectic High Entropy Alloy with NbC Additions. Acta Metallurgica Sinica (English Letters) (2023). https://doi.org/10.1007/s40195-023-01529-4
216. Jain, Reliance; Rahul, M. R.; Jain, Sandeep; Samal, Sumanta; Kumar, Vinod Phase Evolution and Mechanical Behaviour of Co–Fe–Mn–Ni–Ti Eutectic High Entropy Alloys. Transactions of the Indian Institute of Metals (2018). https://doi.org/10.1007/s12666-018-1437-2
217. Shi, Peijian; Ren, Weili; Zheng, Tianxiang; Ren, Zhongming; Hou, Xueling; Peng, Jianchao; Hu, Pengfei; Gao, Yanfei; Zhong, Yunbo; Liaw, Peter K. Enhanced strength–ductility synergy in ultrafine-grained eutectic high-entropy alloys by inheriting microstructural lamellae. Nature Communications (2019). https://doi.org/10.1038/s41467-019-08460-2
218. Sinha, S.; Nene, S. S.; Frank, M.; Liu, K.; Mishra, R. S.; McWilliams, B. A.; Cho, K. C. Microstructural Evolution and Deformation Behavior of Ni-Si- and Co-Si-Containing Metastable High Entropy Alloys. Metallurgical and Materials Transactions A (2018). https://doi.org/10.1007/s11661-018-4968-6
219. Nene, S. S.; Frank, M.; Liu, K.; Mishra, R. S.; McWilliams, B. A.; Cho, K. C. Extremely high strength and work hardening ability in a metastable high entropy alloy. Scientific Reports (2018). https://doi.org/10.1038/s41598-018-28383-0
220. Sinha, Subhasis; Nene, Saurabh S.; Frank, Michael; Liu, Kaimiao; Agrawal, Priyanka; Mishra, Rajiv S. On the evolving nature of c/a ratio in a hexagonal close-packed epsilon martensite phase in transformative high entropy alloys. Scientific Reports (2019). https://doi.org/10.1038/s41598-019-49904-5

221. Frank, M.; Nene, S. S.; Chen, Y.; Gwalani, B.; Kautz, E. J.; Devaraj, A.; An, K.; Mishra, R. S. Correlating work hardening with co-activation of stacking fault strengthening and transformation in a high entropy alloy using in-situ neutron diffraction. Scientific Reports (2020). https://doi.org/10.1038/s41598-020-79492-8
222. Osei-Agyemang, Eric; Balasubramanian, Ganesh Surface oxidation mechanism of a refractory high-entropy alloy. npj Materials Degradation (2019). https://doi.org/10.1038/s41529-019-0082-5
223. Huang, Wenjun; Yin, Shaochun; Wang, Xuejiao; Guo, Ruipeng; Wu, Yucheng; Qiao, Junwei Grain growth and Hall–Petch relationship in $Ti_{37}V_{15}Nb_{22}Hf_{23}W_3$ refractory high-entropy alloys. Journal of Materials Research (2023). https://doi.org/10.1557/s43578-023-00924-y
224. Huang, Shuo; Li, Wei; Holmström, Erik; Vitos, Levente Phase-transition assisted mechanical behavior of $TiZrHfTa_x$ high-entropy alloys. Scientific Reports (2018). https://doi.org/10.1038/s41598-018-30892-x
225. Poulia, Anthoula; Georgatis, Emmanuel; Karantzalis, Alexander Evaluation of the Microstructural Aspects, Mechanical Properties and Dry Sliding Wear Response of MoTaNbVTi Refractory High Entropy Alloy. Metals and Materials International (2019). https://doi.org/10.1007/s12540-019-00283-6
226. Cieślak, Grzegorz; Dąbrowa, Juliusz; Jawańska, Monika; Parzuchowska, Agnieszka; Oleszak, Dariusz Microstructure and Mechanical Properties of the Ductile Al–Ti–Mo–Nb–V Refractory High Entropy Alloys. Metallurgical and Materials Transactions A (2021). https://doi.org/10.1007/s11661-021-06543-8
227. Senkov, O. N.; Pilchak, A. L.; Semiatin, S. L. Effect of Cold Deformation and Annealing on the Microstructure and Tensile Properties of a HfNbTaTiZr Refractory High Entropy Alloy. Metallurgical and Materials Transactions A (2018). https://doi.org/10.1007/s11661-018-4646-8
228. Soni, V.; Senkov, O. N.; Gwalani, B.; Miracle, D. B.; Banerjee, R. Microstructural Design for Improving Ductility of An Initially Brittle Refractory High Entropy Alloy. Scientific Reports (2018). https://doi.org/10.1038/s41598-018-27144-3
229. Dong, Ya-Guang; Chen, Shang; Jia, Nan-Nan; Zhang, Qiu-Hong; Wang, Liang; Xue, Yun-Fei; Jin, Ke Microstructures and mechanical properties of Ta–Nb–Zr–Ti–Al refractory high entropy alloys with varying Ta/Ti ratios. Tungsten (2021). https://doi.org/10.1007/s42864-021-00111-8
230. Zhang, Fengqi; Xiang, Chao; Han, En-Hou; Zhang, Zijian Effect of Nb Content on Microstructure and Mechanical Properties of $Mo_{0.25}V_{0.25}Ti_{1.5}Zr_{0.5}Nb_x$ High-Entropy Alloys. Acta Metallurgica Sinica (English Letters) (2022). https://doi.org/10.1007/s40195-022-01399-2
231. Hu, Yong; Liu, Fei; Zhao, Longzhi; Tang, Yanchuan; Jiao, Haitao Microstructural Evolution and Mechanical Properties of Semi-Solid $Al_{15}Mg_{45}Li_{39}Ca_{0.5}Si_{0.5}$ Light-Weight High Entropy Alloys Fabricated by Isothermal Heat Treatment. Metals and Materials International (2022). https://doi.org/10.1007/s12540-022-01303-8

Chapter 5
Fatigue/Fracture Properties of High Entropy Alloys

5.1 Introduction by Editor

A major factor in the attraction of high-entropy alloys is the investigation of their fatigue and fracture properties, an area that holds the key to maximizing their potential in demanding applications. These properties of materials provide invaluable information about their durability and reliability under cyclic loading and extreme conditions. Owing to unique compositions and resulting microstructures, high entropy alloys challenge conventional understanding of fracture and fatigue studies. The intricate interplay between microstructure and mechanical response under cyclic stress is a focus of intense research, encouraging the pursuit of HEAs for structural applications in aerospace, energy, and beyond. This chapter aims to shed light on fatigue/fracture behavior by investigating the most recent advances in experimental techniques, computational modeling, and experimental frameworks.

A promising new class of materials called High Entropy Alloys (HEAs) is transforming material science and structural engineering. These alloys have exceptional mechanical properties and are appealing candidates for a variety of applications due to their unusual composition of numerous primary elements in equimolar and non-equimolar ratios. Among the numerous factors influencing the performance of HEAs, fatigue and fracture properties stand out as critical aspects that require thorough investigation and understanding.

In this chapter, we set out to investigate the complex realm of fatigue and fracture behavior in High Entropy Alloys. In order to give readers a thorough grasp of the fundamental ideas guiding the mechanical response of materials under cyclic loading circumstances, we start by outlining the fundamental concepts of fatigue and fracture mechanics. Expanding on this understanding, we examine the unique characteristics of HEAs, emphasizing their composition, microstructure, and mechanical attributes that distinguish them from traditional alloys. Fatigue and fracture mechanics are crucial factors in the design and assessment of structural materials, guaranteeing their dependability and safety in diverse engineering applications. Fatigue is the process

S. S. Nene (ed.), *High Entropy Alloys*, https://doi.org/10.1007/978-981-99-7173-2_5

of gradual and localized structural deterioration that happens when a material is repeatedly exposed to cyclic loading. This can eventually cause the material to fail, even when the stress levels are far lower than its static strength. Fracture mechanics focuses on the examination of crack propagation and failure behavior in materials, offering an understanding of the mechanisms that control brittle and ductile fracture processes.

The unique mechanical characteristics of HEAs result from the combination of their intricate microstructure and elemental interactions. Superior strength, ductility, and resistance to deformation are conferred upon HEAs by the creation of a single-phase solid solution in conjunction with the suppression of grain boundaries and dislocation movement. Furthermore, these alloys' high configurational entropy encourages the development of intricate atomic structures, which improves their mechanical performance even more.

The chapter explores the enhancement of fatigue life in HEAs through the incorporation of ductile-transformable multicomponent B2 precipitates. By investigating the role of precipitate phases, this study sheds light on strategies for improving the fatigue resistance of HEAs. The presence of these B2 precipitates contributes to the overall toughness of the alloy by absorbing and redistributing stress, thereby delaying the initiation and propagation of fatigue cracks.

Moving forward, the chapter delves into the fracture toughness and fatigue crack growth behavior of as-cast HEAs, providing insights into the intrinsic toughness and crack propagation characteristics of these alloys. This section looks at how the alloy's resistance to crack initiation and growth under cyclic loading circumstances is affected by intrinsic microstructural properties including grain size and phase distribution. Understanding these factors is crucial for predicting the performance of HEAs in real-world applications where mechanical reliability is paramount.

The impact of microstructural characteristics on the high-cycle fatigue behavior of cryogenic and room-temperature deformed CoCrFeMnNi HEAs is examined next. This investigation highlights how different deformation temperatures can alter the microstructure and, consequently, the fatigue properties of HEAs. The study reveals that cryogenic deformation can enhance dislocation density and refine grain structures, leading to improved fatigue resistance compared to room-temperature deformed counterparts.

Next, the development of grain-orientation-dependent residual stresses in cyclically deformed alloys is scrutinized. Through the clarification of the impact of grain orientation on the creation of residual stress, this study advances our comprehension of fatigue-induced damage buildup in heat-exposed assemblies. Residual stresses, which develop due to cyclic deformation, play a significant role in the overall fatigue performance of HEAs. By examining how different grain orientations affect residual stress distribution, the chapter provides valuable insights into optimizing alloy processing techniques to mitigate fatigue damage.

Finally, the chapter concludes with an exploration of the nanoscale origins of damage tolerance in the CrMnFeCoNi HEA. Through advanced characterization techniques, such as transmission electron microscopy (TEM) and atom probe tomography (APT), this research unveils the mechanisms underlying the exceptional damage tolerance exhibited by HEAs at the nanoscale level. These techniques allow for a detailed analysis of dislocation interactions, phase transformations, and atomic-scale features that contribute to the alloy's ability to withstand high levels of stress and strain without catastrophic failure.

Collectively, the articles presented in this chapter offer a comprehensive overview of the fatigue and fracture properties of HEAs, paving the way for future advancements in material design and engineering applications. By integrating insights from microstructural characterization, mechanical testing, and advanced modeling, this body of work enhances our understanding of how to tailor HEA compositions and processing methods to achieve superior fatigue and fracture performance. These findings are instrumental in guiding the development of next-generation materials for critical applications in the aerospace, automotive, and energy sectors, where reliability and durability are paramount.

5.2 Machine Generated Summaries

Disclaimer: The summaries in this chapter were generated from Springer Nature publications using extractive AI auto-summarization: An extraction-based summarizer aims to identify the most important sentences of a text using an algorithm and uses those original sentences to create the auto-summary (unlike generative AI). As the constituted sentences are machine selected, they may not fully reflect the body of the work, so we strongly advise that the original content is read and cited. The auto generated summaries were curated by the editor to meet Springer Nature publication standards. To cite this content, please refer to the original papers.

Machine generated keywords: fatigue, deformation, crack, neutron, damage, toughness, dislocation, plastic, grain, strain, deformation mechanism, cocrfemnni highentropy, diffraction, failure, stress

Enhancing Fatigue Life by Ductile-Transformable Multicomponent B2 Precipitates in a High-Entropy Alloy [32]

This is a machine-generated summary of:

Feng, Rui; Rao, You; Liu, Chuhao; Xie, Xie; Yu, Dunji; Chen, Yan; Ghazisaeidi, Maryam; Ungar, Tamas; Wang, Huamiao; An, Ke; Liaw, Peter. K.: Enhancing fatigue life by ductile-transformable multicomponent B2 precipitates in a high-entropy alloy [32]

Published in: Nature Communications (2021)

Link to original: https://doi.org/10.1038/s41467-021-23689-6

If you want to cite the papers, please refer to the original.

For technical reasons we could not place the page where the original quote is coming from.

Abstract-Summary

"A fundamental understanding of cyclic-deformation and fatigue-failure mechanisms is critical for the development of fatigue-resistant structural materials."

"Multiple cyclic-deformation mechanisms, including dislocation slips, precipitation strengthening, deformation twinning, and reversible martensitic phase transformation, are observed in the studied high-entropy alloy."

"Its improved fatigue performance at low strain amplitudes, i.e., the high fatigue-crack-initiation resistance, is attributed to the high elasticity, plastic deformability, and martensitic transformation of the B2-strengthening phase."

"This study shows that fatigue-resistant alloys can be developed by incorporating strengthening ductile-transformable multicomponent intermetallic phases."

Introduction

"The conventional alloy-design strategies encounter the dilemma of improving both fatigue strength and fatigue-crack-initiation resistance."

"A new alloy-design concept, called high-entropy alloys (HEAs), shows great potential in enhancing materials' mechanical performance [1–8]."

"Inspired by this idea, here we design a multicomponent B2 precipitates-strengthened HEA to improve the fatigue performance of structural materials."

"We find that the designed alloy shows at least four times longer fatigue life than other conventional alloys at a low plastic strain amplitude of ~ 0.03% by incorporating the ductile-transformable multicomponent B2 precipitates, exhibiting enhanced fatigue-crack-initiation resistance."

"We show that the design idea by integrating ductile-transformable multicomponent intermetallic precipitates and providing various beneficial cyclic-deformation mechanisms leads to a new direction of designing advanced fatigue-resistant alloys."

Results

"At the low strain amplitude of 0.5%, $FWHM_G/d$ of the FCC-{111} and B2-{110} peaks exhibited a stable stage at the first 250 cycles, and then, quickly increased to a higher level until 500 cycles."

"The B2 phase is plastically deformable at all the three strain amplitudes, as demonstrated by the greatly increased $FWHM_G/d$, which is caused by the stored dislocations, as presented by later TEM observations."

"The plastically-deformable B2 phase implies that stress relaxation and strain partitioning occur between the FCC matrix and B2 precipitates, which can relieve the stress concentration during cyclic loading."

"At all strain amplitudes, dense dislocations piled-up around the B2-phase boundaries, due to the strain incompatibility of the FCC and B2 phases."

"The hard B2 precipitates acted as obstacles and resisted the plastic deformation, which hindered the further movements of dislocations, and the internal interphase stress developed, contributing to the cyclic strain hardening."

Discussion

"The ductile-transformable character of the multicomponent B2 phase plays a critical role in strengthening the fatigue-crack-initiation resistance, leading to the enhanced fatigue life at low strain amplitudes."

"At high strain amplitudes (>1%), the deformation twinning occurred, which acted as barriers to hinder the further dislocation movement, and then notable cyclic hardening can be achieved to overcome the softening caused by the microcracks and SRO effect."

"Our results demonstrate that fatigue life can be improved at low strain amplitudes by incorporating multicomponent ductile and transformable intermetallic precipitates, superior to other reported traditional intermetallic precipitates-strengthened alloys, and meanwhile, maintaining comparable fatigue life at high strain amplitudes."

"Our work provides a full picture to understand the cyclic-deformation mechanisms in the multicomponent B2 precipitates-strengthened HEA, and guides the design of fatigue-resistant alloys by introducing ductile-transformable multicomponent intermetallic precipitates that can be easily achieved by tuning HEAs' compositions and thermomechanical processing."

Methods

"The monotonic tension and fully-reversed strain-controlled LCF tests were performed at RT, using an MTS Model 810 servohydraulic machine."

"The monotonic tension was performed at RT with a strain rate of 1×10^{-3} s^{-1}."

"The LCF tests were performed at the strain rate of 1×10^{-2} s^{-1}, and the initial loading is tension."

"In situ neutron-diffraction experiments subjected to tension and fully-reversed strain cycling were performed, using an MTS load-frame on the VULCAN Engineering Materials Diffractometer at the Spallation Neutron Source (SNS), Oak Ridge National Laboratory (ORNL) [9, 10]."

"Low, intermediate, and high strain amplitudes of $\pm$ 0.5%, $\pm$ 1%, and $\pm$ 1.75% were chosen for the in situ neutron-diffraction continuous measurements, respectively."

"The selected cycles for neutron diffraction were recorded with a strain rate of ~ $7.4 \times 10^{-6}\ s^{-1}$, while other fatigue cycles were with a strain rate of $1 \times 10^{-2}\ s^{-1}$."

Fracture Toughness and Fatigue Crack Growth Behavior of As-Cast High-Entropy Alloys [33]

This is a machine-generated summary of:

Seifi, Mohsen; Li, Dongyue; Yong, Zhang; Liaw, Peter K.; Lewandowski, John J.: Fracture Toughness and Fatigue Crack Growth Behavior of As-Cast High-Entropy Alloys [33]

Published in: JOM (2015)

Link to original: https://doi.org/10.1007/s11837-015-1563-9

If you want to cite the papers, please refer to the original.

For technical reasons we could not place the page where the original quote is coming from.

Abstract-Summary

"The fracture toughness and fatigue crack growth behavior of two as-vacuum arc cast high-entropy alloys (HEAs) ($Al_{0.2}CrFeNiTi_{0.2}$ and $AlCrFeNi_2Cu$) were determined."

"The notched and fatigue precracked toughness values were in the range of those reported in the literature for two-phase alloys but significantly less than recent reports on a single phase fcc-HEA that was deformation processed."

Introduction

"High-entropy alloys (HEAs) are a relatively new class of materials that typically combine five principal elements or more and are known to exhibit a number of unique combinations of flow and fracture behavior."

"Some preliminary information is available for fatigue properties [11], but none exists for fatigue-crack-growth behavior, and only limited reports on fracture toughness of HEAs exist."

"[12, 13] In that regard, recent work [14] has documented the microstructure and high compression strength of $Al_xCrFeNiTi_{0.25}$ alloys, with fracture strengths exceeding 1800 MPa and yield strengths in excess of 1300 MPa."

"[14] In particular, when $x = 0.5$ in $Al_xCrFeNiTi_{0.25}$, the fracture strength and plastic strain at failure achieved their maximum values of 3474 MPa and 40%, respectively."

"The present work was conducted to determine the fracture toughness and fatigue-crack-growth behavior of $Al_{0.2}CrFeNiTi_{0.2}$ and $AlCrFeNi_2Cu$ alloys to provide a more complete characterization of microstructure/property relationships in these particular HEA materials."

Materials and Methods

"The hardness was measured by a HXD-1000TM Vickers microhardness tester with a load of 500 g and a dwell time of 15 s. Fracture toughness was determined in three-point bending using single-edge notch (SEN) samples tested on a Model 810 MTS servohydraulic machine in general accordance with ASTM E399 standard."

"Fatigue crack growth tests were performed at a cyclic frequency of 20 Hz in room temperature air with relative humidity of 40% in accordance with ASTM E647 standard (Ref."

"Fracture Technology Associates (FTA™) software was used to monitor crack growth continuously in both the fatigue and fracture toughness tests in order to also comply with existing ASTM requirements."

"Fatigue crack growth tests at load ratios $R = 0.1$, 0.3, and 0.7 were started at a low stress intensity range ΔK, using constant cyclic loading to produce rising ΔK conditions until catastrophic fracture."

Results

"There are minor unidentified peaks at small angles and the structure corresponding to those peaks are being determined by a TEM study as part of future work."

"The matrix in $Al_{0.2}CrFeNiTi_{0.2}$ is enriched in Cr and Fe, whereas the particles and small, elongated precipitates are enriched in Al, Ni, and Ti."

"The matrix in $AlCrFeNi_2Cu$ is enriched in Al, Ni, and Cu, whereas particles are enriched in Fe and Cr."

"These increases in R produced significant reductions in ΔK_{th} and increases in Paris law slope, somewhat similar to a recent report on an as-cast γ-TiAl [15], but much greater than what was reported in as-deposited Ti-6Al-4V (Ref."

Discussion

"Increasing the test temperature to 200 °C increases the toughness, while producing a nonlinear load versus displacement trace although the fracture surface features appeared similar to those of the room-temperature tests in that there appeared to be similar combinations of ductile/dimpled and brittle fracture."

"Future work will examine the effects of testing at higher temperatures in order to induce a brittle-to-ductile transition in the bcc constituent in order to determine its effect on toughness of these two-phase HEA alloys."

"The combinations of brittle and ductile/dimpled features present on the fracture surfaces of the presently tested materials suggest that a similar mechanism may contribute to the significantly higher fatigue-crack growth rates and higher Paris law slopes obtained at higher R. Future work will examine the influence of changing the test temperature in order to induce more ductile behavior of the bcc constituent, as test temperature should have a significant effect on the fatigue-crack growth rate (and Paris law slope)."

Conclusion

"$Al_{0.2}CrFeNiTi_{0.2}$ and $AlCrFeNi_2Cu$ HEA alloys produced via arc casting exhibited a combination of bcc and fcc phases."

"Fracture-surface examinations of the toughness tests revealed a combination of ductile/dimpled and brittle features that roughly corresponded to the fcc and bcc phases, respectively, in both HEA alloys presently tested."

"Fatigue-crack growth experiments conducted on as-cast $Al_{0.2}CrFeNiTi_{0.2}$ and $AlCrFeNi_2Cu$ exhibited high fatigue thresholds that decreased significantly with an increase in R, while Paris law slopes exhibited metallic-like values (e.g., 3–5) at low R with significant increases at high R. Fracture-surface examinations revealed combinations of brittle and ductile/dimpled regions in the overload region, with limited observations of fatigue striations in the Paris law regime."

Effect of Microstructural Features on the High-Cycle Fatigue Behavior of CoCrFeMnNi High-Entropy Alloys Deformed at Room and Cryogenic Temperatures [34]

This is a machine-generated summary of:

Lee, Gyung Tae; Won, Jong Woo; Lim, Ka Ram; Kang, Minju; Kwon, Heoun Jun; Na, Young Sang; Choi, Yoon Suk: Effect of Microstructural Features on the High-Cycle Fatigue Behavior of CoCrFeMnNi High-Entropy Alloys Deformed at Room and Cryogenic Temperatures [34]

Published in: Metals and Materials International (2020)

Link to original: https://doi.org/10.1007/s12540-020-00786-7

If you want to cite the papers, please refer to the original.

For technical reasons we could not place the page where the original quote is coming from.

Abstract-Summary

"The dislocation cell structures and deformation twins were generated by prestraining the CoCrFeMnNi high-entropy alloy at room temperature (27 °C) and a cryogenic temperature (− 196 °C), respectively."

"The results of the rotational bending fatigue tests revealed that the CoCr-FeMnNi high-entropy alloy prestrained at room temperature exhibited higher fatigue resistance and fatigue limit than the specimen prestrained at a cryogenic temperature."

"It is concluded that the different aspects of micro-void formation affected the crack initiation and, consequently, the fatigue properties of the room and cryogenic temperature-prestrained alloys."

Introduction

"The CoCrFeMnNi high-entropy alloy is a representative FCC alloy and has high strength and toughness at cryogenic temperatures due to the generation of deformation twins at low stacking fault energy (SFE)."

"The research team of this study previously reported the efficient improvement of the strength of the CoCrFeMnNi high-entropy alloy by activating deformation twins through the rolling of the alloy at cryogenic temperatures [16]."

"Few studies have been conducted on the fatigue properties of the CoCrFeMnNi high-entropy alloy, and no studies have been conducted on the effects of different deformation mechanisms associated with temperature and the resulting microstructures on fatigue behavior."

"The CoCrFeMnNi high-entropy alloy used in this study exhibits different SFEs depending on its temperature."

"In this study, the effects of initial microstructural differences in a CoCrFeMnNi high-entropy alloy obtained through different deformation mechanisms on its high-cycle fatigue properties were examined."

Experimental Method

"The internal homogeneity of the prepared ingot was also increased by homogenizing heat treatment at 1200 °C for 24 h, followed by furnace cooling to 900 °C, and then air cooling to room temperature."

"To analyze the microstructures of the specimens, electron backscatter diffraction (EBSD) analyses were conducted on their transverse-direction surfaces using an SU-6600 field emission scanning electron microscope (FE-SEM)."

"The resulting fatigue fracture surfaces were observed using a JSM-6610LV scanning electron microscope (SEM) focused on the fatigue crack initiation site, which is an important location affecting fatigue failure under high-cycle fatigue."

"After performing electrolytic polishing on the vertical surface of the fatigue fracture, microstructural analysis of the fatigue crack initiation site was also conducted using an SU-6600 FE-SEM and EBSD analysis."

Experiment Results

"This study then attempted to observe differences in the fatigue failure and crack formation behavior of the RT specimen, which contained a large quantity of dislocation cell structures and LABs, and the CT specimen, which contained a large quantity of deformation twins."

"As with the fatigue fracture surface of the RT specimen, the fracture surface of the CT specimen can be divided into the fatigue crack initiation site, crack propagation site, and final fracture site."

"The fatigue crack initiation site and crack propagation site observations of the CT specimen, however, differ slightly from those of the RT specimen."

"The differences in the fatigue crack initiation sites and crack propagation sites of the RT and CT specimens appear to have been caused by the microstructural differences between the two specimens."

"The spacings indicate that the fatigue crack growth rate of the CT specimen was faster than that of the RT specimen."

Discussion

"The interactions between different microstructures in the RT and CT specimens and the dislocation movements due to cyclic stress were analyzed to explain the observed differences in fatigue crack behavior."

"In relation to the fatigue crack behavior, deformation twins can delay dislocation accumulation at grain boundaries by accepting shifting dislocations under cyclic stress [17]."

"In most fatigue-related studies, deformation twins have been observed to accelerate crack nucleation by acting as dislocation accumulation locations, similar to grain boundaries [18–21]."

"Micro-voids were observed at the grain boundary triple junctions, indicating that these points acted as crack nucleation regions in the RT specimen."

"Micro-voids were found to be concentrated in regions where the grain and twin boundaries intersect, indicating that these boundaries acted as crack nucleation regions in the CT specimen."

"In this instance, the twin boundaries act as fatigue crack nucleation regions due to the occurrence of dislocation accumulation and stress concentration."

Conclusions

"Experiments were performed to examine the effects of different microstructures formed under different prestraining temperatures on the high-cycle fatigue behavior of the CoCrFeMnNi high-entropy alloy."

"These characteristics increased the yield and tensile strengths of the CoCrFeMnNi high-entropy alloy. (2) While LABs do not interfere with the movement of dislocations under cyclic stress, grain and twin boundaries can act as obstacles to limit such movement."

"Twin boundaries act as crack nucleation regions due to the local stress concentrations. (3) In the specimen prestrained at − 196 °C, a large quantity of deformation twins generated numerous small micro-voids."

"These micro-voids adversely affected the fatigue properties of the specimen as they promoted the growth of cracks. (4) The CoCrFeMnNi high-entropy alloy exhibited higher fatigue strength than stainless steels, TWIP steels, TRIP steels, and pure titanium."

The Development of Grain-Orientation-Dependent Residual Stresses in a Cyclically Deformed Alloy [35]

This is a machine-generated summary of:

Wang, Yan-Dong; Tian, Hongbo; Stoica, Alexandru D.; Wang, Xun-Li; Liaw, Peter K.; Richardson, James W.: The development of grain-orientation-dependent residual stresses in a cyclically deformed alloy [35]

Published in: Nature Materials (2003)

Link to original: https://doi.org/10.1038/nmat812

If you want to cite the papers, please refer to the original.

For technical reasons we could not place the page where the original quote is coming from.

Abstract-Summary

"The micromechanical behaviours, particularly the distributions of stresses on the scale of grain size during or after mechanical or electrical fatigue, are crucial to a full understanding of the damage mechanisms in these materials."

"We report a neutron diffraction investigation of the development of intergranular stresses, which vary as a function of grain orientations, in 316 stainless steel during high-cycle fatigue."

"With further increase of fatigue cycles, the intergranular stress decreased, while the elastic intragranular stored energy continued to grow."

"One implication of our findings is that the ratio between the intergranular and intragranular stored energies during various stages of fatigue deformation may validate the damage mechanism and can be used as a fingerprint for monitoring the state of fatigue damage in materials."

Main

"The existence of intergranular stresses in single-phase materials is due to the discontinuity of the stress distribution from grain to grain, caused by so-called elastic or/and plastic incompatibility, which has been investigated by 3D plasticity modelling [22–24] in combination with lattice strain measurements by neutron diffraction."

"These observations gave unambiguous evidence that a large grain-orientation-dependent intergranular stress exists in the 316 stainless steel after high-cycle fatigue."

"Because the stress or strain distributions of fatigued samples with random grain orientations display axial symmetry around the LD, the SODF representation can be conveniently reduced to a two-dimensional fibre-symmetry group."

"Although it is not yet clear whether the observed large intergranular stresses are responsible for the initiation of cracks in fatigued samples, the development of intergranular stress and stored energy determined by neutron diffraction can be directly connected to the microscopic mechanisms of materials damage during fatigue."

Methods

"The average lattice strain is determined from the diffraction line displacement in neutron time-of-flight."

"Individual diffraction peaks in the time-of-flight spectrum were fitted separately by convoluting a Voigt sample broadening function with a pulse-shape instrument function [25]."

"The peak position in time-of-flight is proportional to the average value of the lattice spacing corresponding to a family of {hkl} reflections."

"The experimental lattice strain was calculated from the change in the peak position relative to the virgin (stress-free) sample."

"It is adequate to mention our specific approximation: $W_S = (K/2)(\beta_{hkl}{}^2/C_{hkl})\,F$, where $K = C_{44}/\sqrt{A_i} = 65$ GPa for the stainless steel, $A_i = 2C_{44}/(C_{11} - C_{12}) = 3.77$ is the elastic anisotropy factor, β_{hkl} is the integral breadth for the {hkl} peak, C_{hkl} is the contrast factor, and F is a numerical factor close to 1."

Nanoscale Origins of the Damage Tolerance of the High-Entropy Alloy CrMnFeCoNi [36]

This is a machine-generated summary of:

Zhang, ZiJiao; Mao, M. M.; Wang, Jiangwei; Gludovatz, Bernd; Zhang, Ze; Mao, Scott X.; George, Easo P.; Yu, Qian; Ritchie, Robert O.: Nanoscale origins of the damage tolerance of the high-entropy alloy CrMnFeCoNi [36]

Published in: Nature Communications (2015)

Link to original: https://doi.org/10.1038/ncomms10143

If you want to cite the papers, please refer to the original.

For technical reasons we could not place the page where the original quote is coming from.

Abstract-Summary

"Damage tolerance can be an elusive characteristic of structural materials requiring both high strength and ductility, properties that are often mutually exclusive."

"The single-phase CrMnFeCoNi alloy displays tensile strength levels of ~1 GPa, excellent ductility (~60–70%) and exceptional fracture toughness (K_{JIc} > 200 MPa$\sqrt{m}$)."

"We identify a synergy of multiple deformation mechanisms, rarely achieved in metallic alloys, which generates high strength, work hardening and ductility, including the easy motion of Shockley partials, their interactions to form stacking-fault parallelepipeds, and arrest at planar slip bands of undissociated dislocations."

Introduction

"Intrinsic toughness is promoted by extensive plastic deformation, which requires the easy motion and multiplication of dislocations (that can compromise strength), yet contradictorily requires significant hardening to generate strength and ductility by resisting dislocation motion and delaying plastic (necking) instabilities [26–28]."

"'conflict', the ability to resist cracking in metals and alloys under mechanical loading can be affected by the prevailing modes of plastic deformation and microstructure at the crack tip."

"The solute atoms locally introduce different crystal bonds with varied strengths and also cause local lattice distortion, thereby offering resistance to dislocation motion; as a result, strength is generally enhanced but ductility can be reduced."

"The detailed mechanisms and crack tip processes responsible for the high toughness of this high-entropy alloy have remained unclear."

"By imaging as close as within a few hundred nanometres of the crack tip, we identify multiple deformation mechanisms that are activated at different stages of deformation and act synergistically to contribute to the ultrahigh toughness."

Results

"We observed slip of the leading partial dislocations, formation of corresponding SFs and glide of the trailing partial dislocations."

"These dynamic dislocation sequences clearly demonstrate that because of the low SFE in this alloy [29–31], the slip of partial dislocations and the corresponding formation of SFs have a critical role in accommodating the strain at the early stages of plastic deformation."

"The interactions of partial dislocations also generate other obstacles to dislocation motion that lead to further hardening."

"Although the movement of partial dislocations is relatively easy and provides initial deformability in this alloy, their interaction can lead to significant strain hardening through the formation of such three-dimensional stacking-fault defects that are both stable and a formidable obstacle for other dislocations to overcome."

Discussion

"Toughness or fracture resistance can be considered to result from two classes of mechanisms: 'intrinsic toughening' mechanisms, which are primarily associated with plasticity (and the generation of ductility) and operate ahead of the crack tip to provide resistance to microstructural damage, and 'extrinsic toughening' mechanisms, which operate principally in the wake of the crack tip to inhibit fracture by 'shielding' the crack from the applied driving force [27]."

"One of the unique aspects of this high-entropy alloy is that it can generate fracture resistance from both these classes of mechanisms; the numerous plasticity and hardening mechanisms creating both strength and ductility by intrinsic means, and the near-tip crack bridging, which is extremely rare in a ductile metallic material, providing additional toughening extrinsically."

Methods

"To produce electron-transparent regions for observation and analyses during the in situ TEM tensile tests, the mechanically polished specimens were further thinned using ion milling or jet polishing until a hole appeared in the middle of the foils."

"The in situ TEM tensile tests were conducted at room temperature with a Gatan model 654 single-tilt straining holder, in a FEI Tecnai G2 F20 TEM operating at 200 kV. Only samples that were well attached to the substrate, about 12, free of contamination from the sample preparation procedure and did not rotate or bend during deformation were selected for detailed TEM investigations."

"The ion milling and jet polishing processes produced many tiny cracks in the electron-transparent region around the hole, but our analyses focused mainly on the microstructure near those cracks that were roughly perpendicular to the loading direction."

"The regions in front of the crack tip were monitored during the straining process."

"Since the thickness of the foil increases gradually, the cracks grew from the thin area (electron-transparent region) to the thick area (non-electron transparent)."

References

1. J.W. Yeh, S.K. Chen, S.J. Lin, J.Y. Gan, T.S. Chin, T.T. Shun, C.H. Tsau, and S.Y. Chang, Adv. Eng. Mater. 6, 299 (2004).
2. B. Cantor, I. T. H. Chang, P. Knight, et al., "Microstructural development in equiatomic multicomponent alloys," Mater. Sci. Eng. A, 375, 213–218 (2004).
3. Zhang, Y. et al. Microstructures and properties of high-entropy alloys. Prog. Mater. Sci. 61, 1–93 (2014).
4. Gludovatz, B. et al. A fracture-resistant high-entropy alloy for cryogenic applications. Science 345, 1153–1158 (2014).
5. H.Y. Diao, R. Feng, K.A. Dahmen, and P.K. Liaw, Fundamental Deformation Behavior in High-Entropy Alloys: An Overview, Curr. Opin. Solid State Mater. Sci., 2017, 21(5), p 252–266. https://doi.org/10.1016/j.cossms.2017.08.003
6. T. Yang, Y.L. Zhao, Y. Tong, Z.B. Jiao, J. Wei, J.X. Cai, X.D. Han, D. Chen, A. Hu, J.J. Kai, K. Lu, Y. Liu, C.T. Liu, Science 362, 933 (2018)
7. Z. Lei, X. Liu, Y. Wu, H. Wang, S. Jiang, S. Wang, X. Hui, Y. Wu, B. Gault, P. Kontis, D. Raabe, L. Gu, Q. Zhang, H. Chen, H. Wang, J. Liu, K. An, Q. Zeng, T.-G. Nieh, Z. Lu, Enhanced strength and ductility in a high-entropy alloy via ordered oxygen complexes. Nature 563(7732), 546+ (2018)
8. Li, Z., Pradeep, K. G., Deng, Y., Raabe, D. & Tasan, C. C. Metastable high-entropy dual-phase alloys overcome the strength–ductility trade-off. Nature 534, 227–230 (2016).
9. Wang, X. L. et al. VULCAN-The engineering diffractometer at the SNS. Phys. B Condens. Matter 385-386, 673–675 (2006).
10. An, K. et al. First in situ lattice strains measurements under load at VULCAN. Metall. Mater. Trans. A Phys. Metall. Mater. Sci. 42, 95–99 (2011).
11. M.A. Hemphill, T. Yuan, G.Y. Wang, J.W. Yeh, C.W. Tsai, A. Chuang, P.K. Liaw, Acta Mater., 2012, vol. 60, pp. 5723 – 5734.
12. U. Roy, H. Roy, H. Daoud, U. Glatzel, and K.K. Ray, Mater. Lett. 132, 186 (2014).
13. H. Zhang, Y. He, and Y. Pan, Scripta Mater. 69, 342 (2013).
14. S. Liu, M.C. Gao, P.K. Liaw and Y. Zhang: J. Alloys Compd., 2015, vol. 619, pp. 610-15.
15. M.S. Dahar, S.M. Seifi, B.P. Bewlay, and J.J. Lewandowski, Intermetallics 57, 73 (2015).
16. J.W. Won, S. Lee, S.H. Park, M. Kang, K.R. Lim, C.H. Park, Y.S. Na, J. Alloys Compd. 742, 290–295 (2018)
17. B. Wang, P. Zhang, Q.Q. Duan, Z.J. Zhang, H.J. Yang, J.C. Pang, Y.Z. Tian, X.W. Li, Z.F. Zhang, Mater. Sci. Eng. A 679, 258–271 (2017)
18. S. Hashimoto, H. Ikehata, A. Kato, H. Kato, Y. Kaneko, Interface Sci. 7, 159–171 (1999)

19. L. Llanes, C. Laird, Mater. Sci. Eng. A 157, 21–27 (1992)
20. S. Qu, P. Zhang, S.D. Wu, Q.S. Zang, Z.F. Zhang, Scr. Mater. 59, 1131–1134 (2008)
21. M.D. Sangid, H.J. Maier, H. Sehitoglu, Int. J. Plast. 27, 801–821 (2011)
22. Dawson, P., Boyce, D., MacEven, S. & Rogge, R., Residual strain in HY100 polycrystals: comparisons of experiments and simulations. Metall. Trans. A 31, 1543–1555 (2000).
23. Clausen, B., Lorentz, T. & Leffers, T. Self-consistent modelling of the plastic deformation of f.c.c polycrystals and its implications for diffraction measurements of internal stresses. Acta Mater. 46, 3087–3098 (1998).
24. Pang, J.W.L., Holden, T.M. & Mason, T.E. In situ generation of intergranular strains in an Al7050 alloy. Acta Mater. 46, 1503–1518 (1998).
25. Wang, Y.D., Wang, X.-L., Stoica, A.D., Richardson, J.W. & Lin Peng, R. Determination of the stress orientation distribution function using pulsed neutron sources. J. Appl. Cryst. 36, 14–22 (2003).
26. Lu, K., Lu, L. & Suresh, S. Strengthening materials by engineering coherent internal boundaries at the nanoscale. Science 324, 349–352 (2009).
27. Ritchie, R. O. The conflicts between strength and toughness. Nat. Mater. 10, 817–822 (2011).
28. Zhu, Y. T. & Liao, X. Nanostructured metals: retaining ductility. Nat. Mater. 3, 351–352 (2004).
29. A. Gali and E.P. George, Tensile Properties of High- and Medium-Entropy Alloys, Intermetallics, 2013, 39, p 74-78
30. F. Otto, A. Dlouhý, C. Somsen, et al., "The influences of temperature and microstructure on the tensile properties of a CoCrFeMnNi high-entropy alloy," Acta Mater., 61, No. 15, 5743–5755 (2013).
31. A.J. Zaddach, C. Niu, C.C. Koch, and D.L. Irving, Mechanical Properties and Stacking Fault Energies of NiFeCrCoMn High-Entropy Alloy, JOM, 2013, 65, p 1780–1789.
32. Feng, Rui; Rao, You; Liu, Chuhao; Xie, Xie; Yu, Dunji; Chen, Yan; Ghazisaeidi, Maryam; Ungar, Tamas; Wang, Huamiao; An, Ke; Liaw, Peter. K. Enhancing fatigue life by ductile-transformable multicomponent B2 precipitates in a high-entropy alloy. Nature Communications (2021). https://doi.org/10.1038/s41467-021-23689-6
33. Seifi, Mohsen; Li, Dongyue; Yong, Zhang; Liaw, Peter K.; Lewandowski, John J. Fracture Toughness and Fatigue Crack Growth Behavior of As-Cast High-Entropy Alloys. JOM (2015). https://doi.org/10.1007/s11837-015-1563-9
34. Lee, Gyung Tae; Won, Jong Woo; Lim, Ka Ram; Kang, Minju; Kwon, Heoun Jun; Na, Young Sang; Choi, Yoon Suk Effect of Microstructural Features on the High-Cycle Fatigue Behavior of CoCrFeMnNi High-Entropy Alloys Deformed at Room and Cryogenic Temperatures. Metals and Materials International (2020). https://doi.org/10.1007/s12540-020-00786-7
35. Wang, Yan-Dong; Tian, Hongbo; Stoica, Alexandru D.; Wang, Xun-Li; Liaw, Peter K.; Richardson, James W. The development of grain-orientation-dependent residual stresses in a cyclically deformed alloy. Nature Materials (2003). https://doi.org/10.1038/nmat812
36. Zhang, ZiJiao; Mao, M. M.; Wang, Jiangwei; Gludovatz, Bernd; Zhang, Ze; Mao, Scott X.; George, Easo P.; Yu, Qian; Ritchie, Robert O. Nanoscale origins of the damage tolerance of the high-entropy alloy CrMnFeCoNi. Nature Communications (2015). https://doi.org/10.1038/ncomms10143

Chapter 6
Creep and Super-Plastic Deformation in High Entropy Alloys

6.1 Introduction by Editor

Creep, a time-dependent deformation phenomenon that occurs at high temperatures and constant stress, and superplastic deformation, which is characterized by exceptional plasticity at high temperatures and low strain rates, are important aspects of material behavior for applications ranging from aerospace to energy generation. Exploration of these properties in high-entropy alloys holds enormous promise, as their complex composition and resulting microstructures introduce the possibility of novel creep and superplastic responses that differ from those found in traditional alloys. This chapter takes the reader on a journey through the fascinating world of creep (long-term, nanoindentation, and elevated temperature) and superplastic deformation (ultrahigh strain rate and differential) properties in high-entropy alloys, providing a thorough examination of their behavior under extreme conditions.

Creep, and superplastic deformation, are important aspects of material behavior for applications ranging from aerospace entities to the energy generation sector. Conventionally, alloys in this class include well-studied Ni-rich alloys including Inconel series and CMSX series alloys that are prominently used in gas turbine applications. These alloys are seen to limit their application due to the possible occurrence of high-temperature undesired phase leading to brittleness or limited high-temperature strengthening due to low antiphase boundary energies. In such cases, the metallurgy of the component used proves to be very critical as the undesired phase or morphology that is chemically favorable can prove to be fatal. Moreover, with the increase in demand for these entities in application, it becomes essential that these components be fabricated easily. However, these alloys' high strength and low ductility have been a primary issue limiting its formability. The low strain rate deformation aspect limits their potential for mass production of extensive time-based fabrication techniques. This demands deformation at a high strain rate, which is impossible in the pre-existing alloys as far as the literature suggests.

S. S. Nene (ed.), *High Entropy Alloys*, https://doi.org/10.1007/978-981-99-7173-2_6

Exploration of these properties in high-entropy alloys holds enormous promise, as their complex composition ultimately results in microstructures that introduce a possibility of a novel creep and superplastic responses that differ from those found in traditional alloys. The study of HEAs in this instance has shown that the alloys tend to enhance the properties for deformation under creep in wrought conditions after rolling and annealing or processed conditions like friction stir processing (FSP). Few HEAs on this occasion showed significant plasticity at comparatively higher strain rates as compared to conventional superalloys. These alloys are limited for now and are specific to the processing implied prior to deformation concluding to a fact that the grain morphology plays a crucial role in deformation at high temperature. In literature, these HEAs are still limited. However, the alloy designing approach and processing conditions for similar microstructure and phase distribution, can be implemented to develop new alloys for high strain rate deformation with a similar extent of plasticity. This indeed is an important requirement in the conventional superalloys by considering it for rapid production of a component for high temperature targeted application.

The enhancement of the creep-resistance or super-plasticity in HEAs creates a new pathway in evolving 3d-transitional series HEAs and extending their role in high-temperature applications.

It is important to know that understanding the creep behavior or a deformation mechanism under creep for these newly developed alloys can be a tedious process. Understanding the deformation of these alloys for a condition of varying temperature alone can be severely time-consuming for not less than months or years. These also demand large investments in terms of equipment which practically is impossible for large data set generation in a short period of time. For this, it becomes necessary to develop a technique that can give a basic idea of the creep deformation behavior of an alloy. One of these techniques is nano-indentation. The deformation behavior of an alloy under nanoindentation and the subsequent behavior of the material under creep show a very promising correlation leading to understanding of creep properties of the newly designed HEAs and are very feasible to be implemented to pre-existing HEAs as well. The correlation of nanoindentation and the creep behavior is still in study and cannot be directly represent the creep behavior of an alloy. However, it gives a pathway in reducing the experimentations to the minimal level to find the optimum temperature or deformation rate for obtaining creep or superplastic behavior of the HEA.

Overall, this chapter has selected articles that summarize and provide the reader a basic essence on creep, correlating it with nanoindentation, superplastic deformation (ultrahigh strain rate and differential) of HEAs, providing a thorough examination of their behavior under high temperatures and at varied deformation conditions.

6.2 Machine Generated Summaries

Disclaimer: The summaries in this chapter were generated from Springer Nature publications using extractive AI auto-summarization: An extraction-based summarizer aims to identify the most important sentences of a text using an algorithm and uses those original sentences to create the auto-summary (unlike generative AI). As the constituted sentences are machine selected, they may not fully reflect the body of the work, so we strongly advise that the original content is read and cited. The auto generated summaries were curated by the editor to meet Springer Nature publication standards. To cite this content, please refer to the original papers.

Machine generated keywords: creep, creep behavior, nanoindentation, strain rate, rate, strain, loading, testing, exponent, grain growth, sensitivity, load, superalloy, hold, precipitate

Development and Microstructural Characterization of a New Wrought High Entropy Superalloy [66]

This is a machine-generated summary of:

Shafiee, Ahad; Nili-Ahmadabadi, Mahmoud; Kim, Hyoung Seop; Jahazi, Mohammad: Development and Microstructural Characterization of a New Wrought High Entropy Superalloy [66]

Published in: Metals and Materials International (2019)

Link to original: https://doi.org/10.1007/s12540-019-00360-w

If you want to cite the papers, please refer to the original.

For technical reasons we could not place the page where the original quote is coming from.

Abstract-Summary

"A newly developed γ' precipitation hardened high entropy superalloy (designated as HES-A1 hereafter) was produced and its microstructural features were characterized."

"The new alloy composition is based on the major elements of conventional IN718 alloy, however other elements were added or removed to minimize possible presence of δ (or γ'') phase and increase the volume fraction of very fine γ' precipitates instead."

"The microstructure of the HES-A1 alloy in as-cast, homogenized, hot rolled, annealed and aged conditions were characterized using optical, scanning, and transmission electron microscopes as well as X-ray diffraction technique."

"The as-cast microstructure of HES-A1 contained 5.5% of Laves phase, which was reduced to less than 0.3% through the development of a homogenization treatment."

"The application of double aging treatment, similar to the one for IN718 alloy, showed no presence of γ'' or formation of δ phase in the microstructure."

"The average size of monomodal γ' precipitates was smaller than that observed in conventional alloys after similar treatments."

Introduction

"Some new HEAs containing γ' phase, as their main strengthening phase, have been introduced as high entropy superalloys (HESAs)."

"The microstructure of the developed alloy remained stable without any detrimental phases after isothermal aging at 1050 °C for 500 h. Daoud and others [1] used Thermo-Calc to develop a Ni-rich high-entropy alloy, $Al_8Co_{17}Cr_{17}Cu_8Fe_{17}Ni_{33}$ in which two morphology of γ' precipitates observed after aging at 700 °C; one with spherical γ' precipitates less than 20 nm in diameter and the other with elongated morphology less than 350 nm long."

"The high entropy alloy concept could be used to develop new superalloys with improved microstructural and mechanical properties."

"This paper deals with the alloy design scheme and microstructural characterization of a new wrought high entropy superalloy (HES-A1) in the as-cast, homogenized, hot rolled, annealed, and aged conditions following our previous paper, which deals with mechanical properties of this alloy [2]."

Alloy Design Approach

"The hot working window of wrought superalloys is normally between the solvus temperature of δ or γ' and the solidus of the alloy and often this window is very small."

"While it has commonly been assumed that increasing Al and Ti contents deteriorates the hot formability of superalloys, AD730 alloy with as high as 9 at% Al + Ti content has shown very good formability [3]."

"Using thermodynamic analysis, the proportions of the elements were adjusted for lower γ' solvus temperature and a higher melting temperature in order to broaden the hot working window of the alloy."

"It can be seen that HES-A1 has similar γ' volume fraction to AD730 but significantly higher than Waspaloy, IN718, and 718Plus; suggesting better mechanical properties compared to the last three alloys."

Experimental

"The homogenized ingots with an initial thickness of 26 mm were successfully hot rolled at 1015 °C using a speed of 3 rpm to a final thickness of 16 mm."

"For the first pass, the ingot was hot rolled with 20% reduction in thickness and for the subsequent six passes; about 15% reduction in thickness was applied, per pass."

"The microstructure of hot rolled and annealed samples were characterized using a field emission scanning electron microscope (FE-SEM, Quanta 3D FEG, FEI) equipped with an EDAX."

"To prepare specimens for transmission electron microscope (TEM), after an initial thinning of the specimens by SiC papers up to 100 μm, 3 mm diameter discs were punched out from the 100 μm foil using a Gatan punch and were electropolished using a Struers TenuPol-5 in an electrolyte of 10% Perchloric acid and 90% Acetic acid."

Results and Discussion

"Image analysis of SEM images indicates that the Laves phase volume fraction in the as-cast microstructure was 5.5%."

"Comparing to IN718, the volume fraction of Laves phase changed related to the cooling rate [4], but similar networks of intermetallic Laves phases has been observed and its volume fraction varies from 3.2 [5] to 5.4% [6] for the as-cast condition in different works."

"Earlier in [7] when the Laves phase goes into solution, some porosity may result that are not interconnected and are quickly healed during hot working."

"Spherical shape γ' precipitates with uniform distribution and the average size of 18.20 ± 3.09 nm after about 18 h of aging could be seen in these images."

"Based on the microstructural characterizations of HES-A1, it could be concluded that similar to the other Nb-bearing superalloys, such as IN718, and 718Plus, the as-cast microstructures of HES-A1 has the same level of the Laves phase."

Conclusions

"A new wrought high entropy superalloy (designated as HES-A1) was developed using high entropy alloy concept, CALPHAD, and Phacomp methods."

"The alloy is characterized with monomodal nano-size γ' precipitates, with a high degree of γ/γ' interface coherency."

"The successful hot rolling, with significant grain refinement, of the developed alloy, confirm its potential for application where rolled or forged products are needed."

Long-Term Creep Behavior of a CoCrFeNiMn High-Entropy Alloy [67]

This is a machine-generated summary of:

Rozman, K. A.; Detrois, M.; Liu, T.; Gao, M. C.; Jablonski, P. D.; Hawk, J. A.: Long-Term Creep Behavior of a CoCrFeNiMn High-Entropy Alloy [67]

Published in: Journal of Materials Engineering and Performance (2020)

Link to original: https://doi.org/10.1007/s11665-020-05103-2

If you want to cite the papers, please refer to the original.

For technical reasons we could not place the page where the original quote is coming from.

Abstract-Summary

"While shorter-term creep tests have been reported in the literature on HEAs, not all methodologies utilize repeatable techniques."

"This manuscript reports on over 23,500 accumulated hours of tensile creep testing with adherence to ASTM standards on a melt-solidified ingot of CoCrFeNiMn HEA converted to wrought plate using conventional thermo-mechanical processing techniques."

"The typical standard creep analyses are reported, i.e., Larson–Miller parameter, Monkman–Grant relationship, activation energy for creep, and creep stress exponents were calculated and compared to previously reported short-term creep tests."

Introduction

"Such research has documented the relative stability of the FCC phase in CoCrFeNiMn systems during isothermal aging; however, much is unknown as to the stability of alloys of this type due to the compounding effect of stress in addition to the thermal load, i.e., long-term creep."

"The CoCrFeNiMn alloy features a single-phase FCC microstructure without precipitates. (However, the particular variant reported within this manuscript had some Y-rich precipitates and some alumina inclusions which are discussed later.) The tensile results for the CoCrFeNiMn alloy have been reported previously, see Licavoli and others (Ref. [8]), where tensile testing was carried out from ambient room conditions up to 800 °C, using standard ASTM E-8 practice."

"This manuscript builds upon the tensile results of the HEAs (Ref. [8]), reporting on the intermediate to long-term creep behavior."

Materials and Methods

"Part of the feedstock consisted of a Ni22Cr master alloy previously VIM manufactured to reduce S. Slices of the master alloy plus compressed compacts of balancing feedstock and loose material were VIM processed at 1345 °C with additions of NiCrY to further decrease the concentration of S and O (Ref. [9])."

"Hot rolling into plate progressed with reheats between each rolling pass (calculated based on the varying thickness of the plate throughout the process). (Ref Note: The stainless steel pouches were removed prior to forging.) After the final hot rolling pass, the plate was returned to the furnace and held for 7 min to provide a full solution anneal (from a mechanical strength perspective) and to reduce residual stress due to TMP."

"Specimens for creep testing were machined from the 10-mm-thick plate, parallel with the rolling direction."

"XEDS maps were generated for 2-h intervals at 20 keV. For TEM three-mm-diameter rods were electrodischarged machined (EDM'd) from one half of each failed creep specimen."

Results

"CoCrFeNiMn shows increasing strain with increasing creep time for both constant temperature and constant stress tests, but the general curve features are slightly different."

"About half the CoCrFeNiMn creep specimens exceeded 34% strain, and as a consequence, the creep displacement versus time curves are truncated at long times where the overall creep strain exceeded 34%."

"At constant creep stress with varying temperature, the MCR occurred at approximately the same time fraction (1.7% to 6.3% of lifetime) and strain (0.24 to 0.64%)."

"At constant temperature with varying creep stress, the MCR showed both stress and time dependence with the MCR occurring at ~ 6% strain and 25% of life at 207 MPa, while at 103 MPa, the MCR occurred at ~ 0.4% strain and ~ 5% of creep life."

Discussion

"Pilloni and others (Ref. [10]) reported an apparent activation energy of 441 to 462 kJ/mol for alloy 347 at 127 and 196 MPa stress levels, respectively."

"For other publications, He and others (Ref. [11]) reported an apparent activation energy of 333 kJ/mol under slow strain rate testing of CoCrNiFeMn using dogbone-shaped tensile specimens 4 mm by 1.5 mm in cross-sectional area."

"Cao and others (Ref. [12]) measured an apparent activation energy of 385 kJ/mol for $Al_{0.15}CoCrNiFe$ alloy, using stress relaxation tests with specimen diameter of 5 mm."

"By measuring the hardness of the gage sections from the creep specimens and comparing to the initial hardness, and from strain hardening in room temperature tensile specimen, the relative effect of σ phase precipitation and growth can be assessed."

"This may suggest σ phase offers some strength increase in addition to strain hardening due to creep (while much less than was observed in the room temperature gage section)."

Summary

"The initial microstructure prior to elevated temperature creep testing was single phase at the exception of some large Ni- and Y-rich precipitates which were incoherent with the matrix."

"Was evidence of cavitation damage surrounding the σ phase when examining the creep fracture surfaces."

"The σ phase measured in the final microstructure was significantly larger than that reported elsewhere."

"Utilization of the creep damage tolerance factor showed that while the alloy cavitated due to σ phase, fracture by micro-void initiation, growth, and coalescence did not occur until very late in tertiary creep."

"The approach used is insufficient in understanding whether the increase in strength during creep arises from strain hardening or σ phase acting as a hardening feature."

"It provided insight as to deformation behavior, especially in creep where the microstructure evolution into σ phase was an important observation."

Nanoindentation Creep Behavior of an $Al_{0.3}$CoCrFeNi High-Entropy Alloy [68]

This is a machine-generated summary of:

Zhang, Lijun; Yu, Pengfei; Cheng, Hu; Zhang, Huan; Diao, Haoyan; Shi, Yunzhu; Chen, Bilin; Chen, Peiyong; Feng, Rui; Bai, Jie; Jing, Qin; Ma, Mingzhen; Liaw, P. K.; Li, Gong; Liu, Riping: Nanoindentation Creep Behavior of an $Al_{0.3}$CoCrFeNi High-Entropy Alloy [68]

Published in: Metallurgical and Materials Transactions A (2016)

Link to original: https://doi.org/10.1007/s11661-016-3469-8

If you want to cite the papers, please refer to the original.

For technical reasons we could not place the page where the original quote is coming from.

Abstract-Summary

"The effects of the indentation size and loading rate on creep behavior were investigated."

"The experimental results show that the hardness, creep depth, creep strain rate, and stress exponent are all dependent on the holding load and loading rate."

"The creep behavior shows a remarkable indentation size effect at different maximum indentation loads."

"An obvious loading rate sensitivity of creep behavior is found under different loading rates for the alloy."

Introduction

"The high-entropy alloys (HEAs) with unique compositions, microstructures, and adjustable properties [13, 14] have attracted increasing attentions to the materials science community."

"The addition of other alloying elements, such as V [15], Nb [16], Ti [17], and Mo [18], were investigated to improve the microstructures and mechanical properties of different kinds of HEA systems."

"We use the nanoindentation experiment, which is different from the traditional tensile test to probe the creep behavior of the CG $Al_{0.3}$CoCrFeNi HEA."

"Nanoindentation creep tests have been carried out at different indentation loads to study the indentation size effects (ISEs) on the creep behavior of nanocrystalline materials [19] and bulk metallic glasses [20]. For the polycrystalline Ni films, a significant indentation loading rate sensitivity (LRS) on the creep stress exponent and hardness have been reported recently [21]. In order to understand the influence of the indentation size and loading rate on the CG HEA, it is necessary that the creep properties of the CG $Al_{0.3}$CoCrFeNi HEA be examined by the nanoindentation technique."

Experimental

"The material was prepared by arc-melting a mixture of constituent metals [purity > 99.9 weight percent (wt pct)] with a nominal composition of $Al_{0.3}$CoCrFeNi (atomic ratio) using the direct-current-powered button melter."

"The alloy was re-melted three times under a high-purity argon atmosphere in order to obtain homogeneity."

"For the polished $Al_{0.3}$CoCrFeNi alloy, the creep properties were characterized under a constant loading rate of 1000 $\mu N\ s^{-1}$ with the holding loads ranging from 500 to 9000 μN and the holding time of 40 s."

"For nanoindentations under changing loading rates, the loading rates of 5000, 1000, 500, 250, 165, 100, and 60 $\mu N\ s^{-1}$ with a maximum load of 5000 μN and the holding time of 40 s were applied."

"The hardness was investigated by means of the Oliver–Pharr method [22]. The indentation at each position was repeated at least 10 times."

Results

"For Berkovich tips, the strain rate and the hardness could be described in the following equations [23] with the depth-sensing indentation technique: where h is the displacement rate of the Berkovich tip, h is the instantaneous indenter displacement, and P and H are the indentation load and hardness, respectively."

"The creep behavior and hardness present strong load dependency."

"From the holding stage, we can see that the creep depth increases from 7.7 to 9.3 nm with the increase in the loading rate."

"The nanoindentation creep properties and the hardness show a strong LRS of the $Al_{0.3}$CoCrFeNi HEA."

Discussion

"Self-diffusion, dislocation, and grain boundary are the most effective diffusion paths under creep deformation."

"N increased to about 90 at 9000 μN, which is approximately ten times higher than that at 500 μN. This trend suggests a rapid change of creep mechanisms from the diffusion flow to dislocation climb and slip with increasing indentation loads or indentation depths."

"The LRS in the values of stress exponents and hardnesses may suggest that there is a difference in the deformation mechanism in the grain interior of the $Al_{0.3}$CoCrFeNi HEA during the nanoindentation creep."

"With a longer loading time, the creep deformation will occur mainly at the loading stage, and the creep behaviors at the holding stage will be affected by the smaller creep depth, the higher creep strain rate, and the lower stress exponent."

Conclusion

"The obvious ISEs on the hardness, creep depth, creep strain rate, and stress exponent are found at different maximum indentation loads."

"Self-diffusion along the indenter/specimen interface and along the free surface of the specimen may play an important role at a low indentation load of 500 μN. The nanoindentation creep behavior exhibits a remarkable LRS under different loading rates."

"When the loading rate is very low, creep deformation can run out partially or completely, and a small amount of dislocations dominate the creep behavior with the low stress exponent."

Theoretical Characterization of Indentation Depth-Dependent Creep Behavior of $CoCrFeNiAl_{0.3}$ High-Entropy Alloy [69]

This is a machine-generated summary of:

Meng, Lu; Cui, Weiling; Su, Buyun; Shu, Xuefeng; Xiao, Gesheng: Theoretical Characterization of Indentation Depth-Dependent Creep Behavior of $CoCrFeNiAl_{0.3}$ High-Entropy Alloy [69]

Published in: Acta Mechanica Solida Sinica (2022)

Link to original: https://doi.org/10.1007/s10338-022-00354-z

If you want to cite the papers, please refer to the original.

For technical reasons we could not place the page where the original quote is coming from.

Abstract-Summary

"The effects of indentation loading depth and dynamic pre-compression on the creep behavior of $CoCrFeNiAl_{0.3}$ high-entropy alloy (HEA) were studied through a series of indentation creep tests."

"A phenomenological model based on the holding time and loading depth was established by studying the characteristic relation between the loading depth and the creep displacement of $CoCrFeNiAl_{0.3}$ HEA."

Introduction

"The existing research focuses on the effect of strain rate on creep behavior, for instance, Hu and others [24] showed that the creep behavior is sensitive to loading rate."

"Few dislocations can be stored at low loading rates, and creep behavior is mainly dominated by grain boundary sliding."

"The creep behavior under different loads or depths has rarely been studied in detail."

"It is vital to clearly understand the effect of dynamic deformation on indentation creep behavior."

"The purpose of this work is to establish a phenomenological model to determine the creep curve via given holding time and loading depth, and quantitatively predict and characterize the indentation creep behavior of $CoCrFeNiAl_{0.3}$ HEA at different loading depths."

"The phenomenological model is applied to the creep behavior of $CoCrFeNiAl_{0.3}$ HEA subjected to dynamic loading."

Experimental Procedure and Characterization Method

"Micro-indentation creep tests were carried out using the Agilent Nano Indenter G200 test system with force and displacement being 50 nN and 0.01 nm, respectively."

"On account of the non-destructive nature of micro-indentation test, its creep curve shows two stages [25]."

"The latter creep displacement increases linearly, and the creep rate is stable, which is the quasi-steady-state creep."

"For the creep behavior of materials, h_0 represents the total depth of indentation loading stage."

"The hardening is balanced by dynamic recovery (e.g., dislocation annihilation), which is manifested as the creep displacement increases slowly with the increase in the holding time [26]."

"1) with all parameters determined can predict the creep curves of the materials at different loading depths."

"To further investigate the creep mechanism of materials under different loading depths, the creep stress exponents were calculated."

Results and Discussion

"As the loading depth increases to 1500 nm and the load increases, the high stress causes more dislocation nucleation, dislocation multiplication and dislocation slip in the local area, dislocation motion may be the operative mechanism [25]."

"When the depth continues to increase to 3000 nm, the holding load reaches its maximum, and the dislocation mechanism under high stress is considered to be dominant."

"Grain boundary slip and dislocation motion are the creep mechanisms at the loading depth of 3000 nm."

"After applying pre-compression, the creep displacement decreases at all loading depths."

"The interference of the pre-compression on creep behavior is weakened with the increase in loading depth, which indicates that the pre-compression effect originates form the pre-compression-induced plasticity, scilicet, the dislocation [27, 28]."

"It is found that the pre-compression leads to numerous pre-dislocations in the alloy, and the preexisting dislocations increase the resistance to dislocation motion during the indentation creep test, thus improving the creep resistance of the $CoCrFeNiAl_{0.3}$ HEA."

Conclusions

"Micro-indentation creep tests are performed on $CoCrFeNiAl_{0.3}$ HEA without and with dynamic pre-compression under different loading depths."

"Using the proposed phenomenological model to analyze the effect of dynamic pre-compression on the indentation creep behavior of $CoCrFeNiAl_{0.3}$ HEA, the creep behavior shows sensitive to loading depth."

"Pre-compression promotes the creep resistance of $CoCrFeNiAl_{0.3}$ HEA, and creep behaviors of the two planes are distinct due to different grain deformations of the compression plane and transverse plane under dynamic loading."

"The phenomenological model proposed under the micro-indentation creep test of the alloy without pre-compression is reasonable and reliable, and can be applied to predict and analyze the creep behavior of dynamic pre-compression alloy, which lays a foundation for the application of micro-indentation to obtain the influence of pre-deformation on the mechanical properties of $CoCrFeNiAl_{0.3}$ HEA."

The High Temperature Tensile and Creep Behaviors of High Entropy Superalloy [70]

This is a machine-generated summary of:

Tsao, Te-Kang; Yeh, An-Chou; Kuo, Chen-Ming; Kakehi, Koji; Murakami, Hideyuki; Yeh, Jien-Wei; Jian, Sheng-Rui: The High Temperature Tensile and Creep Behaviors of High Entropy Superalloy [70]

Published in: Scientific Reports (2017)

Link to original: https://doi.org/10.1038/s41598-017-13026-7

If you want to cite the papers, please refer to the original.

For technical reasons we could not place the page where the original quote is coming from.

Abstract-Summary

"This article presents the high temperature tensile and creep behaviors of a novel high entropy alloy (HEA)."

"The microstructure of this HEA resembles that of advanced superalloys with a high entropy FCC matrix and $L1_2$ ordered precipitates, so it is also named as "high entropy superalloy (HESA)"."

"To the best of authors' knowledge, this article is the first to present the elevated temperature tensile creep study on full scale specimens of a high entropy alloy, and the potential of HESA for high temperature structural application is discussed."

Introduction

"Recent studies have shown attractive mechanical properties of HEAs at cryogenic temperature [30] and room temperature [31] with excellent combinations of high strength and toughness, indicating their potential as the structural materials."

"Precipitation strengthening should be considered in developing HEAs for high temperature applications."

"The high temperature tensile and creep behaviors of a novel alloy named high entropy superalloy (HESA) [32] have been investigated."

"HESA is strengthened by the coherent $L1_2$ γ' precipitates, while the FCC matrix remains high entropy ($\Delta S_{mix} > 1.5$ R) [33]."

"To the best of authors' knowledge, there is no reported creep studies on the full scale sample of HEAs at elevated temperature such as 982 °C."

"This article presents the first tensile creep studies on full size specimen of a high entropy alloy, which can benefit the community of materials science and engineering whose interest is in the development of HEAs for elevated temperature applications."

Results

"It also indicates the two phases microstructure, and the γ' solvus and solidus temperature are 1199 and 1276 °C, respectively."

"According to JMatPro calculations, the APB energies of γ' phase in HESA and CMSX-2 are 0.22 and 0.19 J/m^2, respectively."

"Higher APB energy of γ' may contribute to some degree of strengthening in HESA [34]."

"Microstructure evolution during high temperature creep is known to influence creep behaviour, such as the directional coarsening of γ' precipitates—"rafting" [35]."

"The rafting of γ' in HESA has shown to coarsen along the direction of the applied stress, which can be related to the positive lattice misfit (δ) between γ and γ' [36]."

"The microstructure can be stable with high entropy γ matrix and $L1_2$ γ', while the size of γ' has increased from 0.3 μm to 0.4 and 0.6 μm after 750 and 850 °C creep tests, respectively."

Discussion

"Owing to higher volume fractions and higher APB energy of γ' precipitates, HESA can show the superior tensile yield strength to those of the reported high entropy alloys from room to elevated temperatures."

"The high temperature strength of HESA can approach that of the Ni-based superalloy CMSX-2, and the optimizations in larger γ' size, higher degree of misfit value may contribute to further strengthening."

"Regarding to creep resistances, HESA can be strengthened by the lower stacking fault energy, higher creep activation energy and stable γ' phases; the sluggish diffusion of the high entropy γ matrix may have attributed to the higher creep activation energy."

"At higher temperature, although the P-raft γ' at 982 °C has led to a more rapid increase in creep strain rate, the creep life of HESA can be compared to those of some conventional superalloys and with less refractory contents."

Methods

"The solution-heat-treatment (SHT) was then conducted at 1210 °C for 10 h to homogenize chemical segregations, and followed by a primary aging at 1000 °C for 3 h and a secondary aging at 800 °C for 20 h to grow and refine the morphology of γ' precipitates."

"A Pyris Diamond TG/DTA was used to determine the γ' solvus temperature of HESA; 30 mg sample was put in an Al_2O_3 crucible with a heating rate of 10 °C/min to 1350 °C and with 400 sccm of Ar flow."

"A strain aging method was conducted to assist measuring the compositions of γ and γ' phase; samples were subjected to indentation after the solution-heat-treatment process stated above, and then followed by an annealing at 900 °C for 50 h. Therefore, cellular growth of γ' would occur [37], and resulted in the coarsened γ/γ' morphology."

Elevated-Temperature Creep of High-Entropy Alloys via Nanoindentation [71]

This is a machine-generated summary of:

Lin, P. H.; Chou, H. S.; Huang, J. C.; Chuang, W. S.; Jang, J. S. C.; Nieh, T. G.: Elevated-temperature creep of high-entropy alloys via nanoindentation [71]

Published in: MRS Bulletin (2019)

Link to original: https://doi.org/10.1557/mrs.2019.258

If you want to cite the papers, please refer to the original.

For technical reasons we could not place the page where the original quote is coming from.

Abstract-Summary

"For high-temperature industrial applications, creep behavior of structural materials is critical."

"With the high accuracy of depth-sensing technology and a stable temperature-controlling stage, elevated-temperature time-dependent mechanical behavior of

HEAs can be investigated, especially at localized regions without the limitations of the standard specimen size used for traditional creep testing."

"This article overviews current progress in studying creep behavior in HEAs via nanoindentation technology."

Introduction

"[38] For the latter, with the high accuracy of depth-sensing technology and stable temperature-controlling stages in current instruments, the time-dependent mechanical behavior of HEAs can be conducted on localized regions, ignoring the limit of standard-sized specimens, especially in traditional creep testing."

"Several studies have reported the time-dependent mechanical properties of HEAs via nanoindentation, including the elevated-temperature creep of FeCoNiCrMn and FeCoNiCrMn-Al HEAs at 300–600 °C [39], room-temperature creep of as-deposited and annealed CoCrFeCuNi HEA thin films [40], room-temperature creep of as-cast and severely plastic-deformed CoCrFeMnNi [41], initial creep behavior of CoCrNi HEA [42], characteristics of plastic deformation in CoCrFeMnNi HEA from room temperature to 300 °C [43], strain rate sensitivity of AlCoCrFeNi HEA at room temperature [44], room-temperature creep in CoCrFeNiCu HEA with a face-centered-cubic (fcc) structure [45], and room-temperature creep in $CoCrFeNiCuAl_{2.5}$ HEA with a body-centered-cubic (bcc) structure."

"[45] In this article, we focus on creep properties, explored using nanoindentation, on several well-known HEAs, such as FeCoNiCrMn (also known as the Cantor alloy) and extended alloy systems."

"This article summarizes the current findings in HEAs based on nanoindentation creep."

"It is hoped that more exploration using this new method of nanoindentation can provide new scientific findings on the high-temperature properties of HEAs."

Creep behavior of HEAs at elevated temperatures

"[39] The creep strength of the Al-modified Cantor alloy, FeCoNiCrMn-Al, is comparable to that in the standard Cantor alloy, suggesting that the solid-solution effect (only ~ 4 at.% Al remaining in the fcc grains and remaining Al present in the bcc grains) is insignificant for the fcc {111} grains at elevated temperatures."

"[46] Ma and others45 studied the strain-rate sensitivity, m, and activation volume, V_c, of a CoCrFeNiCu (zero Al, namely, Al-0) HEA thin film with a fcc structure and $CoCrFeNiCuAl_{2.5}$ (2.5 at.% Al, namely, Al-2.5) HEA thin film with a bcc structure at room temperature."

"Tsai and others [39] reported the elevated-temperature creep of fcc {100} and {111} FeCoNiCrMn HEAs."

"The activation volumes for {100} at elevated temperatures are also higher than those for {111}, suggesting that the creep resistance of the fcc {111} grains is higher than that of the fcc {100} grains."

Comparison and Discussion

"The creep mechanism at elevated temperatures is controlled by several critical parameters such as strain rate, stress, grain size, and creep temperature."

"When comparing the results of elevated temperature nanoindentation creep response with traditional creep tension tests for the FeCoNiCrMn alloy subjected to a normal lower stress level in the range of 10–150 MPa, the stress exponents (approximately 3–8), activation energies (around 300–400 kJ/mol), and activation volumes (around 40–100 b^3) are similar, suggesting that the mechanisms should be the same, likely due to dislocation climb-controlled creep."

"In order to prevent strong thermal drift, nanoindentation creep testing at elevated temperatures from 400–600 °C usually adopts a much shorter creep testing duration by raising the creep stress level to around 200 to 500 MPa, or even higher."

"The slightly lower activation energy could be a result of grain-boundary and dislocation pipe diffusion that have been involved in the nanoindentation creep testing adopting much higher creep stress levels."

Summary and future directions

"Current findings on nanoscale time-dependent creep behavior of HEAs, explored using the nanoindentation technique, are reviewed."

"HEA creep response related to minor solute addition effects, structure effects, grain-size effects, and orientation effects are reviewed and discussed."

"Traditional creep testing on bulk samples will not show individual creep response from each crystalline orientation."

"The region where the nanoindentation probe interacts with the sample is quite small, which could be an advantage in using nanoindentation creep testing."

Strain Rate Sensitivity of Nanoindentation Creep in an AlCoCrFeNi High-Entropy Alloy [72]

This is a machine-generated summary of:

Jiao, Z. M.; Wang, Z. H.; Wu, R. F.; Qiao, J. W.: Strain rate sensitivity of nanoindentation creep in an AlCoCrFeNi high-entropy alloy [72]

Published in: Applied Physics A (2016)

Link to original: https://doi.org/10.1007/s00339-016-0339-6

If you want to cite the papers, please refer to the original.

For technical reasons we could not place the page where the original quote is coming from.

Abstract-Summary

"Creep behaviors of an AlCoCrFeNi high-entropy alloy with the body-centered cubic structure were investigated by nanoindentation."

Introduction

"Ma and others [40] have reported that the as-deposited CoCrFeCuNi HEA with a face-centered cubic (fcc) structure exhibited the positive strain rate sensitivity (SRS) during creep, and the value of SRS increased with increasing loading rates under nanoindentation."

"For a coarse-grained $Al_{0.3}$CoCrFeNi HEA with a single fcc structure under nanoindentation, the value of SRS decreased with the increase in loading rates during creep [47]."

"These inconsistent results indicate that creep behaviors of HEAs are influenced by many factors like the compositions and microstructures, which stimulates further investigation to uncover the effect of strain rate on the nanoindentation creep properties."

"According to the previous investigations [48, 49], the as-cast AlCoCrFeNi HEA possesses the bcc crystal structure and exhibits excellent mechanical properties."

"The creep behaviors of the as-cast AlCoCrFeNi HEA at room temperature are investigated using nanoindentation."

Experimental

"To understand the effect of strain rate on creep behavior of the present alloy, indentation tests were performed based on the continuous stiffness measurement (CSM) technique under five strain rates of 0.01, 0.02, 0.05, 0.1, and 0.2 s^{-1}."

"In Oliver and Pharr's nanoindentation model [22], the contact stiffness can be obtained as: where P_0 is the amplitude of the harmonic excitation force, Z_0 is response to the displacement amplitude ($\approx$2 nm), Φ is the phase shift between the harmonic displacement and the harmonic excitation force, $\omega = 2\pi f$ is the angular frequency (45 Hz), and K_s, K_f, and m are the spring constant in the vertical direction, frame stiffness, and mass of the indenter, respectively."

"The hardness is defined by: The reduced modulus, a combined response of the indenter tip and the sample, can be computed by: where $\beta = 1.034$ is the shape constant."

Results and Discussion

"The creep displacement is larger under high strain rate due to the release of plastic flow accumulated during the loading stage."

"Since there is a longer time to reach the depth limit at low strain rate, creep deformation will occur during the loading stage, which weakens the subsequent creep behavior during the holding stage."

"It is noteworthy that the creep strain rate during the holding stage increases with increasing strain rates."

"The creep strain rate during the transient stage is roughly two orders of magnitude higher than that during the steady-state stage, which agrees well with the previous studies [40, 47]."

"The larger strain gradient induced by higher strain rate contributes to the occurrence and entanglement of more dislocations, which enhances the creep resistance and reduces the SRS values during creep process [47]."

Conclusions

"The strain rate sensitivity of nanoindentation creep in an AlCoCrFeNi HEA with the bcc crystal structure was investigated."

"The strain gradient becomes larger under higher strain rates, resulting in the enhanced hardness."

"The creep displacement is larger at high strain rate, and creep rate drops rapidly at initial holding stage and then declines steadily."

Ultrahigh High-Strain-Rate Superplasticity in a Nanostructured High-Entropy Alloy [73]

This is a machine-generated summary of:

Nguyen, Nhung Thi-Cam; Asghari-Rad, Peyman; Sathiyamoorthi, Praveen; Zargaran, Alireza; Lee, Chong Soo; Kim, Hyoung Seop: Ultrahigh high-strain-rate superplasticity in a nanostructured high-entropy alloy [73]

Published in: Nature Communications (2020)

Link to original: https://doi.org/10.1038/s41467-020-16601-1

If you want to cite the papers, please refer to the original.

For technical reasons we could not place the page where the original quote is coming from.

Abstract-Summary

"We present a superplastic elongation to 2000% of the original length at a high strain rate of 5×10^{-2} s^{-1} in an $Al_9(CoCrFeMnNi)_{91}$ (at%) high-entropy alloy nanostructured using high-pressure torsion."

"The high-pressure torsion induced grain refinement in the multi-phase alloy combined with limited grain growth during hot plastic deformation enables high strain rate superplasticity through grain boundary sliding accommodated by dislocation activity."

Introduction

"Superplasticity is generally observed in fine-grained materials at low strain rates (10^{-4}–10^{-3} s^{-1}) and high homologous temperatures (>0.5 T_m), where T_m is the melting temperature [50–52]."

"High Strain-Rate Superplasticity (HSRS) is extremely rare in high-strength materials, and in particular, the lack of superplastic forming capability in a recently emerged class of high-strength materials, the so-called High-Entropy Alloys (HEAs), is a serious obstacle for its potential use in engineering complex structures."

"We fabricate an $Al_9(CoCrFeMnNi)_{91}$ HEA with nano-sized FCC grains and B2 phase in the range of a few hundred nanometers to micrometers through thermo-mechanical treatment followed by high-pressure torsion (HPT) processing."

"The presence of B2 phase and the newly formed sigma phase during superplastic testing assist in preventing significant grain growth, favoring GBS with high-strain-rate sensitivity value."

"The fine grain size accompanied by high-strain-rate sensitivity and GBS accommodation by plastic deformation lead to an ultra-high superplasticity of 2000% at a high-strain rate of 5×10^{-2} s^{-1} in the present alloy."

Results and Discussion

"From the domain of superplasticity research that utilizes HPT processing, the present HEA still has the best elongation result compared to the highest previously reported elongation of 1800% for illustrious Zn-22% Al alloy [53]."

"The high-strain-rate and presence of B2 and sigma phases limit grain growth during superplastic deformation, leading to a quasi-theoretical m value."

"Elimination of OR between FCC and B2 points toward the occurrence of GBS that accommodates grain rotation during the superplasticity test [54]."

"Looking at the above evidence, we can say the different phases in the present multi-phase HEA demonstrate quite a heterogeneous mechanical response at the scale of individual grains during the superplasticity test."

"The present multi-phase microstructure is able to achieve an ultra-high HSRS with a the record of 2000% elongation that is significantly greater than anything seen in previous investigations from the severe plastic deformation society and HEA community."

Methods

"The 10 mm discs were processed by five revolutions of HPT at room temperature under an applied pressure of 6 GPa with an anvil rotation rate of 1 revolution per minute."

"The HPT sample was loaded into a ceramic pan and then heated from room temperature to melting temperature with a heating rate of 10 K/min in a nitrogen gas atmosphere."

"The specimens were heated from room temperature to the target superplastic testing temperatures of 873, 973, and 1073 K by halogen furnace (DF-60HG, DaeHeung Co., Korea) in the air atmosphere."

"The Transmission Electron Microscope (TEM) specimens of the as-HPT and grip of the superplastic specimens were prepared by mechanical grinding to a thickness of ~ 60 μm, and they were then thinned using electro-polishing with a solution of 10% $HClO_4$ and 90% CH_3COOH at an applied voltage of 25 V at room temperature."

Phase Composition and Superplastic Behavior of a Wrought AlCoCrCuFeNi High-Entropy Alloy [74]

This is a machine-generated summary of:

Shaysultanov, D. G.; Stepanov, N. D.; Kuznetsov, A. V.; Salishchev, G. A.; Senkov, O. N.: Phase Composition and Superplastic Behavior of a Wrought AlCoCrCuFeNi High-Entropy Alloy [74]

Published in: JOM (2013)

Link to original: https://doi.org/10.1007/s11837-013-0754-5

If you want to cite the papers, please refer to the original.

For technical reasons we could not place the page where the original quote is coming from.

Abstract-Summary

"The forged alloy exhibited unusual superplastic behavior accompanied by a pronounced softening stage, followed by a steady-state flow stage, during tensile deformation at temperatures of 800 °C–1000 °C and at strain rates of 10^{-4}–10^{-1} s^{-1}."

"A detailed analysis of the phase composition and microstructure of the alloy before and after superplastic deformation was conducted, the strain rate and temperature dependences of the flow stress were determined at different stages of the superplastic deformation, and the relationships between the microstructure and properties were identified and discussed."

Introduction

"High-entropy alloys (HEAs) are a relatively new class of metallic materials developed during recent decade."

"[13, 55, 56] Unfortunately, in many cases these alloys were studied in the as-cast condition, and very low room temperature ductility, or even brittle behavior, was often reported."

"The alloy was hot worked by multiaxial forging to produce a fine-grained structure, which resulted in significantly improved ductility and strength at room temperature."

"At high temperatures, the fine-grained alloy exhibited superplastic-like behavior."

"Contributions of different phases presented in the AlCoCrCuFeNi alloy in the superplastic behavior, as well as flow stress controlling mechanisms, have not yet been identified."

"Detailed phase and microstructure analyses of the fine-grained AlCoCrCuFeNi HEA have been conducted before and after superplastic deformation, a constitutive equation for the superplastic flow has been determined, and the relationships between the microstructure and properties have been outlined and discussed."

Experimental Procedures

"The produced ingot was trimmed by cutting out about 10-mm-thick pieces from the top and bottom of the ingot, and a sample of 40 mm in diameter and 35 mm in length was machined from the remaining part of the ingot."

"Uniaxial tensile tests were conducted in air using an Instron 5882 testing machine (Instron Corporation, Norwood, MA, USA) equipped with a radiant furnace."

"The microstructures of polished cross-sections of the nondeformed and deformed specimens were studied using scanning electron microscopes (SEMs) Quanta 200 3D and Quanta 600 (FEI Company, Hillsboro, OR, USA) equipped with backscattered electron detectors, energy-dispersive spectrometers (EDS), and electron backscattered diffraction (EBSD) detectors."

"Additional microstructural investigations were performed using transmission electron microscope (TEM) JEOL-2100 (JEOL Ltd., Tokyo, Japan) equipped with an EDS detector."

"A thorough phase analysis was conducted with the use of X-ray diffraction, SEM, EDS, EBSD, and TEM techniques."

Results

"The lattice parameter of the Cu-rich phase slightly increases, while that of the Co-Cr-Fe-rich phase slightly decreases, probably due to modified compositions of these phases after the forging."

"Relative to the as-cast condition, the volume fraction of the Cu-rich phase increases from 12 to 17%, while the volume fraction of the Co-Cr-Fe-rich phase decreases from 45 to 30%."

"Only the concentration of Cu in the Cu-rich phase in the deformed sections of the specimens noticeable varied; e.g., after the tensile testing at 800 °C it increased to 76%, and after testing at 1000 °C it decreased to 65%, relative to 72% in the forged condition, before tensile testing."

"The volume fractions of the Cu-rich, Al-Ni-rich, and Co-Cr-Fe-rich phases remained nearly unchanged in the nondeformed sections of the tensile samples."

"In the deformed section, after testing at 1000 °C, the volume fraction of the Al-Ni-rich phase decreased to 41%–43%, and the volume fraction of the Co-Cr-Fe-rich phase increased to 30%–37%."

Discussion

"Transformation of the ordered B2 Al-Ni-rich phase into a disordered bcc solid-solution phase at temperatures above 850 °C has recently been reported for an $Al_{0.5}$CoCrCuFeNi HEA [57]. Thermodynamic modeling of the phase diagrams of similar HEA systems, Al_xCoCrFeNi [58] and $Al_{0.5}$CoCrCuFeNi [57], predicts that the hard-to-deform and brittle σ phase is thermodynamically unstable at $T \geq 800$ °C."

"[59, 60] Using this comparison and taking into account that the rates of diffusion in the HEAs are slower than in pure metals [61], one can suggest that the superplastic deformation of the AlCoCrCuFeNi HEA is realized through the GBS accommodated by the following activation processes: (i) at the beginning of deformation by the plastic deformation in the Cu-rich phase; (ii) in the vicinity of the peak stress and at the softening stage by the Fe- and/or Ni- diffusion-controlled transformations and plastic deformation in the Co-Cr-Fe-rich and/or Al-Ni-rich phases; and (iii) during the steady-state plastic flow by the plastic deformation of equilibrium high-temperature phases, controlled by complex diffusion with a high activation energy."

"As soon as the phase transformations are complete, the steady-state flow stage occurs at which GBS is accommodated by the plastic deformation inside the equilibrium high-temperature phases."

Summary and Conclusions

"The hot forging destroyed ordering of the Cu-rich and Co-Cr-Fe-rich phases and transformed the bcc Cr-Fe-rich phase to a tetragonal σ phase."

"The Cu-rich phase had the lowest nanohardness of 3.8 GPa and 3.0 GPa in as-cast and forged states, respectively."

"The remaining Co-Cr-Fe-rich, Al-Ni-rich, and σ phases had nanohardness of 4.1 GPa, 6.8 GPa, and 9.9 GPa, and elastic modulus of 205 GPa, 190 GPa, and 233 GPa, respectively, in hot forged condition."

"It was proposed that GBS was accommodated by (i) the thermally activated deformation of a soft Cu-rich phase at the beginning of the plastic deformation, (ii) the transformations of the low-temperature phases into equilibrium high-temperature phases during the strain-softening stage, and (iii) dislocation glide in the equilibrium high-temperature phases, which was controlled by complex solute–dislocation interactions during a steady-state flow stage."

Differential Superplasticity in a Multi-phase Multi-principal Element Alloy by Initial Annealing [75]

This is a machine-generated summary of:

Nguyen, Nhung Thi-Cam; Asghari-Rad, Peyman; Park, Hyojin; Kim, Hyoung Seop: Differential superplasticity in a multi-phase multi-principal element alloy by initial annealing [75]

Published in: Journal of Materials Science (2022)

Link to original: https://doi.org/10.1007/s10853-022-07616-8

Copyright of the summarized publication:

If you want to cite the papers, please refer to the original.

For technical reasons we could not place the page where the original quote is coming from.

Abstract-Summary

"Multi-phase structure alloys have been widely used in superplasticity deformation due to their ability to inhibit grain growth."

"The current study on multi-phase structure alloys has mainly indicated static grain growth than dynamic grain growth."

"It is suggested that the initial B2, formed during the annealing stage before the HPT process, effectually limits the dynamic grain growth, resulting in remarkably enhanced superplasticity."

"This investigation introduces the new microstructural evolution to uplift superplasticity in multi-phase structures with dynamic grain growth elimination."

Introduction

"In these researches, the alloy with a fine dispersion of a secondary phase or multiphase structure is utilized to limit grain growth during superplastic deformation."

"More understanding of alloy design is required to promote the grain size stability in multi-phase structures under superplastic deformation."

"These multiphase microstructures limit grain growth, particularly dynamic grain growth, at the high homologous temperature and generate excellent superplastic performance [62]."

"Among many MPEAs comprised of FCC, B2, and sigma phase, $Al_{0.5}$CoCrFeMnNi HEA can be considered a prototypal alloy by its excellent superplasticity at high strain rates [62]."

"The high-temperature deformation behaviors of $Al_{0.5}$CoCrFeMnNi HEA, such as superplasticity, creep, and hot compression, were investigated in a wide range of strain rates from 10^{-1} to 10^{-8} s^{-1} [46, 62–64]."

"This research suggests a new path to enhance superplasticity in multi-phase structures by engineering the initial microstructure to suppress dynamic grain growth."

Experimental procedure

"The cold-rolled alloy plate was annealed at 1473 K for 2 h under the Argon gas, followed by water quenching."

"To distinguish the two investigations from a similar alloy, we labeled the present research ($Al_{0.5}CoCrFeMnNi$ HEA with initial single-phase (SP) FCC structure resulting from annealing condition at 1473 K for 2 h) as Al-HEA-SP."

"The published research on $Al_{0.5}CoCrFeMnNi$ HEA with initial dual-phase (DP) structure of FCC-matrix and B2 precipitates (with an average size of 800 nm^{-1} μm) processed by annealing at 1273 K for 15 min was labeled as Al-HEA-DP [62]."

"The testing temperatures were over half the alloy's melting temperature (T_m = 1567 K) [62]."

"The previous investigation suggested the sigma phase formed at 673–1073 K; therefore, the testing temperatures were selected in the range that $Al_{0.5}CoCrFeMnNi$ have a tri-phase mixture of FCC, B2, and sigma phases and above the half of melting temperature [62, 65]."

Result and Discussion

"These results reveal that the Al-HEA-SP attains superplastic elongation of 440% at the high-strain rate of 10^{-2} s^{-1} at 873 K. With an increase in the testing temperature, the elongation rises to 626 and 695% at 923 and 973 K, respectively, and it reaches 936% at 1073 K. At the lowest testing temperature of 873 K (corresponding to $0.56T_m$), HSRS is achieved."

"This elongation is higher than the elongation of 430% reported in Al-HEA-DP at an equivalent testing condition (strain rate of 10^{-2} s^{-1} at 873 K) [62]."

"Al-HEA-SP accomplishes significantly higher elongation in the range of the HSRS at temperatures of 973 and 1073 K. Then, superplastic elongation gradually decreases when the superplastic test is conducted at medium strain rates."

"It is visible that the multiphase structure in this present Al-HEA-SP has a better performance in restricting dynamic grain growth than TiCoCrFeMnNi at the same temperature and strain rate."

Conclusions

"The $Al_{0.5}CoCrFeMnNi$ HEA was annealed at 1472 K for 2 h to attain fully-FCC structure, then subjected to the HPT process for grain refinement."

"Superplasticity with elongation of 440% is recorded at 873 K and a strain rate of 10^{-2} s^{-1}."

"Superplasticity is achieved at 923, 973, and 1073 K with an increment in elongation proportional to the testing temperature."

"The highest elongation of 1100% was achieved at 1073 K and a strain rate of 10^{-1} s^{-1}."

"The dynamic grain coarsening is observed in the sample of 1073 K at the strain rate of 10^{-2} s^{-1}."

References

1. H.M. Daoud, A. Manzoni, R. Völkl, N. Wanderka, U. Glatzel, Microstructure and tensile behavior of Al8Co17Cr17Cu8Fe17Ni33 (at.%) high-entropy alloy. JOM 65, 1805–1814 (2013). https://doi.org/10.1007/s11837-013-0756-3
2. M.C. Gao, J.W. Yeh, P.K. Liaw, Y. Zhang, High-entropy alloys (Springer, New York, 2016)
3. A. Devaux, B. Picque, M.F. Gervais, E. Georges, T. Poulain, P. Heritier, AD730TM—A new Nickel-Based Superalloy for High Temperature Engine Rotative Parts, in Superalloys 2012 12th Int. Symp. Superalloys (2012) pp. 911–919
4. S.G.K. Manikandan, D. Sivakumar, K.P. Rao, M. Kamaraj, Effect of weld cooling rate on Laves phase formation in Inconel 718 fusion zone. J. Mater. Process. Technol. 214, 358–364 (2014). https://doi.org/10.1016/j.jmatprotec.2013.09.006
5. Z. Miao, A. Shan, W. Wang, J. Lu, W. Xu, H. Song, Quantitative characterization of two-stage homogenization treatment of alloy 718, in 7th Int. Symp. Superalloy 718 Deriv. (2010), pp. 107–115. https://doi.org/10.1002/9781118495223.ch7
6. M.J. Cieslak, G.A. Knorovsky, T.J. Headley, A.D. Romig Jr, The solidification metallurgy of alloy 718 and other Nb-containing superalloys. Superalloys 718, 59–68 (1989). https://doi.org/10.7449/1989/superalloys_1989_59_68
7. R.M. Forbes Jones, L.A. Jackman, The structural evolution of superalloy ingots during hot working. JOM 51, 27–31 (1999). https://doi.org/10.1007/s11837-999-0007-9
8. J.J. Licavoli, M.C. Gao, J.S. Sears, P.D. Jablonski, and J.A. Hawk, Microstructure and Mechanical Behavior of High-Entropy Alloys, J. Mater. Eng. Perform., 2015, 24, p 3685–3698
9. M. Detrois, P.D. Jablonski, Trace Element Control in Binary Ni-25Cr and Ternary Ni-30C0-30Cr Master Alloy Castings, in ed. by M.J.M. Krane RMW, S. Rudoler, A.J. Elliott, A. Pate. Proceeding of the Liquid Metal Processing & Casting Conference (2017), pp. 75–84.
10. G. Pilloni, E. Quadrini, and S. Spigarelli, Interpretation of the Role of Forest Dislocations and Precipitates in High-Temperature Creep in a Nb-Stabilised Austenitic Stainless Steel, Mater. Sci. Eng. A, 2000, 279, p 52–60
11. D. Li, M.C. Gao, J.A. Hawk, and Y. Zhang, Annealing Effect for the Al0.3CoCrFeNi High-Entropy Alloy Fibers, J. Alloys Compd., 2019, 778, p 23–29
12. Ö.N. Doğan, B.C. Nielsen, and J.A. Hawk, Elevated-Temperature Corrosion of CoCrCuFe-NiAl0.5Bx High-Entropy Alloys in Simulated Syngas Containing H2S, Oxid. Met., 2013, 80, p 177–190
13. J.W. Yeh, S.K. Chen, S.J. Lin, J.Y. Gan, T.S. Chin, T.T. Shun, C.H. Tsau, and S.Y. Chang, Adv. Eng. Mater. 6, 299 (2004).
14. Zhang, Y. et al. Microstructures and properties of high-entropy alloys. Prog. Mater. Sci. 61, 1–93 (2014).
15. N.D. Stepanov, D.G. Shaysultanov, G.A. Salishchev, M.A. Tikhonovsky, E.E. Oleynik, A.S. Tortika, O.N. Senkov, Effect of v content on microstructure and mechanical properties of the CoCrFeMnNiVx high entropy alloys. J. Alloys Compd. (2015). https://doi.org/10.1016/j.jallcom.2014.12.157
16. W. Liu, J. He, H. Huang, H. Wang, Z. Lu, and C. Liu, Effects of Nb Additions on the Microstructure and Mechanical Property of CoCrFeNi High-Entropy Alloys, Intermetallics, 2015, 60, p 1-8
17. Y.J. Zhou, Y. Zhang, Y.L. Wang, and G.L. Chen, Appl. Phys. Lett. 90, 181904 (2007).
18. C.Y. Hsu, C.C. Juan, W.R. Wang, T.S. Sheu, J.W. Yeh, and S.K. Chen, Mater. Sci. Eng. A 528, 3581 (2011).
19. Z. H. Cao, P. Y. Li, H. M. Lu, Y. L. Huang, Y. C. Zhou, and X. K. Meng: Scripta Materialia, 2009, vol. 60, pp. 415-418.
20. Y. J. Huang, J. Shen, Y. L. Chiu, J. J. J. Chen, and J. F. Sun: Intermetallics, 2009, vol. 17, pp. 190–194.
21. Z. S. Ma, S. G. Long, Y. Pan, Y. C. Zhou: J Mater Sci, 2008, vol. 43, pp. 5952–5955.

22. Oliver WC, Pharr GM (1992) An improved technique for determining hardness and elastic modulus using load and displacement sensing indentation experiments. J Mater Res 7:1564–1583
23. C. L. Wang, M. Zhang, and T. G. Nieh: J. Phys. D: Appl. Phys, 2009, vol. 42, pp. 115405.
24. Hu J, Sun G, Zhang X, Wang G, Jiang Z, Han S, et al. Effects of loading strain rate and stacking fault energy on nanoindentation creep behaviors of nanocrystalline Cu, Ni-20 wt.%Fe and Ni. J Alloys Compd. 2015;647(1):670–80.
25. Kassner ME. Fundamentals of creep in metals and alloys. London: Butterworth-Heinemann, an imprint of Elsevier; 2009.
26. Mayo MJ, Nix WDA. micro-indentation study of superplasticity in Pb, Sn, and Sn-38 wt% Pb. Acta Metall. 1988;36(8):2183–92.
27. Dirras G, Couque H, Lilensten L, Heczel A, Tingaud D, Couzinié JP, et al. Mechanical behavior and microstructure of Ti20Hf20Zr20Ta20Nb20 high-entropy alloy loaded under quasi-static and dynamic compression conditions. Mater Charact. 2016;111:106–13.
28. Song H, Yavas H, Van der Giessen R, Papanikolaou S. Discrete dislocation dynamics simulations of nanoindentation with pre-stress: hardness and statistics of abrupt plastic events. J Mech Phys Solids. 2018;123:332–47.
29. Murakami, H., Harada, H. & Bhadeshia, H. The Location of Atoms in Re-Containing and V-Containing Multicomponent Nickel-Base Single-Crystal Superalloys. Appl. Surf. Sci. 76, 177–183 (1994).
30. Gludovatz, B. et al. A fracture-resistant high-entropy alloy for cryogenic applications. Science 345, 1153–1158 (2014).
31. He, J. Y. et al. A precipitation-hardened high-entropy alloy with outstanding tensile properties. Acta Mater. 102, 187–196 (2016).
32. Yeh, A. C. et al. Developing new type of high temperature alloys – High Entropy Superalloys, Int. J. Met. Mater. Eng. 1 (2015).
33. Yeh, J. W. Alloy design strategies and future trends in high-entropy alloys. Jom 65, 1759–1771 (2013).
34. Tsao, T. K. & Yeh, A. C. The thermal stability and strength of highly alloyed Ni3Al. Mater. Trans. 56, 1905–1910 (2015).
35. Veron, M., Brechet, Y. & Louchet, F. Directional coarsening of nickel based superalloys: Driving force and kinetics, Superalloys 181–190 (1996).
36. Ratel, N., Bruno, G., Bastie, P. & Mori, T. Plastic strain-induced rafting of γ' precipitates in Ni superalloys: Elasticity analysis. Acta Mater. 54, 5087–5093 (2006).
37. Oblak, J. M. & Owczarsk, W. Cellular Recrystalization in Nickle-Base Superalloy. T. Metall. Soc. Aime. 242, 1563–1568 (1968).
38. S.Z. Chavoshi, S. Xu, MRS Commun. 8, 15 (2018).
39. M.T. Tsai, J.C. Huang, P.H. Lin, T.Y. Liu, Y.C. Liao, J.S.C. Jang, S.X. Song, T.G. Nieh, Intermetallics 103, 88 (2018).
40. Y. Ma, G.J. Peng, D.H. Wen, T.H. Zhang, Mater. Sci. Eng. A 621, 111 (2015).
41. D.H. Lee, M.Y. Seok, Y. Zhao, I.C. Choi, J. He, Z. Lu, J.Y. Suh, U. Ramamurty, M. Kawasaki, T.G. Langdon, Spherical nanoindentation creep behavior of nanocrystalline and coarse-grained CoCrFeMnNi high-entropy alloys. Acta Mater. 109, 314–322 (2016)
42. Z. Wang, S. Guo, Q. Wang, Z. Liu, J. Wang, Y. Yang, C.T. Liu, Intermetallics 53, 183 (2014).
43. D.-H. Lee, I.-C. Choi, G. Yang, Z. Lu, M. Kawasaki, U. Ramamurty, R. Schwaiger, J.-i. Jang, Scr. Mater. 156, 129 (2018).
44. Z.M. Jiao, Z.H. Wang, R.F. Wu, J.W. Qiao, Appl. Phys. A 122, 794 (2016).
45. Y. Ma, Y.H. Feng, T.T. Debela, G.J. Peng, T.H. Zhang, Int. J. Refract. Metals Hard Mater. 54, 395 (2016).
46. W.J. Kim, H.T. Jeong, H.K. Park, K. Park, T.W. Na, E. Choi, J. Alloys Compd. 802, 152 (2019).
47. Zhang L, Yu P, Cheng H, Zhang H, Diao H, Shi Y, et al. Nanoindentation creep behavior of an Al0.3CoCrFeNi high-entropy alloy. Metall Mater Trans A. 2016;47(12):5871–5.
48. Z. Tang, O.N. Senkov, C.M. Parish, C. Zhang, F. Zhang, L.J. Santodonato, G.Y. Wang, G.F. Zhao, F.Q. Yang, P.K. Liaw, Mater. Sci. Eng. A 647, 229 (2015)

49. Y.P. Wang, B.S. Li, M.X. Ren, C. Yang, and H.Z. Fu: Mater. Sci. Eng. A, 2008, vol. 491, pp. 154–58.
50. Langdon, T. G. Seventy-five years of superplasticity: historic developments and new opportunities. J. Mater. Sci. 44, 5998–6010 (2009).
51. McFadden, S. X., Mishra, R. S., Valiev, R. Z., Zhilyaev, A. P. & Mukherjee, A. K. Low-temperature superplasticity in nanostructured nickel and metal alloys. Nature 398, 684–686 (1999).
52. Berdichevskij, E. G. Use of the superplasticity effect of materials in engineering technology. IOP Conf. Ser. Mater. Sci. Eng. 441, 012009 (2018).
53. Kawasaki, M. & Langdon, T. G. Developing superplasticity and a deformation mechanism map for the Zn–Al eutectoid alloy processed by high-pressure torsion. Mater. Sci. Eng. A 528, 6140–6145 (2011).
54. Clark, M. A. & Alden, T. H. Deformation enhanced grain growth in a superplastic Sn-1% Bi alloy. Acta Met. 21, 1195–1206 (1973).
55. C.W. Tsai, Y.L. Chen, M.H. Tsai, J.W. Yeh, T.T. Shun, and S.K. Chen, J. Alloys Compd. 486, 427 (2009).
56. O.N. Senkov, G.B. Wilks, D.B. Miracle, C.P. Chuang, and P.K. Liaw, Intermetallics 18, 1758 (2010).
57. K.Y. Tsai, M.H. Tsai, and J.W. Yeh, Acta Mater. 61, 4887 (2013).
58. Ng, C., Guo, S., Luan, J., Shi, S. & Liu, C. T. Entropy-driven phase stability and slow diffusion kinetics in an Al 0.5 CoCrCuFeNi high entropy alloy. Intermetallics 31, 165–172 (2012).
59. S. Meagher, R.S. Borch, J. Groza, A.K. Mukherjee, and H.W. Green II, Acta Metall. Mater. 40, 159 (1992).
60. J.L. Campbell and C.W. Schulte, J. Appl. Phys. 19, 149 (1979).
61. W. Ren, J. Guo, G. Li, and J. Wu, Mater. Trans. 42, 1731 (2004).
62. Nguyen TCN, Asghari-Rad P, Sathiyamoorthi P, Zargaran A, Lee CS, Kim HS (2020) Ultrahigh high-strain-rate superplasticity in a nanostructured high-entropy alloy. Nat Commun 11:1–7. https://doi.org/10.1038/s41467-020-16601-1
63. Jeong HT, Kim WJ (2021) Calculation and construction of deformation mechanism maps and processing maps for CoCrFeMnNi and Al0.5CoCrFeMnNi high-entropy alloys. J Alloys Compd 869:159256. https://doi.org/10.1016/j.jallcom.2021.159256
64. Jeong HT, Park HK, Kim WJ (2022) Class I type creep behavior of coarse-grained Al0.5CoCrFeMnNi high entropy alloy. Mater Sci Eng A 845:143239. https://doi.org/10.1016/j.msea.2022.143239
65. Uporova SA, Ryltseva RE, Bykov VA, Estemirova SKh, Zamyatin DA (2020) Microstructure, phase formation and physical properties of AlCoCrFeNiMn high-entropy alloy. J Alloys Compd 820:153228. https://doi.org/10.1016/j.jallcom.2019.153228
66. Shafiee, Ahad; Nili-Ahmadabadi, Mahmoud; Kim, Hyoung Seop; Jahazi, Mohammad Development and Microstructural Characterization of a New Wrought High Entropy Superalloy. Metals and Materials International (2019). https://doi.org/10.1007/s12540-019-00360-w
67. Rozman, K. A.; Detrois, M.; Liu, T.; Gao, M. C.; Jablonski, P. D.; Hawk, J. A. Long-Term Creep Behavior of a CoCrFeNiMn High-Entropy Alloy. Journal of Materials Engineering and Performance (2020). https://doi.org/10.1007/s11665-020-05103-2
68. Zhang, Lijun; Yu, Pengfei; Cheng, Hu; Zhang, Huan; Diao, Haoyan; Shi, Yunzhu; Chen, Bilin; Chen, Peiyong; Feng, Rui; Bai, Jie; Jing, Qin; Ma, Mingzhen; Liaw, P. K.; Li, Gong; Liu, Riping Nanoindentation Creep Behavior of an $Al_{0.3}$CoCrFeNi High-Entropy Alloy. Metallurgical and Materials Transactions A (2016). https://doi.org/10.1007/s11661-016-3469-8
69. Meng, Lu; Cui, Weiling; Su, Buyun; Shu, Xuefeng; Xiao, Gesheng Theoretical Characterization of Indentation Depth-Dependent Creep Behavior of CoCrFeNi$Al_{0.3}$ High-Entropy Alloy. Acta Mechanica Solida Sinica (2022). https://doi.org/10.1007/s10338-022-00354-z
70. Tsao, Te-Kang; Yeh, An-Chou; Kuo, Chen-Ming; Kakehi, Koji; Murakami, Hideyuki; Yeh, Jien-Wei; Jian, Sheng-Rui The High Temperature Tensile and Creep Behaviors of High Entropy Superalloy. Scientific Reports (2017). https://doi.org/10.1038/s41598-017-13026-7

71. Lin, P. H.; Chou, H. S.; Huang, J. C.; Chuang, W. S.; Jang, J. S. C.; Nieh, T. G. Elevated-temperature creep of high-entropy alloys via nanoindentation. MRS Bulletin (2019). https://doi.org/10.1557/mrs.2019.258
72. Jiao, Z. M.; Wang, Z. H.; Wu, R. F.; Qiao, J. W. Strain rate sensitivity of nanoindentation creep in an AlCoCrFeNi high-entropy alloy. Applied Physics A (2016). https://doi.org/10.1007/s00339-016-0339-6
73. Nguyen, Nhung Thi-Cam; Asghari-Rad, Peyman; Sathiyamoorthi, Praveen; Zargaran, Alireza; Lee, Chong Soo; Kim, Hyoung Seop Ultrahigh high-strain-rate superplasticity in a nanostructured high-entropy alloy. Nature Communications (2020). https://doi.org/10.1038/s41467-020-16601-1
74. Shaysultanov, D. G.; Stepanov, N. D.; Kuznetsov, A. V.; Salishchev, G. A.; Senkov, O. N. Phase Composition and Superplastic Behavior of a Wrought AlCoCrCuFeNi High-Entropy Alloy. JOM (2013). https://doi.org/10.1007/s11837-013-0754-5
75. Nguyen, Nhung Thi-Cam; Asghari-Rad, Peyman; Park, Hyojin; Kim, Hyoung Seop Differential superplasticity in a multi-phase multi-principal element alloy by initial annealing. Journal of Materials Science (2022). https://doi.org/10.1007/s10853-022-07616-8

Chapter 7
Welding/Joining of High Entropy Alloys

7.1 Introduction by Editor

Welding, a fundamental fabrication process, joins components to form cohesive structures. The unique compositions and microstructures of high-entropy alloys, on the other hand, introduce complexities that necessitate advances in welding techniques. The intricate connection among the composition, crystal structures, and thermodynamic properties of multiple elements within HEAs demands novel techniques to ensure strong welds, maintain material integrity, and exploit the enhanced properties of these alloys. Researchers and practitioners are interested in learning more about the complexities of phase stability, solidification behavior, and microstructural evolution during the welding process. This chapter is dedicated to the exploration of welding strategies that optimize joint quality, minimize defects, produce different types of welds (similar and dissimilar), and unlock the full potential of high-entropy alloys in welded structures.

Welding of metals has always been considered a critical cause of material failure starting from a small automotive entity to giant structures used in nuclear industries. These failures ultimately prove to be fatal, leading to severe casualties and deaths. This makes the study of new welding techniques an area of interest and developing new materials as a challenging topic for research. Welding studies in metallurgy primarily include the welding technique and the metals to be welded. Welding techniques are categorized into two sections which include fusion welding and solid-state welding techniques. In fusion welding the adhesion of two components includes the liquefication of either base metals or the metal added. In solid-state welding, the metal adhesion is without metal liquefication, usually at the temperatures at which the metal begins to plasticize.

Conventionally, welding is not much matter of concern with a defined welding technique for similar metal/alloy joining. A defined welding technique is a feasible welding methodology that can weld the metal with minimal or no defects. The

S. S. Nene (ed.), *High Entropy Alloys*, https://doi.org/10.1007/978-981-99-7173-2_7

issue arises when the weld requirement is between two dissimilar metals. Dissimilar metals in this case are the ones with base metals of different categories like steel with aluminum, or aluminum with titanium, etc. Here, both, the alloy and the welding technique prove to be crucial factors in determining the quality of the weld, microstructure, and grain morphology, affecting the mechanical properties altogether. In most cases, the weld regions are prone to form intermetallic compounds (IMCs) jointly deteriorating the welding efficiency. IMCs usually form in metals or alloys that have limited or no solubility within themselves. Many of these structural metals possess a tendency to form an IMC.

High entropy alloys have been studied for welding and are proving to be a desirable way of reducing the potential of IMC formation. This is due to their high entropy behavior and the tendency to undergo sluggish diffusion. These ultimately resist the chances for compound formation leading to an alloy in the form of a solid solution. In combination with the alloy designing philosophy with HEA design, advanced welding techniques like friction stir welding (FSW) have proved to be an effective way of limiting the IMC formation, tending to almost minimal IMC formation. This enhancement of weld qualities ultimately gives an edge for designing more HEAs over conventional alloys.

Welding overall has shown an improvement in metals and alloy characteristics and has extended the possibility of its use over a broader range. To understand this, the following chapters cover the welding of similar metals (HEAs in this case), and dissimilar metal joints (conventional metal to a HEA), HEA powders as a binder between two dissimilar conventional metals. Not only the categories of metals, the summary also covers welding techniques from conventional fusion welding like gas tungsten arc welding, metal inert gas welding, and laser welding to advanced solid-state welding techniques like friction stir welding and relating it with the microstructure and material properties and performance. The summary covers a wide range of parameters and processes to minimize the defects and the quality of weld.

Overall, this chapter also covers the uniquely designed compositions by high entropy approach, their subsequent microstructures and, on the other hand, introduces the requirement of advanced welding techniques focusing enhanced weld quality. Researchers and practitioners are interested in learning more about the complexities of phase stability, solidification behavior, and microstructural evolution during the welding process when HEAs are part of the welding. This chapter explores welding strategies that optimize joint quality, minimize defects, produce different types of welds (similar and dissimilar), and unlock the full potential of high-entropy alloys in welded structures.

7.2 Machine Generated Summaries

Disclaimer: The summaries in this chapter were generated from Springer Nature publications using extractive AI auto-summarization: An extraction-based summarizer aims to identify the most important sentences of a text using an algorithm and uses those original sentences to create the auto-summary (unlike generative AI). As the constituted sentences are machine selected, they may not fully reflect the body of the work, so we strongly advise that the original content is read and cited. The auto generated summaries were curated by the editor to meet Springer Nature publication standards. To cite this content, please refer to the original papers.

Machine generated keywords: weld, welding, joint, zone, weldability, dissimilar, stainless, stainless steel, laser, stir, friction stir, friction, join, cocrfemnni, fusion

Dissimilar Welding of High-Entropy Alloy to Inconel 718 Superalloy for Structural Applications [40]

This is a machine-generated summary of:

Sokkalingam, R.; Pravallika, B.; Sivaprasad, K.; Muthupandi, V.; Prashanth, K. G.: Dissimilar welding of high-entropy alloy to Inconel 718 superalloy for structural applications [40]

Published in: Journal of Materials Research (2021)

Link to original: https://doi.org/10.1557/s43578-021-00352-w

If you want to cite the papers, please refer to the original.

For technical reasons we could not place the page where the original quote is coming from.

Abstract-Summary

"For high-temperature applications, where dissimilar materials are in demand, HEAs may be joined with commercially available structural materials to improve their performance-life ratio."

"A dissimilar joint was fabricated by gas tungsten arc welding between $Al_{0.1}$CoCrFeNi-HEA and Inconel 718."

"The columnar dendritic grains are growing epitaxially at the $Al_{0.1}$CoCrFeNi-HEA/weld metal interface, where their compositions are matching."

"The results demonstrate the applicability of GTAW in fabricating the dissimilar weld joints between HEA and Inconel 718 for structural applications."

Introduction

"Due to this exceptional mechanical properties, $Al_{0.1}CoCrFeNi$-HEAs is believed to be a substitute to the currently available commercial high-temperature alloys in some critical high-temperature applications [1]."

"Study on dissimilar welding between HEA and commercially available structural alloys (stainless steels, nickel-base superalloys, titanium alloys, etc.) is vital for preparing most economic and reliable structures."

"Nam and others has welded HEAs with different processing conditions (as-cast and as-rolled conditions) by laser welding and found evolution of columnar dentritic grains epitaxially and strength of the dissimilar weld joint is comparable with the as-cast HEA as it fails at as-cast HEA [2]."

"Sokkalingam and others have examined dissimilar weld joint between $Al_{0.1}CoCrFeNi$-HEA and AISI304 stainless steel by gas tungsten arc welding (GTAW) [3] and electron beam welding (EBW) [4] processes."

"With the view of using HEA in high-temperature structural applications like aero-engine structural components, an investigation has made on the dissimilar joint fabricated between high-entropy alloy and Ni-based superalloy (i.e., Inconel 718) by gas tungsten arc (GTA) welding process."

Results and discussion

"The presence of NbC and Laves phases could increase the possibility of hot cracking susceptibility and may reduce the ductility of the weld [5]."

"Grain coarsening coupled with segregation could deteriorate the mechanical properties as a consequence of a reduction in grain boundary strengthening and an early crack initiation at pores, NbC, and the brittle Laves phases during plastic deformation."

"The reduction in the strength of the dissimilar $Al_{0.1}CoCrFeNi$-HEA/IN718 weld joint is attributed to the coarse-grained structure at the FZ compared to the BMs."

"Further, the possible earlier crack initiation at pores, SGBs and, brittle NbC and Laves phases in the coarse weld metal grains could have caused the fracture in the weld, associated with a reduction in ductility [6]."

"The quasi-cleavage mode of fracture in the weld metal is due to the presence of hard secondary phases like NbC and Laves phases along the SGBs and interdendrites in the coarse columnar grains."

Conclusions

"In GTA welding of $Al_{0.1}CoCrFeNi$—HEA to IN718, the epitaxial growth of columnar dendritic grains occurs at the HEA-weld interface as the chemical composition of the weld is close to $Al_{0.1}CoCrFeNi$—HEA composition."

"Unmixed zones with a dispersion of large fractions of NbC followed by columnar dendritic coarse grain, are observed along with the IN718 weld interface."

"The weld center reveals the presence of discrete NbC/Laves phases."

"The presence of a high fraction of random high-angle boundary (RGBs) with high grain boundary energy encourages the segregation of the NbC/Laves phases."

"Fracture occurrence at the weld fusion zone is attributed to the coarse grains and the presence of the hard phases like NbC and laves along with SGBs in the weld fusion zone."

Materials and Methods

"$Al_{0.1}$CoCrFeNi-HEA and IN718 alloys with a thickness of 2 mm were considered as the base metals (BM) for the present study."

"$Al_{0.1}$CoCrFeNi-HEA was fabricated through a vacuum arc melting technique, as mentioned in earlier studies [7–9]."

"The resultant button type sample (~ 7 mm thick) was homogenized (at 1050 °C for 60 h with subsequent furnace cooling) [7, 8] and drop forged (at ~ 700 °C) to a thickness of 2.5 mm followed by surface grinding to reduce the thickness to 2 mm."

"The 2 mm thick $Al_{0.1}$CoCrFeNi-HEA was annealed (at 1050 °C for 2 h with subsequent furnace cooling) and was used as one of the base materials."

"A coupon of dimension 25 mm × 250 mm × 2 mm was extracted using wire-EDM cutting from the as-received IN718 alloy and was heat-treated at 900 °C for one hour and subsequently cooled in the atmospheric air."

Welding Dissimilar Alloys of CoCrFeMnNi High-Entropy Alloy and 304 Stainless Steel Using Gas Tungsten Arc Welding [41]

This is a machine-generated summary of:

Zhang, Penglin; Qi, Yongfeng; Cheng, Qianqian; Sun, Xuemin: Welding Dissimilar Alloys of CoCrFeMnNi High-Entropy Alloy and 304 Stainless Steel Using Gas Tungsten Arc Welding [41]

Published in: Journal of Materials Engineering and Performance (2023)

Link to original: https://doi.org/10.1007/s11665-023-08229-1

If you want to cite the papers, please refer to the original.

For technical reasons we could not place the page where the original quote is coming from.

Abstract-Summary

"The weldability of a CoCrFeMnNi high-entropy alloy (HEA) and 304 stainless steel (304 SS) was investigated to determine their potential for application in the nuclear and aerospace fields."

"There is an unmixed zone near the 304 SS side, which can be attributed to subcooling of the composition caused by the lower liquidus temperature of the bulk weld metal than that of the base metal."

"A small increase in the hardness of the fusion zone compared with that of the CoCrFeMnNi HEA was observed."

"The joints exhibited a tensile strength of ~ 465 MPa and ductility of 38%, where the strength was comparable to that of the CoCrFeMnNi HEA, and a joint fracture was found at the side of the CoCrFeMnNi HEA."

Introduction

"Kashaev and others (Ref. [10]) prepared CoCrFeMnNi HEA butt joints using a laser beam welding method and found that laser welding did not significantly affect the static mechanical and fatigue properties of the alloy."

"Adomako and others (Ref. [11]) laser welded CoCrFeMnNi onto duplex stainless steel and found that the tensile strength and ductility of the welded joint were 584 MPa and 8.2%, respectively."

"Oliveira and others (Ref. [12]) obtained defect-free joints by laser welding a CoCrFeMnNi HEA and 316 stainless steel and found an increase in hardness in the fusion zone (FZ) compared with the parent material, with a tensile strength of 449 MPa."

"To apply HEA to nuclear reactor structural parts, the heterogeneous welding of CoCrFeMnNi-HEA prepared by the conventional gas tungsten arc welding process with 304 stainless steel was investigated."

Experimental Procedures

"Sheets of isoatomic CoCrFeMnNi HEA and commercially available 304 stainless steel were used as base materials."

"Available 304 SS and HEA were cut into dimensions of 50 × 30 × 1.5 mm."

"Self-heterogeneous butt welding was performed between the CoCrFeMnNi HEA and 304 SS using a six-axis robotic arm with a conventional GTAW torch."

"After welding, the welded joint cross section was ground, polished, and etched."

"A tensile force was applied perpendicular to the welding direction at a strain rate of 1 mm min^{-1}, and the fracture morphology was observed by SEM."

Results and discussion

"Soysal and others (Ref. [13]) investigated the mechanism of UZ formation in the fusion welding of dissimilar metals; that is, generation of the UZ was related to the liquid-phase line temperature of 304 SS (T_{L1}) and the bulk weld metal (T_{LW})."

"The CoCrFeMnNi HEA base material grains exhibit coarse austenite grains, and relevant studies on CoCrFeMnNi have shown the presence of coarse grains under both as-cast and annealed conditions (Refs. [11, 14, 15])."

"Similar results were obtained in studies on dissimilar welded joints of CoCrFeMnNi HEA with 316 stainless steel and duplex stainless steel (Refs. [11, 12])."

"The average $HV_{0.2}$ values in the FZ were higher than those in the CoCrFeMnNi HEA owing to the increased hardness of the refined equiaxed grains in the FZ."

"Despite the presence of a certain range of UZ near the 304 SS side, the higher hardness of the FZ compared to the CoCrFeMnNi HEA allowed the fusion zone to still obtain superior strength compared to the HEA."

Conclusions

"Defect-free joints between a CoCrFeMnNi high-entropy alloy (HEA) and 304 stainless steel (SS) were fabricated using gas tungsten arc welding (GTAW), which exhibited good weldability."

"Because of the composition undercooling caused by the relatively low liquid-phase temperature of the welded metal compared to that of the base metal, an unmixed zone (UZ) was formed near the 304 SS side."

"The microhardness distribution of the joint was asymmetric, with the highest hardness found in the UZ and the fusion, and it was higher than that of the HEA base metal."

"Failure occurred on the side of the HEA base metal, indicating that the joint is an excellent candidate for future nuclear industry structural components."

Micro-characteristics of High-Entropy Alloy Weld Formed by Friction Stir Lap Welding of 1060 Al and $Al_{0.1}$CoCrFeNi [42]

This is a machine-generated summary of:

Zhang, H. J.; Zhao, L. Y.; Liu, X.; Zhang, H.; Wang, Y. L.; Song, J. L.: Micro-characteristics of High-Entropy Alloy Weld Formed by Friction Stir Lap Welding of 1060 Al and $Al_{0.1}$CoCrFeNi [42]

Published in: Transactions of the Indian Institute of Metals (2020)

Link to original: https://doi.org/10.1007/s12666-020-02110-8

If you want to cite the papers, please refer to the original.

For technical reasons we could not place the page where the original quote is coming from.

Abstract-Summary

"1060 Al and $Al_{0.1}$CoCrFeNi are friction stir lap welded in this study."

"Coarsening of large-size Al(Cr)-rich particles and dissolution of small-size Al(Cr)-rich particles simultaneously take place in the heat-affected zone of high-entropy alloy during the welding."

"The plastic deformation of high-entropy alloy induced by welding tool refines and homogenizes the microstructures in thermo-mechanically affected zone and stir zone."

"Friction stir lap weld of high-entropy alloy presents higher hardness than the base metal in each zone."

Introduction

"Different from arc, laser and electron beam welding, friction stir welding (FSW) is a solid-state joining technique, in which two facing workpieces are joined through the heat generated by friction between a rotating tool and the workpiece material."

"FSW has become an important and hotspot technology for the welding of HEAs in recent years."

"Jo and others [16] investigated the microstructures and mechanical properties of friction stir welded 2-mm-thick CrMnFeCoNi HEA."

"Another of their study [17] performed the FSW of a $Co_{16}Fe_{28}Ni_{28}Cr_{28}$ HEA, which revealed a similar trend of hardness increase in the SZ compared to the BM."

"It has been observed that the previous investigations on the FSW of HEAs were mainly related to the butt or bead welding."

"In light of this, friction stir lap welding (FSLW) between aluminum and a $Al_xCoCrFeNi$ HEA characterized by excellent specific strength, low temperature mechanical properties and radiation resistance [18, 19] is conducted in this paper."

Experimental Procedure

"1060 Al sheet was placed over $Al_{0.1}CoCrFeNi$ alloy and had stiff contact to form a lap joint configuration, after which an FSW machine and a welding tool were utilized to perform the FSLW experiments."

"The tool shoulder (12 mm in diameter) having three equally spaced scrolls on the end surface was utilized in the study."

"The conical tool pin with the length of 2.85 mm had right-hand threads for counterclockwise rotation."

"During the welding, a shoulder plunge depth of 0.2 mm and a tilting angle of 2° with respect to workpiece were applied to the welding tool, and the rotation speed and welding speed of the tool were kept at 800 rpm and 50 mm/min, respectively."

"Room-temperature tensile shear test was carried out on an electronic universal tensile testing machine (WDW-3100) at a constant testing speed of 1 mm/min."

Results and discussion

"The large-sized elements Al and Cr tend to segregate to form the dark network structures during the preparation of HEA, while the small-sized elements Fe, Ni and Co as the gradually slower diffuser are mainly distributed in the bulk structures."

"The SZ of $Al_{0.1}CoCrFeNi$ HEA here exhibits an average hardness improvement of ~ 300 Hv in contrast to the BM with hardness of ~ 100 Hv, which reflects that the operative strengthening mechanisms are more complicated than pure grain refinement."

"The diffusion of Al from the upper Al sheet toward the lower HEA sheet should be limited due to the slow-diffusion effect of HEA [20], nevertheless, the downward flow of aluminum and its mixing with the HEA during FSW will contribute to the improvement in Al content in the SZ."

"Both advantages of friction stir modification and Al content improvement are integrated by the FSLW of Al and HEA in the present research, leading to superior hardness values in the $Al_{0.1}CoCrFeNi$."

Conclusions

"Based on the investigations above mentioned, the conclusions of significance are drawn as follows: (1) Crescent-shaped deformed zones are formed in the HEA during the FSLW of Al and HEA."

"The enhancement of solid-solution strengthening leads to higher hardness of HAZ relative to the BM. (3) The plastic deformation during FSW refines and homogenizes the microstructures in the TMAZ and SZ of HEA, which contributes to higher hardness in both the zones."

"The hardness of SZ is improved to a maximum of 600 Hv, higher than that of $Al_xCoCrFeNi$ traditionally obtained at various x values. (4) Al content is improved in the SZ of HEA during the Al-HEA FSLW, leading to the enhancement of solid-solution strengthening and the occurrence of BCC structures in the SZ."

Initial Weldability Study on $Al_{0.5}CrCoCu_{0.1}FeNi$ High-Entropy Alloy [43]

This is a machine-generated summary of:

Martin, Alexander C.; Fink, Carolin: Initial weldability study on $Al_{0.5}CrCoCu_{0.1}FeNi$ high-entropy alloy [43]

Published in: Welding in the World (2019)

Link to original: https://doi.org/10.1007/s40194-019-00702-7

If you want to cite the papers, please refer to the original.

For technical reasons we could not place the page where the original quote is coming from.

Abstract-Summary

"The weld solidification behavior, segregation tendency, phase formation, and cracking susceptibility of $Al_{0.5}CoCrCu_{0.1}FeNi$ high-entropy alloy (HEA) were investigated."

"Microstructural changes in the fusion zone and heat-affected zone (HAZ) were investigated by optical microscopy and micro-hardness."

"The fusion zone solidification microstructure of $Al_{0.5}CoCrCu_{0.1}FeNi$ alloy consists of primary face centered cubic dendrites and interdendritic body centered cubic phase that forms at the end of solidification."

"GTA welding results in the partial dissolution of the interdendritic phase in the high-temperature HAZ, adjacent to the fusion boundary."

"Welding on the heat-treated alloy leads to HAZ softening due to the dissolution of strengthening nano-sized precipitates."

"Laser welding did not result in a drop in hardness or significant microstructural changes in the HAZ."

"The CPTT results correspond well with the predicted solidification temperature range (141 K) for $Al_{0.5}CoCrCu_{0.1}FeNi$ alloy and its overall good weldability that was observed in the present work."

Introduction

"The concept of high-entropy alloys was introduced almost 15 years ago as a new route for developing advanced materials with distinctive properties, some of which not achieved by the conventional micro-alloying approach [21–24]."

"The high configurational entropy was believed to stabilize random solid solutions in these alloys against the formation of intermetallic phases, which were considered often detrimental to alloy properties [24]."

"High-entropy alloys are also often referred to as multi-principal-element alloys (MPEA), which might be more suitable for (near) equiatomic alloys with or without minor element additions and composed of solid solution phase(s) without any or with some intermetallic phases [24, 25]."

"Because of their unique multi-principal-element composition, high-entropy alloys can exhibit useful properties, some of them not seen in conventional alloys."

"Welding and weldability become key issues in high-entropy alloys for structural engineering applications."

Experimental procedures

"Three different welding trials were performed using gas tungsten arc (GTA) welding on the alloy in (1) as-melted, (2) heat-treated, and (3) hot-rolled and heat-treated conditions."

"Autogenous GTA spot welding was performed on "button-shaped" ingots (10 g) in as-melted condition using a stationary welding torch at 120 A and 11 V for 3 s at 5-mm standoff distance."

"Autogenous linear GTA welding was performed on a 20-g plate (63 × 13 × 3 mm) in heat-treated condition (650 °C for 2.5 h)."

"Autogenous GTA butt welding was performed to join two plates (63 × 13 × 3 mm) in the hot-rolled and heat-treated condition."

"Laser welding using a Trumpf Nd:YAG pulsed laser system was performed as an autogenous linear weld on a 20-g plate (63 × 13 × 3 mm) in heat-treated condition (650 °C for 2.5 h)."

Results and discussion

"In previous work by the authors [26], the predicted STR of $Al_{0.5}CoCrCu_{0.1}FeNi$ alloy is narrow as compared to compositions with a higher Cu content ($\geq Cu_{0.3}$), which have a calculated STR > 300 K. This is due to strong partitioning of Cu to the interdendritic regions during solidification, which results in the formation of a low-melting Cu-rich phase that can lead to solidification cracking in the fusion zone and heat-affected zone liquation cracking."

"It can be seen that the hardness in the fusion zone is at about the same level for all GTAW welds, i.e., independent from the processing condition of the alloy (as-melted, heat-treated, or hot-rolled and heat-treated)."

"The GTA butt weld (blue line) exhibits the largest drop of hardness in the fusion zone and HAZ as compared to the initial hot-rolled and heat-treated alloy condition."

Conclusion

"Non-equilibrium weld solidification of $Al_{0.5}CoCrCu_{0.1}FeNi$ alloy results in primary FCC dendrites and the formation of interdendritic BCC phase at the end of solidification."

"The GTA welds were free of solidification cracking or heat-affected zone (HAZ) liquation cracking for all alloy conditions and welding setups."

"Partial dissolution of the interdendritic phase of the initial alloy microstructure was observed in the high-temperature HAZ adjacent to the fusion boundary."

"This might explain the absence of HAZ liquation cracking in the welds, since the formation of low-melting liquid might be inhibited in the absence of the interdendritic phase close to the fusion line."

"Welding on the heat-treated alloy resulted in HAZ softening due to the dissolution of the strengthening phase in the dendrite matrix and interdendritic regions."

"No HAZ liquation cracking was observed in keyhole welding mode."

"In conductive welding mode, no solidification or liquation cracking was observed."

Effect of Initial Grain Size on Friction Stir Weldability for Rolled and Cast CoCrFeMnNi High-Entropy Alloys [44]

This is a machine-generated summary of:

Park, Sangwon; Nam, Hyunbin; Na, Youngsang; Kim, Hyoungseop; Moon, Younghoon; Kang, Namhyun: Effect of Initial Grain Size on Friction Stir Weldability for Rolled and Cast CoCrFeMnNi High-Entropy Alloys [44]

Published in: Metals and Materials International (2019)

Link to original: https://doi.org/10.1007/s12540-019-00466-1

If you want to cite the papers, please refer to the original.

For technical reasons we could not place the page where the original quote is coming from.

Abstract-Summary

"The rolled and cast HEAs had a grain size of 2.8 and 308 μm, respectively."

"After friction stir welding, the cast HEA weld had a grain size of 1.8 μm, which was coarser than that of the rolled HEA weld (1.4 μm)."

"The cast HEA weld with a large grain size contained a higher density of high-angle boundaries and twins than the rolled HEA weld with the small grain size."

"The cast HEA weld had a higher tunnel defects ratio than the rolled HEA weld."

"The total unbonded ratios of the rolled and cast HEA welds were 0.2 and 7.2%, respectively, indicating that the rolled HEA had better weldability than the cast HEA."

Introduction

"The low-temperature tensile properties of cryogenic materials have reached their limit and CoCrFeMiNi high entropy alloys (HEAs) have received immense attention recently due to their significant increase of low-temperature strength and elongation [21, 27, 28]."

"The mechanical properties of HEAs have been reported for welding technologies with a low-heat input, such as electron beam welds, laser beam welds (LBWs), and friction stir welds (FSWs) [29–36]."

"For rolled CoCrFeMnNi HEAs, Wu and others and Nam and others [29, 30] reported that the tensile strength of the LBWs was lower than that of the base metal (BM) owing to the coarse grains and decreased hardness of the welded zone."

"For the cast $Al_xCoCrFeMnNi$ HEAs, the LBW increased the hardness of the weld zone owing to the fine grain and B2 phase particles in the nanoscale [31]."

"This study investigates the influence of the initial grain size on the joint defects, microstructure, hardness, and tensile properties for the FSWs of rolled and cast CoCrFeMnNi HEAs."

Experimental Procedure

"Friction stir welding was performed using an FSW machine (Hwacheon Company in South Korea) with a 25 kW spindle servo motor and the following conditions: rotation speed of 800 rpm and welding speed of 100 mm min^{-1}."

"To study the phase evolution of the HEAs and their welds, x-ray diffraction (XRD) using Cu K_α radiation was performed at a scan speed of 2° min^{-1}, with a range of 20–90°, voltage of 40 kV, and current of 30 mA. The microstructure was observed by light optical microscopy (LOM) and electron backscatter diffraction (EBSD) equipped in a field emission scanning electron microscope (FE-SEM)."

"Using a field-emission electron-probe micro analyser (FE-EPMA), an area analysis was conducted with a voltage of 10 kV, current of 100 nA, and step sizes of 0.1–0.5 mm in the non-etched specimen."

Results and discussion

"The cast HEA had an ~ 100-times larger grain size than the rolled HEA."

"The unbonded ratio of the RS region was 0.2 and 6.5% for the rolled and cast HEA welds, respectively."

"Both rolled and cast HEA welds showed grain refinement owing to the DRX."

"The cast HEA with coarse grains could be associated to the high density of the twin boundary in the SZ, resulting in a large SHR, sluggish plastic flow, and large unbonded ratio."

"The cast HEA (coarse grain) produced a large SHR and hardness than the rolled HEA (fine grain) and the cast HEA weld indicated a larger hardness than the rolled HEA weld."

"It was associated to significant wear of the stirring WC tool, sluggish plastic flow during FSW, and large unbounded ratio in the cast HEA weld."

Conclusions

"The cast HEA weld showed a larger grain size than the rolled HEA weld owing to the significantly larger grain size of the cast BM. (3) The SHR of the cast HEA became higher than that of the rolled HEA at large strains above 0.2."

"The cast HEA weld with large grain size had a higher density of HABs and $\sum 3$ twin boundaries than the rolled HEA weld with small grain size. (4) The cast HEA weld had a wider and more significant tungsten carbide region than the rolled HEA weld."

"The cast HEA weld had a three times larger amount of tungsten than the rolled HEA weld. (5) The cast HEA produced a large SHR and hardness than the rolled HEA, therefore inducing significant wear of the stirring WC tool, sluggish plastic flow during FSW, and large unbounded ratio and tunnel defects in the cast HEA weld."

Dissimilar Metal Joining of 304 Stainless Steel to SMA490BW Steel Using the Filler Metal Powders with a High-Entropy Design [45]

This is a machine-generated summary of:

Liu, Dejia; Guo, Rui; Hu, Yong; Shen, Mingxue; Tang, Yanchuan; Zhao, Longzhi; Li, Deying; Wang, Xiangjie: Dissimilar Metal Joining of 304 Stainless Steel to SMA490BW Steel Using the Filler Metal Powders with a High-Entropy Design [45]

Published in: Metals and Materials International (2019)

Link to original: https://doi.org/10.1007/s12540-019-00400-5

If you want to cite the papers, please refer to the original.

For technical reasons we could not place the page where the original quote is coming from.

Abstract-Summary

"Multi-component mixed powders of FeCoCrNiMn and $CrFeNi_{2.4}Al_{0.6}$, based on a high-entropy design, as well as the comparative 316L stainless steel powders, were used as the filler metals, to achieve the dissimilar welding between 304 stainless steel and SMA490BW steel by laser deposition welding."

"By comparative analysis of the microstructure and mechanical properties of three types of joints, the feasibility and weld-ability of the filler metal powders based on a high-entropy design were studied."

"That the melting of base metal (BM) and weld metal dilution had an impact on the set high-entropy component in the weld zone."

"High-entropy structures were achieved in the weld zone by using the powders of $CrFeNi_{2.4}Al_{0.6}$."

"Multi-component mixed powders of FeCoCrNiMn and $CrFeNi_{2.4}Al_{0.6}$, based on a high-entropy design, were used as the filler metals to achieve the dissimilar welding of 304 stainless steel to SMA490BW steel by laser deposition welding."

"A detailed investigation of microstructures and mechanical properties of those dissimilar joints was carried out to explore the feasibility and weld-ability of the filler metal powders based on a high-entropy design."

Introduction

"It should be pointed out that, the high strength and good fracture toughness of HEAs are particularly suitable to apply in welding, especially served as the filler metal during welding process [19, 37, 38]."

"It is meaningful to evaluate the microstructures and mechanical properties of the welded joints using the filler metal of HEAs, so as to improve the strength and toughness of the joints and expand the application of HEAs served as the filler metal during the welding process."

"Two kinds of multi-component mixed powders based on a high-entropy design are used as the filler metals, to achieve the dissimilar welding of 304 stainless steel to SMA490BW steel by laser deposition welding."

"By comparative analysis of the microstructure and mechanical properties of three types of dissimilar joints, the feasibility and weld-ability of the filler metals of multi-component mixed powders based on a high-entropy design were studied."

Experimental Procedures

"The raw materials of iron, cobalt, chromium, nickel, manganese, and aluminum powders were used to manufacture the high-entropy filler metal in the experiment."

"Another low-cost Co-free high-entropy filler metal was prepared by multi-component mixed powders with a nominal Cr:Fe:Ni:Al molar ratio of 1:1:2.4:0.6, named as the $CrFeNi_{2.4}Al_{0.6}$ powders."

"To evaluate the weld-ability of two types of high-entropy filler metal, commercial 316L stainless steel powders are selected as the comparative filler metal, named as 316L powders."

"Three types of filler metals were used in the study."

"The same technological parameters were used during the welding process for three types of filler metals."

"To evaluate the corrosion resistance of the WZs by using three types of filler metals, potentiodynamic polarization tests were carried out in a corrosion cell containing 450 ml of 3.5-wt% NaCl solution at room temperature by using a corrosion measurement system."

Results

"Dissimilar metals of 304 stainless steel to SMA490BW steel are successfully joined by laser deposition welding by using three types of filler metal powders."

"It is measured that the fracture line is located in a distance of 0.32–1.79 mm from the BM/WZ interface in the SMA490BW steel side for the joint using the powders of 316L, and that is about 0.41–1.82 mm for the joint using the powders of FeCoCrNiMn."

"The average measured atomic ratio of Fe:Co:Cr:Ni:Mn is about 54.1:12.1:17:7:9.8, for the WZ near the BM/WZ interface in 304 stainless steel side by using the powders of FeCoCrNiMn."

"For the WZ by using the powders of $CrFeNi_{2.4}Al_{0.6}$, columnar crystals are observed near the fusion zone in the 304 stainless steel side."

Discussion

"Based on the experimental results in the present study, the melting of BM and weld metal dilution during fusion welding may change the set high-entropy components in the WZ."

"To form only simple ordered phase, the ranges of those parameters are more strict: $-15 \leq \Delta H_{mix} \leq 5$ kJ/mol, $\delta \leq 4.3$, and $12 \leq \Delta S_{mix} \leq 17.5$ J/(K mol) [39]. In order to analyze the effects of Fe element content on the phase structures in high-entropy welded joints, some physicochemical parameters of $Fe_xCoCrNiMn$ alloys are calculated to predict whether sole simple ordered phase or their mixtures can be formed in those alloys."

"As a summary, in the present study, the multi-component mixed powders of FeCoCrNiMn and $CrFeNi_{2.4}Al_{0.6}$, based on a high-entropy design, were used to achieve the dissimilar welding of 304 stainless steel to SMA490BW steel by laser deposition welding."

Conclusions

"High-entropy structures were achieved in the WZ by using the powders of $CrFeNi_{2.4}Al_{0.6}$. (3) All the dissimilar joints welded with three types of filler metals presented a higher notched tensile strength compared to the BM of SMA490BW steel."

"All those WZs had a better corrosion resistance than the BM of SMA490BW steel. (4) The dissimilar joint welded using the powders of $CrFeNi_{2.4}Al_{0.6}$ had the lowest hardness value in the WZ, in which the joint was fractured during the notched tensile tests."

"The other two joints fractured at the regions near the notch in the SMA490BW steel side."

"Transgranular fracture and a typical dimple fracture were observed in the fractured joints."

References

1. R. Sokkalingam, K. Sivaprasad, V. Muthupandi, Welding of high entropy alloys, in High entropy alloys: innovations, advances, and applications. ed. by T.S. Srivatsan, M. Gupta (CRC Press, Boca Raton, 2020), p. 598
2. H. Nam, S. Park, E.J. Chun, H. Kim, Y. Na, N. Kang, Laser dissimilar weldability of cast and rolled CoCrFeMnNi high-entropy alloys for cryogenic applications. Sci Technol Weld Join. 25, 127 (2019)
3. R. Sokkalingam, V. Muthupandi, K. Sivaprasad, K.G. Prashanth, Dissimilar welding of Al0.1CoCrFeNi high-entropy alloy and AISI304 stainless steel. J. Mater. Res. 34, 2683 (2019)
4. R. Sokkalingam, P. Mastanaiah, V. Muthupandi, K. Sivaprasad, K.G. Prashanth, Electron beam welding of high entropy alloy and stainless steel: microstructure and mechanical properties. Mater. Manuf. Process. 35, 1885 (2020)
5. H.S. Hosseini, M. Shamanian, A. Kermanpur, Microstructural and weldability analysis of Inconel 617/AISI 310 stainless steel dissimilar welds. Int. J Pres. Ves. Pip. 144, 18 (2016)
6. H. Ming, R. Zhu, Z. Zhang, J. Wang, E.H. Han, W. Ke, M. Su, Microstructure, local mechanical properties and stress corrosion cracking susceptibility of an SA508-52M-316LN safe-end dissimilar metal weld joint by GTAW. Mater. Sci. Eng. A. 669, 279 (2016)
7. R. Sokkalingam, S. Mishra, S.R. Cheethirala, V. Muthupandi, K. Sivaprasad, Enhanced relative slip distance in gas-tungsten-arc-welded Al0.5CoCrFeNi high-entropy alloy. Metall Mater Trans A. 48, 3630 (2017)
8. R. Sokkalingam, K. Sivaprasad, V. Muthupandi, M. Duraiselvam, Characterization of laser beam welded Al0.5CoCrFeNi high-entropy alloy. Key Eng Mater. 775, 448 (2018)
9. R. Sokkalingam, K. Sivaprasad, V. Muthupandi, M. Duraiselvam, K.G. Prashanth, Novel welding of Al0.5CoCrFeNi high-entropy alloy: corrosion behavior. J Alloy Compd. 817, 153163 (2020)
10. N. Kashaev, V. Ventzke, N. Petrov, M. Horstmann, S. Zherebtsov, D. Shaysultanov, V. Sanin, and N. Stepanov, Fatigue Behaviour of a Laser Beam Welded CoCrFeNiMn-Type High Entropy Alloy, Mater. Sci. Eng. A, 2019, 766, 138358.
11. N.K. Adomako, G. Shin, N. Park, K. Park, and J.H. Kim, Laser Dissimilar Welding of CoCrFeMnNi-High Entropy Alloy and Duplex Stainless Steel, J. Mater. Sci. Technol., 2021, 85, p 95–105.
12. J.P. Oliveira, J. Shen, Z. Zeng, J.M. Park, Y.T. Choi, N. Schell, E. Maawad, N. Zhou, and H.S. Kim: Scripta Mater., 2022, vol. 206, 114219.
13. T. Soysal, S. Kou, D. Tat, and T. Pasang, Macrosegregation in Dissimilar-Metal Fusion Welding, Acta Mater., 2016, 110, p 149–160.
14. H. Nam, S. Park, E.-J. Chun, H. Kim, Y. Na, and N. Kang, Laser Dissimilar Weldability of Cast and Rolled CoCrFeMnNi High-Entropy Alloys for Cryogenic Applications, Sci. Technol. Weld. Join., 2020, 25(2), p 127–134.
15. S. Park, H. Nam, Y. Na, H. Kim, Y. Moon, and N. Kang, Effect of Initial Grain Size on Friction Stir Weldability for Rolled and Cast CoCrFeMnNi High-Entropy Alloys, Met. Mater. Int., 2020, 26(5), p 641–649.
16. Jo M G, Kim H J, Kang M, Madakashira P P, Park E S, Suh J Y, Kim D I, Hong S T, and Han H N, Met Mater Int 24 (2018) 73.
17. Zhu Z G, Sun Y F, Ng F L, Goh M H, Liaw P K, Fujii H, Nguyen Q B, Xu Y, Shek C H, Nai S M L, and Wei J, Mater Sci Eng A 711 (2018) 524.
18. Tsai, M. H. Physical properties of high entropy alloys. Entropy 15, 5338–5345 (2013).
19. Zhang, Y. et al. Microstructures and properties of high-entropy alloys. Prog. Mater. Sci. 61, 1–93 (2014).
20. Y. Zhang, Z.P. Lu, S.G. Ma, P.K. Liaw, Z. Tang, Y.Q. Cheng and M.C. Gao: MRS Comm., 2014, vol. 4, pp. 57–62.
21. J.W. Yeh, S.K. Chen, S.J. Lin, J.Y. Gan, T.S. Chin, T.T. Shun, C.H. Tsau, and S.Y. Chang, Adv. Eng. Mater. 6, 299 (2004).

22. B. Cantor, I. T. H. Chang, P. Knight, et al.,"Microstructural development in equiatomic multicomponent alloys," Mater. Sci. Eng. A, 375, 213–218 (2004).
23. C.Y. Hsu, J.W. Yeh, S.K. Chen, and T.T. Shun, Metall. Mater. Trans. A 35A, 1465 (2004).
24. Miracle, D. B. & Senkov, O. N. A critical review of high entropy alloys and related concepts. Acta Mater. 122, 448–511 (2017).
25. Chen H-L, Mao H, Chen Q (2018) Database development and Calphad calculations for high entropy alloys: challenges, strategies, and tips. Mater Chem Phys 210:279–290
26. Martin AC, Oliveira JP, Fink C (2018) Elemental effects on weld cracking susceptibility in AlxCoCrCuyFeNi high entropy alloy. Not yet published
27. F. Otto, A. Dlouhy, Ch. Somsen, H. Bei, G. Eggeler, E.P. George, Acta Mater. 61, 5743 (2013)
28. W. Zhang, P.K. Liaw, Y. Zhang, Sci. China Mater. 61, 2 (2018)
29. Z. Wu, S.A. David, Z. Feng, H. Bei, Scr. Mater. 124, 81 (2016)
30. H. Nam, C. Park, C. Kim, H. Kim, N. Kang, Sci. Technol. Weld. Join. 23, 420 (2018)
31. N. Kashaev, V. Ventzke, N. Stepanov, D. Shaysultanov, V. Sanin, S. Zherebtsov, Intermetallics 96, 63 (2018)
32. M.G. Jo, H.J. Kim, M. Kang, P.P. Madakashira, E.S. Park, J.Y. Suh, D.I. Kim, S.T. Hong, H.N. Han, Met. Mater. Int. 24, 73 (2018)
33. Z.G. Zhu, Y.F. Sun, M.H. Goh, F.L. Ng, Q.B. Nguyen, H. Fujii, S.M.L. Nai, J. Wei, C.H. Shek, Mater. Lett. 205, 142 (2017)
34. B. Bagheri, M. Abbasi, JWJ 37, 243 (2019)
35. S. Park, C. Park, Y. Na, H. Kim, N. Kang, J. Alloy. Compd. 770, 222 (2019)
36. D. Shaysultanov, N. Stepanov, S. Malopheyev, I. Vysotskiy, V. Sanin, S. Mironov, R. Kaibyshev, G. Salishchev, S. Zherebtsov, Mater. Charact. 145, 353 (2018)
37. Z. Wu, S.A. David, Z. Feng, and H. Bei, Weldability of a High Entropy CrMnFeCoNi Alloy, Scr. Mater., 2016, 124, p 81–85.
38. H. Nam, C. Park, C. Kim, H. Kim, N. Kang, Sci. Technol. Weld. Joi. 23, 420 (2017)
39. Y. Zhang, Y.J. Zhou, J.P. Lin, G.L. Chen, and P.K. Liaw, Adv. Eng. Mater. 10, 534 (2008).
40. Sokkalingam, R.; Pravallika, B.; Sivaprasad, K.; Muthupandi, V.; Prashanth, K. G. Dissimilar welding of high-entropy alloy to Inconel 718 superalloy for structural applications. Journal of Materials Research (2021). https://doi.org/10.1557/s43578-021-00352-w
41. Zhang, Penglin; Qi, Yongfeng; Cheng, Qianqian; Sun, Xuemin Welding Dissimilar Alloys of CoCrFeMnNi High-Entropy Alloy and 304 Stainless Steel Using Gas Tungsten Arc Welding. Journal of Materials Engineering and Performance (2023). https://doi.org/10.1007/s11665-023-08229-1
42. Zhang, H. J.; Zhao, L. Y.; Liu, X.; Zhang, H.; Wang, Y. L.; Song, J. L. Micro-characteristics of High-Entropy Alloy Weld Formed by Friction Stir Lap Welding of 1060 Al and $Al_{0.1}$CoCrFeNi. Transactions of the Indian Institute of Metals (2020). https://doi.org/10.1007/s12666-020-02110-8
43. Martin, Alexander C.; Fink, Carolin Initial weldability study on $Al_{0.5}CrCoCu_{0.1}$FeNi high-entropy alloy. Welding in the World (2019). https://doi.org/10.1007/s40194-019-00702-7
44. Park, Sangwon; Nam, Hyunbin; Na, Youngsang; Kim, Hyoungseop; Moon, Younghoon; Kang, Namhyun Effect of Initial Grain Size on Friction Stir Weldability for Rolled and Cast CoCrFeMnNi High-Entropy Alloys. Metals and Materials International (2019). https://doi.org/10.1007/s12540-019-00466-1
45. Liu, Dejia; Guo, Rui; Hu, Yong; Shen, Mingxue; Tang, Yanchuan; Zhao, Longzhi; Li, Deying; Wang, Xiangjie Dissimilar Metal Joining of 304 Stainless Steel to SMA490BW Steel Using the Filler Metal Powders with a High-Entropy Design. Metals and Materials International (2019). https://doi.org/10.1007/s12540-019-00400-5

Chapter 8
Corrosion Behaviour of High Entropy Alloys

8.1 Introduction by Editor

Corrosion, an unavoidable and universal material challenge, can jeopardize the structural integrity, functionality, and lifespan of a crucial component. The investigation of corrosion behavior in high-entropy alloys focuses on how their complex composition and microstructure influence their susceptibility to various types of corrosion, including uniform, localized, and environmentally assisted corrosion. This study of the corrosion behavior of high-entropy alloys requires an integrative approach that incorporates elements of metallurgy, electrochemistry, materials characterization, and environmental science. Researchers strive to design high-entropy alloys with superior corrosion resistance tailored to specific environments ranging from marine and manufacturing processes to biomedical applications as they comprehend the convoluted interaction occurring at material surfaces and interfaces. This chapter consists discussion of the aqueous and chemical corrosion of high entropy alloys, their correlation with mechanical properties, and different processes to prevent corrosion such as coating and uniform phase distribution by annealing.

Corrosion is an unavoidable phenomenon that the metal undergoes to form a chemically stable state that i.e., metal oxide. The extent or the behavior of an alloy to undergo corrosion is material-dependent. Though the corrosion kinetics in a few of the metals or alloys may be low it is practically inevitable. The condition that an alloy undergoes corrosion may vary based on the nature or severity of applications ranging from temperature change, varying pH conditions, the flow rate of the fluid medium, loading conditions, etc. These conditions are not conventional or constant throughout the exposure. However, the extent of corrosion resistance can be predicted experimentally based on the severity of the corrosion medium.

Traditionally, conventional alloys in many instances have proven to be effective in sustaining the corrosion properties, but under constrained conditions, for, e.g. Stainless steel are excellent example of alloys that deliver significant corrosion resistance, however, can be used in limited applications. There exist no counterparts in

S. S. Nene (ed.), *High Entropy Alloys*, https://doi.org/10.1007/978-981-99-7173-2_8

structural alloys for the conventional Al, Ni, or Ti category. To cater to this issue, multi-principal element alloys here have entered this domain of material properties on a high note and have proven to address the issues effectively. High entropy alloys in the form of solid solutions tend to improve the alloy potential as compared to existing conventional alloys. These alloys tend to show improvement in the desired properties along with enhancement in their corrosion behavior.

A variety of HEAs we have seen before this chapter have proven to be an effective structural material surpassing multiple conventional alloy properties. On detailed investigation of corrosion behavior in HEAs, these alloys have successfully proven to display how these complex compositions and microstructures are prominent in resisting various types of corrosion including uniform, localized, and environmentally assisted corrosion in an alloy. Several new mechanisms in these cases were proven to be effective in the improvement of delivering corrosion resistance for the defined condition. In many instances, a small change by effective system selection, or change in processing condition was proven to be highly beneficial in varying the corrosion resistance of the targeted alloy. Limiting to the corrosion resistance alone, HEAs are exciting arena for researchers for developing next-generation materials with prolonged lifetime. This ultimately contributes to a confident approach to developing new alloys on a large scale and implied for large structural components even in actual applications.

This study of the corrosion behavior of high-entropy alloys as a structural component or as a coating requires an integrative approach that incorporates elements of metallurgy, electrochemistry, materials characterization, and environmental science. Researchers strive to design high-entropy alloys with superior corrosion resistance tailored to specific environments ranging from marine and manufacturing processes to biomedical applications as they comprehend the convoluted interaction occurring at material surfaces and interfaces.

To understand this, the following chapter is carefully designed by selecting the research articles from every specified class of alloys with the potential to deliver in prominent selected fields. For a variety of possibilities of corrosion, this review includes 3d-transition series HEAs, multiple Alx-series alloys, and Cu-containing HEAs. The corrosion resistance in the form of coatings or some surface treatments like laser claddings on these alloys are also reviewed here which shows the progressive potential of HEAs. This chapter primarily consists of a glimpse of the aqueous and chemical corrosion in high entropy alloys, and their correlation with alloy composition, phases formed, their grain morphology and its correlation with corrosion resistance. Well-defined techniques such as coating, cladding, and uniform phase distribution by deformation and annealing for improved corrosion resistance in the aqueous as well as acidic conditions are proven to be highly effective and are summarised briefly in the following chapter.

8.2 Machine Generated Summaries

Disclaimer: The summaries in this chapter were generated from Springer Nature publications using extractive AI auto-summarization: An extraction-based summarizer aims to identify the most important sentences of a text using an algorithm and uses those original sentences to create the auto-summary (unlike generative AI). As the constituted sentences are machine selected, they may not fully reflect the body of the work, so we strongly advise that the original content is read and cited. The auto generated summaries were curated by the editor to meet Springer Nature publication standards. To cite this content, please refer to the original papers.

Machine generated keywords: corrosion, ref, corrosion resistance, coating, corrosion behavior, nacl solution, layer, sample, film, surface, cocrfeni, laser, clad, nacl, increase content

Corrosion of High Entropy Alloys [91]

This is a machine-generated summary of:

Qiu, Yao; Thomas, Sebastian; Gibson, Mark A.; Fraser, Hamish L.; Birbilis, Nick: Corrosion of high entropy alloys [91]

Published in: npj Materials Degradation (2017)

Link to original: https://doi.org/10.1038/s41529-017-0009-y

If you want to cite the papers, please refer to the original.

For technical reasons we could not place the page where the original quote is coming from.

Abstract-Summary

"High entropy alloys represent a unique class of metal alloys, comprising nominally five or more elements in near equiatomic proportions."

"While such a single-phase structure can occur in unique systems, it is now appreciated that the definition of high entropy alloys can be broader, with systems comprising only four elements possible of forming single phases, and most five (or more) element systems actually being multi (> 2) phases."

"The notion of compositionally complex alloys is a more general description, with the concise review herein focusing on the corrosion of compositionally complex alloys (inclusive of high entropy alloys)."

Introduction

"High entropy alloys (HEAs) are systems that comprise five (but possibly more) principal metallic elements in near equiatomic ratios, forming one or more solid solution phases."

"[1] These alloys are termed HEAs because they have a high entropy of mixing compared to conventional alloys, thus notionally favouring the formation of solid solution phases."

"The five transition metal elements used to manufacture that HEA were found to exhibit a high degree of intersolubility to form a single FCC solid solution."

"The microstructures that may evolve in CCAs, and the accompanying elemental segregation is unlike that seen in most other engineering alloys, and the influence of such microstructures on passivity (and uniformity of dissolution and/or surface films) is largely unknown."

"Specifically the influence of alloying elements such as aluminium, titanium, chromium, molybdenum and nickel on corrosion of CCAs are examined."

Aqueous corrosion of HEAs

"The corrosion rate of CCAs with Al may therefore be expected to increase with increasing Al content in depassivating conditions or conditions that promote dissolution; however, the beneficial role of Al when incorporated into surface films for passive alloys (in passive conditions) is well documented in the case of Ni-based corrosion-resistant alloys (CRAs)."

"[2] The Cr concentration in Cr-containing CCAs is typically higher than 12%, and it is therefore plausible that Cr-containing CCAs have corrosion resistance (akin to Fe–Cr–Ni and Ni–Cr alloys)—should the hitherto unknown threshold for stimulation of a chromium oxide passive film in CCAs be fulfilled."

"Al and Cu (among other elements such as Co, Fe and Ni) tend to lower the corrosion resistance of Cr-containing CCAs."

"It is recognised that Ni may combine with Al to form the Al⁻Ni-rich B2 phase, which may also deteriorate the corrosion resistance of Ni-containing $Al_2CrFeCoCuTiNi_x$ alloys."

General discussion

"Undoubtedly, the present literature review has been able to identify some of the most, and least, corrosion-resistant CCAs and to relate their corrosion characteristics to their elemental composition and microstructure."

"The literature surveyed in the present review tended to support the fact that elements such as Mo, Cr and Ti serve to lower the corrosion rate of CCAs."

"The addition of elements such as Al, Cu and B was detrimental to the corrosion resistance of CCAs; however, the level of corrosion resistance alteration arising from Al additions was considered minor and not conclusive."

"Cr has a low solubility in Al (maximum solubility is around 0.77 wt.%) [3]; therefore, it is possible that several secondary phases or intermetallic compounds form in CCAs containing both Al and Cr, which could undermine the corrosion resistance of such CCAs."

Future Work

"Such local probe methods would enable one to monitor the influence of the heterogeneous microstructure on the corrosion of the CCAs, as revealing the distribution of anodes and cathodes at micrometre scale on the alloy surface."

"A structural alloy is usually typically tested for its resistance to atmospheric corrosion, pitting, crevice corrosion, intergranular corrosion, hydrogen embrittlement and stress corrosion cracking following standards such as the ASTM-G50 (practise for conducting atmospheric corrosion tests), ASTM-G48 (test of evaluating the pitting resistance of austenitic stainless steel), ASTM-A262 (for intergranular corrosion of stainless steel) etc. A similar matrix of tests for CCAs has not been presented and will be an important step in further highlighting the corrosion characteristics and for the qualification for any industrial applications."

"A systematic study on the corrosion of CoCrFeNi-based CCAs including $Al_xCoCrFeNi$, $CoCrFeNiCu_x$, $AlCoFeNiCr_x$, $CoCrFeNiTi_y$, $CoCrFeNiMo_x$ and $AlCoCrFeNi_x$ using rigorous electrochemical and surface characterisation could be explored as a first approach towards unravelling the corrosion characteristics of CCAs."

Conclusions

"The apparently good corrosion resistance of CCAs may be aided by most CCAs having Cr as one of their constituent (and often major) alloying elements (with > 15 wt.%".

"The possibility of Cr-based passive film forming upon CCAs akin to that formed upon stainless steels is possible; however, the chemical composition of films upon CCAs remains largely unexplored to date."

"The presence of elements such as Al, Cu and B was shown to possibly compromise corrosion resistance, even in Cr-containing CCAs, but this was dependent on the overall CCA composition—and not general in all cases."

"These complex interactions, while frustrating a general understanding, reveal that optimisation of CCAs for corrosion resistance is both possible and unexplored."

"The corrosion resistance demonstrated thus far for CCAs is promising and undoubtedly CCAs may play several important roles in specific industrial applications."

Influence of WC Particle on the Metallurgical, Mechanical, and Corrosion Behavior of AlFeCuCrCoNi-WC_x High-Entropy Alloy Coatings [92]

This is a machine-generated summary of:

Vyas, Akash; Menghani, Jyoti; Natu, Harshad: Influence of WC Particle on the Metallurgical, Mechanical, and Corrosion Behavior of AlFeCuCrCoNi-WC_x High-Entropy Alloy Coatings [92]

Published in: Journal of Materials Engineering and Performance (2021)

Link to original: https://doi.org/10.1007/s11665-021-05523-8

If you want to cite the papers, please refer to the original.

For technical reasons we could not place the page where the original quote is coming from.

Abstract-Summary

"The high-entropy alloy AlFeCuCrCoNi-WC_x (x = 5, 10, and 15 wt.%) multi-principal element claddings were synthesized with the aid of laser-assisted cladding process on AISI 316 steel substrate."

"Microstructural observations also conclude that increasing WC leads to refinement in the grain size."

"As compared to the substrate AISI 316 (164.5 $HV_{0.5}$), the microhardness of the AlFeCuCrCoNi-WC_5, AlFeCuCrCoNi-WC_{10} and AlFeCuCrCoNi-WC_{15} claddings is enhanced by approximately 49.6%, 73.5% and 79.7%, respectively."

"The corrosion resistance of the AlFeCuCrCoNi-WC_x multi-principal element coating in 3.5 wt.% NaCl corrosive environment is better than that of the substrate AISI 316."

"The corrosion rate of the high-entropy alloy coating increases and corrosion resistance is deteriorated with the rise in WC content."

Introduction

"Among the various compositions of coating materials, the high-entropy alloys (HEA) exhibit outstanding corrosion resistance and mechanical properties."

"The multiple elements with proper composition may offer high-entropy alloys with a wider range of applications due to high hardness and mechanical strength along with excellent wear resistance (Ref. [4]), distinctive electrical and magnetic properties (Ref. [5]), good erosion resistivity as well as excellent thermal stability (Refs. [6–8])."

"In many tribological applications, high-entropy alloy coatings are widely used because of their high hardness and excellent wear resistance."

"There are few investigations on laser-cladded high-entropy alloy coatings strengthened by reinforcing phase elements."

"Considering this as basis for the selection of present system and with aim to have excellent mechanical and corrosion properties simultaneously, WC was added in the high-entropy alloy system (Ref. [9])."

Materials and Methods

"The high input laser energy may lead to distortion of large size substrate during laser-cladding process if clamping is not done properly."

"The variation in the clad microstructure starting from the cladding zone followed by interface zone and substrate was observed by optical microscopy (Leica S8APO) and field emission scanning electron microscope (FESEM-JSM7100F), respectively."

"The microhardness of the cladding layers from the coating surface to the substrate was determined using a micro vickers hardness tester (Future Tech Corporation, Japan – FM 700) applying a load of 0.5 kg with a dwell time of 15 s. The distance between two progressive indentations was 200 μm."

"The surface area of $1 cm^2$ of cladded specimen and substrate AISI 316 was investigated for corrosion study."

"The field emission scanning electron microscopy (FESEM-JSM7100F) is used to observe the surface morphology of the corroded substrate as well as cladded specimen."

Results and Discussion

"With increasing in the amount of WC, there is no change in phase composition of high-entropy alloy, but more content of chromium carbide is formed in high-entropy alloy coating."

"As WC contribution in high-entropy alloy coating increased (10% and 15%), chromium-rich carbide form quite easily due to the availability of a high amount of carbon."

"The formation of chromium carbide is limited to the $AlFeCuCrCoNi\text{-}WC_{10}$, $AlFeCuCrCoNi\text{-}WC_{15}$ high-entropy alloy coatings due to the insufficient presence of carbon content in $AlFeCuCrCoNi\text{-}WC_5$ high-entropy alloy coatings to form chromium carbide."

"The reduction in corrosion resistivity of high-entropy alloy cladding with an increase in the contribution of WC is attributed to one or a combination of the following reasons."

"Increasing the weight fraction of WC in the high-entropy alloy cladding tends to form chromium carbide ($Cr_{23}C_6$) in the inter-dendritic zone."

Conclusion

"The attempt was made to synthesize $AlFeCuCrCoNi\text{-}WC_x$ high-entropy alloy coating on AISI 316 substrate."

"The influence of WC elements on their phase constituent, microstructure, micro-hardness and corrosion performance has been briefly investigated, and the major conclusions can be drawn as follows: a. The crystal structures of AlFeCuCrCoNi-WC_x high-entropy alloy coatings evolved from FCC + BCC + WC phases for X = 5% and FCC + BCC + WC + $Cr_{23}C_6$ phases for X = 10 and 15%, respectively."

"d. The corrosion resistance performance of the AlFeCuCrCoNi-WC_x HEA coating in 3.5 wt.% NaCl corrosive environment is better than that of the substrate AISI 316."

"The corrosion rate of the high-entropy alloy coating increases and corrosion resistance is degraded with the rise in WC contribution in HEA matrix."

Influence of Microstructure and Passive Film on the Corrosion Behavior of CoCrFeMnNi High-Entropy Alloy Prepared by Different Cooling Methods in a Simulated Alkaline Soil Solution [93]

This is a machine-generated summary of:

Zhu, M.; Zhao, B. Z.; Yuan, Y. F.; Guo, S. Y.; Pan, J.: Influence of Microstructure and Passive Film on the Corrosion Behavior of CoCrFeMnNi High-Entropy Alloy Prepared by Different Cooling Methods in a Simulated Alkaline Soil Solution [93]

Published in: Journal of Materials Engineering and Performance (2022)

Link to original: https://doi.org/10.1007/s11665-022-06966-3

If you want to cite the papers, please refer to the original.

For technical reasons we could not place the page where the original quote is coming from.

Abstract-Summary

"The influence of microstructure and passive film on the corrosion behavior of CoCrFeMnNi high-entropy alloy (HEA) prepared by different cooling methods in alkaline soil simulation solution was investigated by electrochemical measurements, immersion test, atomic force microscopy, and x-ray photoelectron spectrometry."

"The results show that the corrosion resistance of the HEA decreases with decreasing cooling rate, and the water-cooled sample exhibits the best anti-corrosion property."

"The corrosion behavior of the HEA is closely related to σ-phase, grain size, oxide, and passive film."

"σ-phase distributed at GBs and the oxide increases the susceptibility to pitting corrosion."

"The increasing number of σ-phase significantly reduces the corrosion resistance of HEA."

"Pits around GB regions and σ-phase are attributed to the selective dissolution of Fe, Co, and Cr elements."

Introduction

"The typical equiatomic CoCrFeMnNi HEA crystallizes into a single-phase face-centered cubic (FCC) solid solution (Refs. [10–13]) and has good mechanical properties and corrosion resistance, as well as excellent plasticity and toughness at room temperature and low temperatures."

"Li and others (Ref. [14]) studied the mechanical properties of an equiatomic CoCrFeMnNi HEA and revealed that the tensile strength increased from 601 to 649 MPa after selective laser melting (SLM) treatment."

"To explore the possible application of HEA in the pipeline engineering field, we studied the corrosion behavior of CoCrFeMnNi HEA interfered with by AC current in a simulated alkaline soil environment, and revealed that AC interference enhanced the corrosion of HEA, and resulted in an evolution of corrosion form (Ref. [15])."

Experimental

"For the Mott-Schottky curve, first, the passive films were grown on the surface of various HEA samples by potentiostatic polarization method at a constant passive potential of 0.65 V for 3 h. Then, the Mott-Schottky curve was tested at a frequency of 1 kHz in the range of − 1.0 V ~ 1.0 V with a step length of 50 mV. By this, the difference in the corrosion resistance of HEA specimens treated with various cooling methods can be clarified to a certain extent with respect to the protective ability of passive films."

"In the two tests, the passive films on the surface of different treated HEA samples were identical to those in the Mott-Schottky curve."

"The HEA samples treated by different cooling methods were immersed in the alkaline soil simulation solution for 336 h. At the end of the test, the corrosion products on the specimens were removed completely."

Result and Discussion

"The water-cooled sample with the lowest values of N_D and N_A suggests that its passive film is stable, dense, and protective (Ref. [16])."

"The high fluctuation at the transect line is slightly larger than the water-cooled sample, which tends to cause the accumulation of solution ions in the low areas, resulting in the breakdown of the passive film and promoting the initiation of pitting corrosion."

"The content of Cr oxide within the passive film of water-cooled sample is far higher than those of the other HEA samples."

"A higher content of Cr oxide was formed within the passive film of water-cooled sample."

"The passive film of water-cooled sample exhibits better protective property due to its minimum Mn oxide content."

"The passive film on the water-cooled sample has higher contents of p-type Cr_2O_3 (Ref. [17]), n-type Fe_2O_3 (Ref. [18]) and bound water, and lower contents of p-type Mn oxides and FeO (Ref. [19]), resulting in better protection of the matrix."

Conclusions

"The corrosion behavior of HEA prepared with different cooling methods is closely related to σ-phase, grain size, oxide, and surface passive film."

"Σ-phase distributed at grain boundaries and the oxide increases the susceptibility to pitting corrosion."

"As the cooling rate decreases, the size and number of Cr-rich σ-phase distributed at GB increase."

"For the water-cooled sample, its excellent corrosion resistance may be related to the stable structure and composition of passive film."

Microstructure, Mechanical Properties, Corrosion and Wear Behavior of High-Entropy Alloy AlCoCrFeNi$_x$ (removenotallowedtagthe) and Medium-Entropy Alloy (removenotallowedtag) [94]

This is a machine-generated summary of:

Sim, R. K.; Xu, Z.; Wu, M. Y.; He, A.; Chen, D. L.; Li, D. Y.: Microstructure, mechanical properties, corrosion and wear behavior of high-entropy alloy AlCoCrFeNi$_x$ (removenotallowedtagthe) and medium-entropy alloy (removenotallowedtag) [94]

Published in: Journal of Materials Science (2022)

Link to original: https://doi.org/10.1007/s10853-022-07256-y

If you want to cite the papers, please refer to the original.

For technical reasons we could not place the page where the original quote is coming from.

Abstract-Summary

"Medium-entropy alloys with fewer elements can also have superior performance with less structural complication."

"The Ni-free four-element medium-entropy alloy, AlCoCrFe, was examined and compared with its five-element high-entropy companion, AlCoCrFeNi."

"It was observed that the medium-entropy alloy was harder and had higher resistances to wear and corrosion than the five-element alloy."

"Detailed microstructural analyses demonstrated that the medium-entropy alloy AlCoCrFe had a spinodal-decomposed nanostructure, which grew into a two-phase microstructure with phase transformations in the five-element high-entropy alloy."

"The medium-entropy alloy showed certain advantages with superior properties while having less complicated microstructure features."

Introduction

"We investigated a medium-entropy alloy, AlCoCrFe with $\Delta S_{conf} = 1.39R$, and compared it to its five-element high-entropy companion, $AlCoCrFeNi_x$ ($x = 0.5$–2.0), to see how the microstructure and corresponding properties of the alloy, $AlCoCrFeNi_x$, change with increasing the Ni content from $x = 0$ to $x = 2.0$."

"The present study is conducted to comprehensively investigate how the Ni element affects various properties that are of importance to industrial applications of the alloy, including mechanical properties, wear, corrosion and corrosive wear."

"Besides, there is a lack of understanding the anti-corrosion and wear properties of this high-entropy alloy with respect to the Ni content and its medium-entropy companion for comparison purposes."

"The medium-entropy AlCoCrFe alloy was investigated and compared to the high-entropy AlCoCrFeNi, in order to make use or effective use of the Ni-free four-element medium-entropy alloy."

Experimental method

"The TEM samples were prepared by slicing the cast alloys, then grinding both sides of each sample to a thickness of 100 μm or less, followed by electro-polishing."

"Compression tests were performed for the alloys under study."

"Compressive stress–strain curves of $AlCoCrFeNi_x$ alloys with different Ni contents were obtained at a strain rate of 0.3 mm/min using a 313 Electromechanical Universal Test Machine (TESTRESOURCES, USA)."

"The samples for the compressive tests had a rectangular cross section of 4×4 mm^2 and 7 mm in height."

"Wear performances of the $AlCoCrFeNi_x$ alloys in dry condition and in the acidic H_2SO_4 ($pH = 3$) solution were evaluated, respectively, using a pin-on-disc tribometer (CSEM Instruments)."

"The dry wear tests were performed under two conditions: one was performed in the ambient condition and the other was performed with an air flow to cool the sample surface in order to minimize oxidation caused by the frictional heat."

Results

"The Ni-free medium-entropy alloy, AlCoCrFe, appears to be homogeneous, while two phases or domains (dark and light) show up and become more distinctive with increasing the Ni content."

"In order to investigate how the AlCoCrFe medium-entropy alloy performs during corrosive wear and how the Ni content influences the performance of high-entropy alloy $AlCoCrFeNi_x$ under the same corrosive wear condition, corrosive wear tests were performed for all the alloys in the dilute acidic solution ($pH = 3$)."

"Different from their performances during the dry wear tests, AlCoCrFe showed the lowest resistance to corrosive wear, while Ni-containing alloys performed better."

"A possible reason for the lower resistance of AlCoCrFe to corrosive wear, compared to the Ni-containing alloys, could be ascribed to corrosion-enhanced local cracking in this less tough alloy."

Conclusions

"The following conclusions can be drawn: (a) The medium-entropy alloy, AlCoCrFe, shows a spinodal-decomposed nanostructure consisting of a Cr-Fe-rich disordered BCC phase and an Al-Ni-rich ordered BCC one (B2)."

"In AlCoCrFeNi2.0, nanoparticles of 10–20 nm in diameter are observed inside the B2 phase, which have a higher Cr content than the surrounding B2 matrix. (b) The medium-entropy alloy is harder than the five-element high-entropy alloy, which should be attributed to its spinodal-decomposed nano-scaled disordered and ordered BCC phases, generating numerous interfacial barriers to the movement of dislocations."

"The wear resistance decreases with Ni added, resulting from the decrease in hardness due to the microstructure coarsening and the transformation of the disordered BCC into the softest FCC phase."

"The correspondingly lowered fraction of passive elements, Cr and Al, in the alloy due to the increased Ni content might also influence the protective role of the passive film. (d) Although the medium-entropy alloy shows the lowest wear and corrosion, it performs the poorest during corrosive wear tests."

Investigation on the Microstructure, Wear and Corrosion Resistance of FeCoNiCrMo$_x$ High-Entropy Alloy Coatings Deposited on 40Cr by Laser Cladding [95]

This is a machine-generated summary of:

Zhou, Chao; Wang, Shouren; Wang, Gaoqi; Zhang, Jianpeng; Liu, Wenlong: Investigation on the microstructure, wear and corrosion resistance of FeCoNiCrMo$_x$ high-entropy alloy coatings deposited on 40Cr by laser cladding [95]

Published in: Journal of Materials Science (2022)

Link to original: https://doi.org/10.1007/s10853-022-07684-w

If you want to cite the papers, please refer to the original.

For technical reasons we could not place the page where the original quote is coming from.

Abstract-Summary

"To improve the wear resistance and corrosion resistance of 40Cr, the FeCoNiCrMo$_x$ ($x = 0, 0.3, 0.6, 0.9, 1.2$) HEAs coatings were deposited on the surface of 40Cr by laser cladding technology, and the phase structure, microstructure and hardness of the coatings were investigated."

"Friction and wear tests, electrochemical tests and neutral salt spray tests were conducted to investigate the effect of Mo content on the wear resistance and corrosion resistance of the coatings."

"With the increasing content of Mo elements, solid solution strengthening, fine grain strengthening, and second phase strengthening occurred in the coatings, which led to the increased hardness and wear resistance of the coatings."

"With the increase of Mo content, the corrosion potential of the coatings drifted positively, the corrosion current density decreased, and the corrosion resistance increased."

"With excessive Mo element content, the σ phase will precipitate, leading to galvanic coupling corrosion of the coatings and reducing the corrosion resistance."

"It had the best corrosion resistance."

"The neutral salt spray test result also showed that the corrosion resistance of the $FeCoNiCrMo_{0.9}$ coating was the best."

Introduction

"The results showed that the surface of FeCoNiCrMo HEAs coatings generates stable MoO_3 and Cr_2O_3, which could effectively enhance the corrosion resistance of TC4 alloy."

"To enhance the corrosion resistance of civil commercial engine blades, Li and others [20] prepared FeCoCrNi and $FeCoNiCrMo_{0.2}$ HEAs coatings on the surface of 304 stainless steel by laser cladding technology."

"It is important to determine the content range of Mo in HEAs coatings to improve the corrosion resistance and wear resistance of the coatings."

"In order to investigate the tribocorrosion performance of the structural engineering parts, the $FeCoNiCrMo_x$ ($x = 0, 0.3, 0.6, 0.9, 1.2$) HEAs coatings with different Mo contents deposited on 40Cr substrate by laser cladding technology were fabricated."

"The effect of Mo element content on the wear and corrosion resistance of the $FeCoNiCrMo_x$ HEAs coatings were analyzed in this paper."

Materials and Methods

"China) was used to test the polarization curves of the substrate and the $FeCoNiCrMo_x$ HEAs coatings in 3.5%NaCl solution and calculated the corresponding corrosion parameters to analyze the corrosion resistance of the substrate and coatings."

"The neutral salt spray test of the substrate and the $FeCoNiCrMo_x$ HEAs coatings were carried out in the salt spray test chamber (YWX/Q-150, Nanjing Jinling Huanshi Instrument Equipment Co., Ltd.) further to analyze the corrosion resistance of the substrate and the coatings."

"To avoid the influence of the non-coatings surface on the experimental results, the non-coatings surface of the specimens should be sealed with the sealant, leaving only the coatings surface for the test."

"The neutral salt spray test process was static at room temperature, and the test time were 72 h and 168 h. After the salt spray test, the specimens were removed from the salt spray chamber."

Results and Discussion

"The improvement of the wear resistance of the coating is mainly due to the increasing content of Mo elements in the coating, which leads to the enhancement of solid solution strengthening, fine grain strengthening and the precipitation of the σ phase as a hard phase in the structure of the coatings, which together further increase the hardness of the coatings and thus improve the wear resistance of the coatings."

"With the addition of the Mo element, the solid solution strengthening, fine grain strengthening and second phase strengthening of the coatings are also enhanced, increasing the hardness and wear resistance of the coatings."

"Due to the addition of a small amount of Mo element, the corrosion resistance of the $FeCoNiCrMo_{0.3}$ coating is improved compared with the substrate and the $FeCoNiCrMo_0$ coating."

Conclusion

"The influence of Mo content on the wear and corrosion resistance of the coatings was investigated."

"With the increase of Mo content, the phase composition of the $FeCoNiCrMo_x$ HEAs coatings changed and consisted of the FCC phase and σ phase. (2) With the increased Mo content, the wear resistance of the $FeCoNiCrMo_x$ HEAs coatings were improved, and the wear mechanism changed."

"The $FeCoNiCrMo_{1.2}$ coating had the best wear resistance. (3) Compared with the substrate, the corrosion resistance of the $FeCoNiCrMo_x$ HEAs coatings was improved with the increased Mo content."

"When the Mo content was too much, galvanic corrosion would occur in the coatings, resulting in the corrosion resistance of the coatings decreased."

"The $FeCoNiCrMo_{0.9}$ coating had the highest self-corrosion potential, the lowest corrosion current density and the best corrosion resistance. (4) With the increase of the experimental time of the neutral salt spray test, the corrosion of the substrate and the $FeCoNiCrMo_x$ HEAs coatings gradually aggravated."

Effects of Nickel Content on the Microstructure, Microhardness and Corrosion Behavior of High-Entropy AlCoCrFeNix Alloys [96]

This is a machine-generated summary of:

López Ríos, M.; Socorro Perdomo, P. P.; Voiculescu, I.; Geanta, V.; Crăciun, V.; Boerasu, I.; Mirza Rosca, J. C.: Effects of nickel content on the microstructure, microhardness and corrosion behavior of high-entropy AlCoCrFeNix alloys [96]

Published in: Scientific Reports (2020)

Link to original: https://doi.org/10.1038/s41598-020-78108-5

Copyright of the summarized publication:

The Author(s) 2020

If you want to cite the papers, please refer to the original.

For technical reasons we could not place the page where the original quote is coming from.

Abstract-Summary

"The effect of three different nickel concentration on the microstructure, hardness and corrosion properties of high entropy alloys (HEAs) from AlCrFeCoNi system as an alternative material for medical instruments fabrication was investigated."

"The analyzed HEAs were AlCrFeCoNix obtained by vacuum arc remelting from high purity raw materials and having nickel atomic ratio x = 1.0, 1.4 and 1.8."

"The corrosion properties of the synthesized HEAs were evaluated using a potentiodynamic polarization method."

"The low corrosion rates, low corrosion currents and high polarization resistance attest the good stability of these HEAs in simulated biological environment indicating their possible use for surgical and dental instruments."

Introduction

"Due to the great advance of science and technology, new metallic alloys containing more than 2 base metals, with a different metallurgical concept, have been recently explored [21, 22]."

"High-entropy alloys (HEAs) are one of the most promising results of the exploration of new chemical compositions for metallic materials with improved performance [23–25]."

"Although many interesting topics have been explored, only few studies deal with corrosion properties of this high-entropy alloy, in general depending of fabrication method: if is synthesized by laser additive [26], by electrospark process [27] and by spark plasma sintering with pre-alloy powders obtained through gas atomization [28]."

"To be able to use the high-entropy alloys for the manufacturing of medical instruments (like cutters, saws, scalpels etc.), their mechanical characteristics and corrosion resistance in physiological fluids which contain 1% wt."

"The effect of three different nickel concentrations on the microstructure, hardness and corrosion properties of high entropy alloys from AlCrFeCoNi system has been investigated."

Experimental

"All the measurements were performed in Ringer Grifols solution (from Grifols Laboratories, Barcelona, Spain) with the following composition in mmol/l: Na^+ 129.9; K^+ 5.4; Ca^{2+} 1.8; Cl^- 111.7 and $C_3H_5O_3^-$ 27.2."

"Open circuit potential measurements during 3 days were performed, followed by potentiodynamic polarization measurements."

"All tests were performed three times and data were processed using EC-LAB software."

"In order to calculate the Tafel slopes for the partial anodic processes (b_a), and the Tafel slopes for the partial cathodic processes (b_c), the linear polarization curves have been shifted from $E_{OCP} - 150$ mV to $E_{OCP} + 150$ mV using a scanning rate of 10 mV/s. The polarization studies to evaluate the passivation process continued with measurements from − 800 mV (vs.SCE) to + 500 mV (vs. SCE), increasing the potential at a scanning rate of 1 mV/s. SP-150 potentiostate was used to perform the tests and data were processed using EC-LAB software, both from BioLOGIC Science Instruments."

Results and Discussions

"In the case of Ni_1 (AlCrFeCoNi) TT sample, the chemical microanalysis on light grey phase indicates that it contains about 5.08 at% Al, 9.14 at% Co, 31.76 at% Cr, 29.80 at% Fe and 20.51 at% Ni."

"In the case of $Ni_{1.8}$ ($AlCrFeCoNi_{1.8}$) TT sample, the light grey phase contains 21.73 at% Al, 5.31 at %Co, 9.85 at% Cr, 12.82 at% Fe and 48.88 at% Ni."

"After the thermal treatment, the elemental composition of HEAs does not change; however, after the corrosion stage, the Al concentration in the surface region significantly dropped, followed by Ni; Co concentration went up, while Fe and Cr did not change much."

"For other HEA system, substitution of Al with different Zr concentrations, determine changes in the microhardness values, attributed to the phase changes in the alloy structure [29]."

Conclusions

"The annealing determined a more uniform distribution of the phases for the three high-entropy alloys and the modification of the morphology of the grain boundaries."

"It also resulted in a significant increase of the grain sizes."

"The formation after the annealing treatment of an Al-Ni compound is hypothesized, which should be corroded faster than the main HEA phase."

"The lattice parameter of the cubic phase varies very slightly, from 2876 to 2870 Å by increasing the Ni content, but the grain size decreases considerably (from 398 to 206 Å) for as-cast samples."

"For the FCC phase, the lattice parameters also vary very slightly but the grain size increases with the nickel content for both as-cast and annealing samples."

Investigation of Microstructure, Mechanical and Corrosion Resistance Properties of Cu–Al–Zn–Sn High-Entropy Alloy for Kitchen Utensils and Other Industrial Applications [97]

This is a machine-generated summary of:

Ayuba, Samuel Ushe; Fayomi, Ojo Sunday Isaac; Omotosho, Olugbenga Adeshola: Investigation of microstructure, mechanical and corrosion resistance properties of Cu–Al–Zn–Sn high-entropy alloy for kitchen utensils and other industrial applications [97]

Published in: Chemical Papers (2022)

Link to original: https://doi.org/10.1007/s11696-022-02336-4

If you want to cite the papers, please refer to the original.

For technical reasons we could not place the page where the original quote is coming from.

Abstract-Summary

"A sample from aluminium (Al) material and samples of Cu–Al–Zn–Sn high entropy alloys were produced using the stir casting process."

"The potentiodynamic polarization method was used to study the Corrosion Rate (CR) of the samples in 3.65 wt.% NaCl solution, while the CETR reciprocating sliding tribometer was used to examine the Wear Rate (WR) following ASTM G99 standard."

"For the aluminium material sample the CR, WR, VH and MTS were 0.9327 mm/year, 1.888×10^{-5} mm^3/N/m, 56.25 mN/mm^2 and 610 °C, respectively."

"The CR of the aluminium sample indicated significant deterioration of the metal's active sites."

"The best performing sample; HEA 1 sample exhibited CR and MTS of 0.3118 mm/year and 720 °C, respectively, relatively lower WR of 1.583×10^{-5} mm^3/N/m and comparatively higher VH of 478.44 mN/mm^2, compared to the other samples."

"The CR, WR, VH and MTS of the HEA samples indicated the passivating and high strengthening of the Cu–Zn–Sn constituent of the HEAs."

Introduction

"There is a paradigm shift to high entropy alloys (HEAs) in recent times, due to their inherent microstructural properties and many other impressive properties, such as high hardness and strength, outstanding corrosion resistance, fatigue, fracture, thermal stability and irradiation resistance, in terms of which they engulf the conventional alloys with several potential applications (Zhang and others [30, 31])."

"The outstanding properties such as improved strength-ductility, enhanced fatigue resistance, high thermal stability and high fracture toughness make HEAs attractive in several manufacturing fields, particularly as advanced structural alloys, which are enormously demanded in intense service environments such as in the aerospace, turbine, and nuclear industries (Santodonato and others [32]; Seifi and others [33]).”

“This present work examination of mechanical, microstructure and corrosion resistance properties of Cu–Al–Zn–Sn high-entropy alloy for kitchen utensils and other industrial applications.”

Materials and Methods

“The stir casting process was used for the melting of the metallic materials.”

“The wear properties of the alloy samples were studied using a tribometer of 8.1.8 version following ASTM G99 standard.”

“The wear properties were investigated using normal loads of 10 N at the laboratory temperature of 19.67 °C, the target temperature of 28 °C, the humidity of 30.24%, frequency of 15 Hz and linear speed of 0.05 cm/s. With the tribometer, parameters such as sample wear rate (mm^3/N/m), friction coefficient (μ) and worn track section (μm^2) were observed.”

“A nanoindenter was used to examine the micro-hardness of the high entropy alloy samples at the frequency of 400.0 Hz following ASTM A-370 procedures.”

Results and Discussion

“The CONTROL 1 sample also exhibited the highest polarization resistance of 522.75 Ω. The unique corrosion properties of this HEA sample could be ascribed to the higher percentage composition of tin (Sn), which synergistically coalesced with the other elements to enhance the passivation potential of the alloy developed.”

“It is also worthy of note that the HEA samples exhibited higher values of corrosion potential (Ecorr) compared to the corrosion potential of CONTROL 2, indicating that the metallic alloys had a more anodic corrosion protection effect due to their ability to form passive layers on their surfaces, which are stable against the anodic dissolution in the 3.65 wt% NaCl medium (Fayomi and others [34]).”

Conclusions

“(i) The corrosion rate of the best performing sample; HEA 1 ($Cu_{55}Al_{35}Zn_5Sn_5$) was 0.3118 mm/year, while that of the aluminium material was 0.9327 mm/year.”

“(ii) The aluminium material sample exhibited a wear rate and Vickers hardness of 1.888×10^{-5} mm^3/N/m and 56.25 mN/mm^2, respectively, while the best performing wear-resisting sample; HEA 1 ($Cu_{55}Al_{35}Zn_5Sn_5$) exhibited the wear rate and Vickers hardness of 1.583×10^{-5} mm^3/N/m and 478.44 mN/mm^2, respectively.”

“(iii) The HEA 1 ($Cu_{55}Al_{35}Zn_5Sn_5$) sample exhibited the maximum temperature of stability of 720 °C, the maximum temperature of stability for the aluminium material sample was 610 °C.”

“The SEM morphology and the unique XRD crystallites in the HEA samples can be attributed to the chemical and microstructural homogeneity, texture quality, stability and high entropy mixing of the constituent elements and low porosity of samples.”

Corrosion Behavior in Nitric Acid Solution and Tensile Properties of $(CuFeNiMn)_{1-x}Cr_x$ HEAs [98]

This is a machine-generated summary of:

Chen, Xu; Hu, Jiaxuan; Liu, Ye; Xiang, Feng: Corrosion Behavior in Nitric Acid Solution and Tensile Properties of $(CuFeNiMn)_{1-x}Cr_x$ HEAs [98]

Published in: Metals and Materials International (2020)

Link to original: https://doi.org/10.1007/s12540-020-00887-3

If you want to cite the papers, please refer to the original.

For technical reasons we could not place the page where the original quote is coming from.

Abstract-Summary

"The empirical prediction on the phase formation in as-cast $(CuFeMnNi)_{1-x}Cr_x$ alloys was consistent with the previous experimental results, which indicated that the BCC phase formed in the alloys was in the thermodynamically unstable state."

"The corrosion resistance of the alloys in 2 wt% HNO_3 solution was increased with the increased Cr content which was mainly related to the formation of large anode and small cathode, increasing total volume fraction of Cr-rich FCC1 and BCC phases, the more uniform distribution of Cu and decreasing relative Cu content."

Introduction

"High-entropy alloys (HEAs) had been investigated owing essentially to the unexpected microstructures and excellent properties since the alloys were reported [35–39]."

"The microstructure characteristic of CuCrFeNiMn HEAs had been studied by Ren and others [40, 41], and it was found that the alloys had FCC phase or FCC + BCC phases."

"Hyun and others [42] revealed that the as-cast CuCrFeNi alloy contained FCC phase, and the microstructure transformed from FCC phase to FCC1 + FCC2 + BCC phases when Mn element was added."

"The phase formation, microstructures, mechanical properties and tensile fracture morphologies of the $Fe_{0.4}Cr_{0.4}NiMn_xCu$ ($x < 1.4$) HEA were investigated systematically [43]."

"The influence of Cr element on corrosion resistance in 2 wt% HNO_3 solution and mechanical properties of the $(CuFeNiMn)_{1-x}Cr_x$ alloys ($x = 0$–0.25) was systematically studied in this work with a purpose to fully understand the corrosion behavior and the tensile properties of CuCrFeMnNi HEAs."

Experimental Process

"The $(CuFeNiMn)_{1-x}Cr_x$ alloy which x from 0 increased to 0.25 (at.%) were prepared by means of standard arc-melting technique."

"The alloys for electrochemical tests were cut with the size of 1 cm × 1 cm × 0.5 cm, and then the specimens were mounted by epoxy."

"The exposed surfaces in solution were ground and then cleaned by alcohol and deionized water carefully before electrochemical experiment."

"Potentiodynamic polarization and electrochemical impedance spectroscopy (EIS) were carried out at room temperature in a typical three-electrode cell, which consisted of reference electrode (saturated calomel electrode (SCE)), working electrode (specimen) and auxiliary electrode (platinum sheet)."

"The morphologies of those surfaces for the specimens were observed at once by SEM (AJSM-6360LV)."

"The morphology of fracture surfaces for tensile samples was characterized by SEM (ZEISS-EVO MA10)."

Results

"Previous experimental researches had found that FCC1 and FCC2 phases were formed when the Cr content was 0%, 5% and 10% (at.%), while FCC1, FCC2 and BCC phases were obtained in 15%Cr, 20%Cr and 25%Cr alloys [44]."

"The values of VEC in the as-cast $(CuFeMnNi)_{1-x}Cr_x$ alloys were higher than 8.0, indicating that FCC1 and FCC2 phases could existed stably."

"The corrosion rate and I_{corr} decreased with the increasing Cr, indicating the corrosion resistance was generally strengthened."

"When the alloy contained 5 at.% Cr, its corrosion resistance was significantly improved, which was almost an order of magnitude higher."

"This indicated that the corrosion resistance was enhanced by increasing Cr, which was consistent with the potentiodynamic polarization curve results."

"The ultimate tensile strength increased dramatically to 1017 MPa and the yield strength increased to 489 MPa in the as-cast 25%Cr alloy."

Discussions

"That Cr together with Fe and Ni were mainly segregated in the DR to form the FCC1 phase in the 5%Cr and 10%Cr alloys, and Cu element coupled with Mn were mainly segregated in the ID to form the FCC2 phase."

"The volume fraction of Cr-rich FCC1 phase decreased while that of the Cr-rich BCC phase increased as the Cr content increased."

"As the Cr content increased, the Cr-rich corrosion-resistant BCC phase appeared and the potential of the cathode (DR) increased [45]."

"When the Cr content increased from 0 to 10%, the volume fraction of FCC1 phase (DR) decreased while the FCC2 phase (ID) increased, indicating that the large anode and the small cathode were obtained and weaken the galvanic corrosion."

"When the content of Cr was more than 10%, the total volume fraction of BCC + FCC1 phases (DR) increased while that of the FCC2 phases (ID) decreased."

Conclusions

"The corrosion behavior of as-cast $(CuFeMnNi)_{1-x}Cr_x$ alloys was investigated by means of electrochemical impedance spectroscopy measurements and potentiodynamic polarization measurements which conducted in 2 wt% HNO_3 solution at room temperature."

"The $(CuFeMnNi)_{0.75}Cr_{0.25}$ alloy displayed a better corrosion resistance, however, the CuFeMnNi alloy without Cr exhibited the worst corrosion resistance in nitric acid solution, indicating the corrosion resistance was enhanced with the increasing Cr content."

"The strength properties for the investigated alloys increased while the elongation-to-fracture decreased as Cr increased."

"The empirical prediction on the phase formation of the as-cast $(CuFeNiMn)_{1-x}Cr_x$ alloys agreed with the previous experimental results."

Effect of Copper on Mechanical Properties and Corrosion Behavior of Powder Metallurgy Processed Ni–Co–Cr–Fe–Mn–Cu_x High Entropy Alloy [99]

This is a machine-generated summary of:

Veerappan, G.; Ravichandran, M.; Mohanavel, V.; Pritima, D.; Rajesh, S.: Effect of Copper on Mechanical Properties and Corrosion Behavior of Powder Metallurgy Processed Ni–Co–Cr–Fe–Mn–Cu_x High Entropy Alloy [99]

Published in: Arabian Journal for Science and Engineering (2022)

Link to original: https://doi.org/10.1007/s13369-022-07041-8

If you want to cite the papers, please refer to the original.

For technical reasons we could not place the page where the original quote is coming from.

Abstract-Summary

"The compacted specimens were subjected to two different sintering temperatures such as 1200° and 1300 °C, and further, alloys were subjected to scanning electron microscopy (SEM) and optical microscopy test."

"The sintered specimens were also subjected to compression test, hardness test and corrosion test."

"$Ni_{5.26}$–$Co_{5.23}$–$Cr_{5.94}$–$Fe_{5.53}$–$Mn_{5.26}$–Cu (4 at.% Cu) alloy sintered at 1300 °C exhibited high compression strength, high corrosion resistance and high hardness."

"Potentiodynamic investigations pointed out that corrosion resistance for $Ni_{3.39}$–$Co_{3.37}$–$Cr_{3.83}$–$Fe_{3.56}$–$Mn_{3.62}$–Cu alloy (6 at.% Cu) sintered at 1300 °C is highest with corrosion rate of 0.22 Mils per year (mpy)."

Introduction

"To typical alloys, HEAs have excellent thermal stability, fatigue, wear, and corrosion resistance, and increased hardness, as well as superior high and cryogenic temperature properties, according to recent studies [46]."

"They serve as high temperature materials, high hardness materials, wear resistance, and corrosion-resistant materials [47]. The constraints in casting route for producing alloys or composites are difficulty in promoting the reinforcement distribution in matrix and achieving uniform microstructure."

"At high operating temperatures and under long exposure, nickel based chromium super alloy exhibits excellent mechanical strength, creep resistance, good corrosion resistance and oxidation resistance."

"Al–Cr–Fe–Ni–Co–Cu high entropy alloy was synthesized using powder metallurgy route to investigate the mechanical properties, microstructure and its corrosion resistance."

"Mammundi Azaath and others synthesized Cu-Ni based alloy reinforced with silicon carbide using PM route and it was found that 4% reinforcement exhibits better corrosion resistance which was sintered at 750 °C [48]."

Materials and Methods

"Elemental powders of nickel, cobalt, chromium, iron, manganese and copper were ball milled in a sealed container containing hardened steel ball (dia = 15 mm; AISI 4140, ball to powder = 10:1) using ball milling for 3 h [49]."

"The green compacts were sintered at two different temperatures such as 1200 °C, and1300 °C."

"The conditions for both the sintering process are same."

"Sintered compacts of Ni–Co–Cr–Fe–Mn–Cu_x HEAs were investigated for morphology and microstructure by using SEM."

"The compressive strength of the sintered specimens was carried out using universal testing machine using ASTM E9 standard [50, 51]."

Results and Discussions

"The change in percentage of density after sintering the alloys, namely HEA C1 (Cu 0 at.%), HEA C2 (2 at.% Cu), HEA C3 (4 at.% Cu), and HEA C4 (6 at.% Cu) which was sintered at 1200 °C are 2.78, 2.89, 2.87 and 3.39, respectively."

"HEA C4 (6 at.% Cu) sintered at 1300 °C able to achieve near density of 7.94, i.e., > 90% $\rho_{th\ de}$."

"Cu%) sintered at 1300 °C has the higher corrosion rate of 0.71MPY and possesses the higher corrosion resistance of HEA C1 (0 at.% Cu) alloy sintered at 1200 °C."

"The corrosion rate for the prepared HEA alloys such as HEA C1 (Cu 0 at.%), HEA C2 (2 at.% Cu), HEA C3 (4 at.% Cu), and HEA C4 (6 at.% Cu) which was sintered at 1200 °C are 0.71, 0.48, 0.37 and 0.22, respectively."

Conclusions

"The change in percentage of density after sintering the alloys, namely HEA C1 (Cu 0 at.%), HEA C2 (2 at.% Cu), HEA C3 (4 at.% Cu), and HEA C4 (6 at.% Cu) which was sintered at 1200 °C are 2.78, 2.89, 2.87 and 3.39, respectively."

"The change in percentage of density after sintering the alloys, namely HEA C1 (0 at.%), HEA C2 (2 at.%), HEA C3 (4 at.%), and HEA C4 (6 at.%) which was sintered at 1300 °C are 3.45, 3.28, 3.13 and 3.79, respectively."

"HEA C4 alloy sintered at 1300 °C was able to achieve near density of 7.94 g cm^{-3}."

"HEA C3 alloy sintered at 1300 °C possesses highest compressive strength of 763 MPa and highest hardness."

"HEA C4 alloy sintered at 1300 °C exhibits better corrosion resistance than 1200 °C sintered alloy."

Addition Al and/or Ti Induced Modifications of Microstructures, Mechanical Properties, and Corrosion Properties in CoCrFeNi High-Entropy Alloy Coatings [100]

This is a machine-generated summary of:

Ma, Guoliang; Zhao, Yong; Cui, Hongzhi; Song, Xiaojie; Wang, Mingliang; Lee, Kwangmin; Gao, Xiaohua; Song, Qiang; Wang, Canming: Addition Al and/or Ti Induced Modifications of Microstructures, Mechanical Properties, and Corrosion Properties in CoCrFeNi High-Entropy Alloy Coatings [100]

Published in: Acta Metallurgica Sinica (English Letters) (2021)

Link to original: https://doi.org/10.1007/s40195-021-01219-z

If you want to cite the papers, please refer to the original.

For technical reasons we could not place the page where the original quote is coming from.

Abstract-Summary

"The purpose of this work is to study the influences of Al and/or Ti addition on the microstructures, mechanical properties and corrosion properties of CoCrFeNi-(Al, Ti) high entropy alloy (HEA) coatings."

"Three coatings, AlCoCrFeNi (I), $CoCrFeNiTi_{0.5}$ (II) and $AlCoCrFeNiTi_{0.5}$ (III), were fabricated by laser cladding successfully."

"The AlCoCrFeNi (I) coating exhibited a simple body-centered cubic (BCC) solid-solution structure, whereas the $CoCrFeNiTi_{0.5}$ (II) alloy exhibited a face-centered cubic (FCC) solid-solution and a small amount of Laves phase."

"The $AlCoCrFeNiTi_{0.5}$ (III) coating with dual-phase BCC structure showed the optimal performance of both mechanical and corrosion properties, which was superior to BCC-based AlCoCrFeNi (I) and FCC-based CoCrFeNi$Ti_{0.5}$ (II) coatings."

Introduction

"The wear resistance of HEAs has large improvement with Al-content increase due to high hardness of Al-rich BCC phase."

"The multi-phase structures of AlCoCuFeNi-based HEAs [52] with Cr, Ti addition are corroded easily, and the Ti element trend to promote the formation of FCC solid-solution phases."

"The lightweight single-phase AlTiVCr HEAs [53] were designed to include more light-weight alloy elements, Al and Ti, and were observed to have a noble pitting corrosion resistance."

"Motivated by above previous reports of the influences of Al and/or Ti additions on the microstructures, mechanical and corrosion properties of (CoCrCuFeNi)-based HEAs, in this study, three HEAs, AlCoCrFeNi (I), CoCrFeNi$Ti_{0.5}$ (II), AlCoCrFeNi$Ti_{0.5}$ (III) prepared by laser cladding were chosen to investigated, aiming at achieving a better understanding of the impact of adding Al and/or Ti elements, and microstructure characteristics on the corrosion and mechanical properties of CoCrFeNi-(Al, Ti) HEA coatings."

Experimental

"The crystal structures of HEA coatings were analyzed by X-ray diffraction (XRD) (Model: D/Max 2500 PC Rigaku, Japan) with a Cu Kα radiation at 40 kV and 100 mA, scanning range 2θ from 20° to 100° with a scanning rate of 4°/min."

"The mechanical properties of the coatings were also characterized via nano-indentation tester (Model: UNHT CSM, Switzerland) equipped with a Berkovich tip in the depth-control mode with a maximum loading of 30 mN at a load rate of 1 mN/s. 15 indentations were carried out to each phase of samples to verify the accuracy of the indentation data."

"The electrochemical tests (potentiodynamic polarization and electrochemical impedance spectroscopy (EIS)) in 3.5 wt% NaCl solution at room temperature were investigated by an electrochemical workstation (Model: REF3000-32,117, Gamry) via a classical three-electrode system, a saturated calomel electrode (SCE) as the reference electrode, a Pt sheet as the counter electrode, and coatings with a working surface of 1 cm^2 as the work electrode."

Results and Discussion

"Compared with the E_{pit} of AlCoCrFeNi (I) alloy, both the Ti-containing CoCrFeNi$Ti_{0.5}$ (II) and AlCoCrFeNi$Ti_{0.5}$ (III) coatings exhibit a higher E_{pit} and a much larger passive region (E_{pit}-E_{corr}), which shows that the addition of Ti element will improve the pitting corrosion resistance of HEA coatings observably."

"The value of R_{ct} for AlCoCrFeNi$Ti_{0.5}$ (III) coating is two orders of magnitude higher than that of AlCoCrFeNi (I) coating, and three orders of magnitude higher than that of CoCrFeNi$Ti_{0.5}$ (II) coating, which indicates that the passive films on the

$AlCoCrFeNiTi_{0.5}$ (III) HEA coatings have less activated sites, further causing the higher corrosion resistance of the coating."

"The addition of Al (in AlCoCrFeNi (I) and $AlCoCrFeNiTi_{0.5}$ (III) HEAs) caused the formation of Al-rich (Cr-depleted) phase, which has a negative influence on the pitting corrosion."

"In consideration of the unavoidable existence of Al-rich (Cr-depleted) phases in this current system, the addition of Ti is an available way to enhance the pitting resistance as observed for the $AlCoCrFeNiTi_{0.5}$ (III) HEA coating in this paper."

Conclusions

"The AlCoCrFeNi (I), $CoCrFeNiTi_{0.5}$ (II) and $AlCoCrFeNiTi_{0.5}$ (III) HEA coatings were fabricated by laser cladding."

"The AlCoCrFeNi (I) coating exhibited a simple BCC phase (BCC2), while the phases of $AlCoCrFeNiTi_{0.5}$ (III) coating were characterized as Fe–Cr rich disordered BCC phases (BCC2) and Al–Ni–Ti rich ordered BCC phases (BCC1), respectively."

"Compared with AlCoCrFeNi (I) and $CoCrFeNiTi_{0.5}$ (II) coatings, the $AlCoCrFeNiTi_{0.5}$ (III) exhibited the highest microhardness."

"The dual-phase BCC-$AlCoCrFeNiTi_{0.5}$ (III) HEA coating was investigated via nanoindentation tests."

A Thermodynamic Study of Corrosion Behaviors for CoCrFeNi-Based High-Entropy Alloys [101]

This is a machine-generated summary of:

Zhang, Boliang; Zhang, Yi; Guo, S. M.: A thermodynamic study of corrosion behaviors for CoCrFeNi-based high-entropy alloys [101]

Published in: Journal of Materials Science (2018)

Link to original: https://doi.org/10.1007/s10853-018-2652-2

If you want to cite the papers, please refer to the original.

For technical reasons we could not place the page where the original quote is coming from.

Abstract-Summary

"To better understand the corrosion behaviors of CoCrFeNi-based high-entropy alloys (HEAs), the CALculation of PHAse Diagrams (CALPHAD) method was used to simulate the Pourbaix diagrams for CoCrFeNi, CoCrFeNiCu and CoCrFeNiAl HEAs."

Introduction

"The standard potentiodynamic polarization tests, together with the corresponding characterization methods, are routinely adopted to experimentally study the general trend of the corrosion behaviors of CoCrFeNi-based high-entropy alloys [54–56]."

"Pourbaix diagram has been prevalently used to provide a basic guidance for the study of corrosion behaviors of alloys [57, 58]."

"In order to generate the Pourbaix diagram for an alloy system, a simple superposition method is usually adopted [59]."

"The simple superposition method would not be adequate to generate the Pourbaix diagrams accurately for complex alloys under nonstandard conditions."

"The CALculation of PHAse Diagrams (CALPHAD) method is used to examine the corrosion behaviors of CoCrFeNi-based high-entropy alloys."

"In specific, the effects of additional alloying of Cu and Al to the corrosion behaviors of baseline CoCrFeNi are examined."

"This study demonstrates that CALPHAD method is a viable tool for the prediction of complex alloy's corrosion behaviors."

The simulation approach

"Before conducting aqueous corrosion simulations, the phase stability of the high-entropy alloy matrixes was modeled using CALPHAD method to evaluate the potential corrosion resistant."

"After knowing the alloys' matrix phases, the corrosion behavior was further modeled by introducing the aqueous phases and the oxides phases into the equilibrium calculations."

"By considering the aqueous phases, together with alloy matrix phases, compound phases, secondary solid phases (oxide films) and gas phases, a Pourbaix diagram can be set up by performing equilibrium calculations under different electric potentials and hydrogen ion concentrations (pH value)."

"The interactions between the alloy matrix structures and an aqueous solution were examined to generate the Pourbaix diagrams, which show all equilibrium phases at different corrosion potentials and environmental pH conditions."

"Similar CALPHAD calculations were performed on CoCrFeNiCu and $CoCrFeNiAl_{0.5}$, to show the effect of additional alloying elements to the corrosion behavior of the CoCrFeNi-based HEAs."

CALPHAD calculations and Discussions

"It has been proved experimentally that the baseline CoCrFeNi HEA is a single-phase alloy with a stable FCC crystal structure [60–64]."

"To demonstrate the impact of additional alloying on the corrosion behaviors of the CoCrFeNi baseline alloy, the pseudo-binary CoCrFeNi–Cu_x phase diagram was calculated with the variation of Cu ratio to show the influence of Cu on the phase stability of CoCrFeNi-based HEA."

"Comparing with the baseline HEA CrCoFeNi, the quinary equimolar HEA CrCoFeNiCu with Cu introduced as an additional principle element shows both new matrix alloy phases and passive oxide phases in the Pourbaix diagram."

"New passive phases like $Cu_2Fe_2O_4$ are found to form on the surface of the alloy during corrosion."

"The diaspore phase is formed spontaneously at the beginning of the corrosion, unlike the cases in CrCoFeNi and CrCuFeNiCu in which $CoCr_2O_4$ and $CoFe_2O_4$ grow gradually in the active-to-passive transition region."

Conclusions

"Using thermodynamically based CALPHAD simulations, the matrix phase structures and the aqueous corrosion behaviors of the CoCrFeNi-based high-entropy alloy systems, namely CoCrFeNi, $CoCrFeNiCu_x$ and $CoCrFeNiAl_x$, were predicted."

"With Pourbaix diagrams, the corrosion regions, the passivation regions and the immunity regions along with the Eh and pH variations for CoCrFeNi-based high-entropy alloys can be determined."

"Assisted by previously published experimental results, CALPHAD simulations provided insights to the corrosion mechanisms for CoCrFeNi-based high-entropy alloys."

"The corrosion behaviors of CoCrFeNi-based high-entropy alloys were found as follows."

"CALPHAD simulations, assisted by experimental results on CoCrFeNi-based HEA systems, indicate that thermodynamic phase calculation is a viable tool to guide the design of specific multi-principle elements alloys with improved corrosion resistant."

Effect of Ti Addition on the Microstructure and Corrosion Resistance of $AlCrCuFeNiNb_{0.2}$ High-Entropy Alloy [102]

This is a machine-generated summary of:

Liang, Xuewei; Su, Yunhai; Yang, Taisen; Dai, Zhiyong; Wang, Yingdi: Effect of Ti Addition on the Microstructure and Corrosion Resistance of $AlCrCuFeNiNb_{0.2}$ High-Entropy Alloy [102]

Published in: JOM (2022)

Link to original: https://doi.org/10.1007/s11837-022-05620-6

If you want to cite the papers, please refer to the original.

For technical reasons we could not place the page where the original quote is coming from.

Abstract-Summary

"The effect of Ti on the phase structure and corrosion resistance of $AlCrCuFeNiNb_{0.2}Ti_x$ HEA surfacing layer is discussed in detail."

"The results demonstrate that the microstructure of the surfacing layer was mainly composed of BCC solid solution with Fe–Cr phase and FCC solid solution with tiny amounts of MC phase."

"It indicates that the addition of Ti can significantly promote the corrosion resistance of the surfacing layer."

"The corrosion resistance effect is stronger than that of 304 stainless steel."

"With the continuous addition of Ti, the self-corrosion current density of polarization curve decreases, the capacitance impedance radius of electrochemical impedance spectroscopy (EIS) increases, and the corrosion resistance of the surfacing layer improves."

"When the addition was $Ti_{0.8}$, the corresponding corrosion resistance value of the surfacing layer impedance was the smallest."

Introduction

"The appearance of high-entropy alloys (HEAs) meets, to a certain extent, the requirements of mechanical equipment surface material properties."

"Through reasonable composition design [65–68], alloy materials with higher hardness, higher wear resistance, higher corrosion resistance, higher temperature oxidation resistance, and excellent mechanical and electromagnetic properties can be obtained."

"Lin and others [69] studied the corrosion resistance of $Cu_{0.5}$CoCrFeNi HEAs as-cast and as-aged at different temperatures in 3.5% NaCl solution, and compared it with 304L stainless steel."

"When the alloy is in the passivation state, a dense oxide film easily forms on the surface of the alloy, which has excellent corrosion resistance."

"The mechanical properties and corrosion resistance of Ti-containing HEAs were studied in references."

"The phase structure and microstructure of the HEA surfacing layers are discussed, and the relationship between phase structure and corrosion resistance is discussed."

Experimental Materials and Methods

"Al, Cr, Cu, Ni, Nb, and Ti metal powders were used as the flux core, and Fe came completely from the steel strip of flux-cored wire."

"At the start of the flux-cored wire control system, the powder falls into the U-shaped steel strip, and after multi-pass rolling and drawing, the flux-cored wire has a diameter of 2.4 mm."

"Wire cutting produced a surfacing layer of 10 mm × 10 mm × 10 mm, and then microstructure observation and electrochemical detection was completed."

"The surfacing layer was polished and the phase composition of the surfacing alloy was investigated by x-ray diffraction."

"The phase composition of the surfacing layer was characterized by a Geminisem 300 field emission scanning electron microscope equipped with electron backscatter diffraction (EBSD) and an energy dispersive spectrometer (EDS)."

Results and Discussion

"[70, 71] The electrochemical results showed that the corrosion resistance of HEA surfacing layers was improved after adding Ti."

"With the continuous addition of Ti element, the corrosion rate of surfacing layers decreases gradually."

"With the continuous addition of Ti element, R_{pass} increases and the corrosion resistance of the $AlCrCuFeNiNb_{0.2}Ti_x$ HEA surfacing layer increases."

"With the increase of the content of Ti, the degree of corrosion of the surfacing layer gradually decreases."

"The Fe–Cr solid solution phase based on the BCC structure occupies most of the volume, and the large amount of corrosion-resistant element Cr on the alloy surface can improve the stability of the passivation film in the corrosive environment, playing a good protective role in it."

"Excessive Ti is added to the BCC solid solution, which leads to an increase in grain size and thus improves the corrosion resistance of the alloy."

Conclusion

"The phase structure of $AlCrCuFeNiNb_{0.2}Ti_x$ HEA surfacing alloy is mainly Fe–Cr solid solution with BCC structure and a tiny amount of (Nb,Ti)C with FCC structure."

"The continuous addition of Ti element is beneficial to the formation of BCC solid solution, and the BCC phase always occupies the dominant position in HEA surfacing alloy."

"When the Ti content was $Ti_{0.8}$, the corrosion resistance of the HEA surfacing alloy was best."

"For the corrosion resistance mechanism of $AlCrCuFeNiNb_{0.2}Ti_x$ HEA surfacing alloy, on the one hand, the addition of corrosion-resistant components Cr and Ti is conducive to the formation of a passivation film on the surface of the alloy, and the cocktail effect results in corrosion resistance of the high-entropy surfacing alloy."

Effect of Cold Rolling on the Microstructure Evolution, Mechanical, and Corrosion Properties of $AlCoCrFeNi_{2.4}$ High-Entropy Alloy [103]

This is a machine-generated summary of:

Bijnavandi, Mehrab Seifpour; Ghaderi, Armin; Dehghani, Kamran: Effect of Cold Rolling on the Microstructure Evolution, Mechanical, and Corrosion Properties of $AlCoCrFeNi_{2.4}$ High-Entropy Alloy [103]

Published in: Journal of Materials Engineering and Performance (2023)

Link to original: https://doi.org/10.1007/s11665-023-08101-2

If you want to cite the papers, please refer to the original.

For technical reasons we could not place the page where the original quote is coming from.

Abstract-Summary

"X-ray diffraction (XRD) analysis revealed that the constituent phases in the as-cast sample were FCC and BCC."

"With the increase in thickness reduction from cold rolling, the microhardness, yield strength, shear strength, and ultimate strength of the alloy all increased, despite a decrease in ductility."

"After the alloy was cold-rolled to 90%, the alloy demonstrated yield strengths of 1791 MPa, ultimate tensile strengths of 1956 MPa, and elongation to fracture values of 10.17%."

Introduction

"High-entropy alloys (HEAs), also known as multi-principal-element alloys, have received considerable attention (Refs. [1, 72])."

"In the early stages of HEA research, single-phase solid solutions with intriguing properties, such as high strength (Refs. [73–75]) and high plasticity at room temperature (Refs. [76–78]), were the primary focus."

"A wide range of HEA systems have been developed and shown to have various exceptional properties, including high strength (Refs. [73, 79]), high plasticity (Ref. [80]), corrosion resistance (Ref. [81]), wear resistance (Refs. [82, 83]), and exceptional cryogenic and elevated temperature properties (Refs. [84, 85])."

"Dual-phase alloys have undergone extensive research and development to obtain high-entropy alloy, which is advantageous for increased strength and excellent ductility (Refs. [86, 87])."

"A study of the as-cast EHEA at room temperature demonstrated high tensile ductility and high fracture strength (Ref. [88])."

Materials and Methods

"The alloy's final thickness of 0.75 mm was achieved through seven passes of cold rolling, i.e., 90% reduction."

"Before tensile testing, the samples were mechanically polished to eliminate the effects of cold rolling due to any possible impact of concentrated surface stress on their tensile strength."

"Sample preparation for Shear punch testing (SPT) was performed in compliance with ASTM D732 standard procedures."

"Each sample's average hardness was determined by measuring it at least five times."

Results and Discussion

"The cold-rolled specimen exhibits peaks similar to those of the as-cast specimen, though the intensity of the peaks differs significantly."

"In the cold rolling condition, the alloy's yield and ultimate strengths increased significantly compared to the as-cast state."

"Following a 90% reduction by cold rolling, the alloy's yield and tensile strengths increased significantly, reaching 1791 and 1956 MPa, respectively."

"The present study indicates that cold rolling is an effective technique for optimizing the mechanical properties of HEAs."

"Of the microstructural modification, decreasing grain size, and work hardening, the cold rolling process generally results in a significant increase in strength."

"This indicates that the cold-rolled sample has a lower ductility."

"When exposed to a NaCl corrosive environment, the as-cast specimen exhibited greater corrosion resistance than the cold-rolled specimen."

"A characteristic dimple is visible on the as-cast specimen, indicating that it has undergone significant plastic deformation."

Conclusions

"The tensile yield strength in the as-cast and cold-rolled states was approximately 831 and 1791 MPa, respectively."

"The shear yield strength increased from 223 MPa for the as-cast state to 331 MPa for the cold-rolled state."

"The alloy's hardness increases from 288 HV when as-cast to 460 HV after cold rolling."

Understanding the Effect of Tungsten on Corrosion Behavior of $AlCuCrFeMnW_x$ High-Entropy Alloys in 3.5 wt.% NaCl Solution [104]

This is a machine-generated summary of:

Kumar, Devesh; Maulik, Ornov; Sharma, V. K.; Prasad, Y. V. S. S.; Kumar, Vinod: Understanding the Effect of Tungsten on Corrosion Behavior of $AlCuCrFeMnW_x$ High-Entropy Alloys in 3.5 wt.% NaCl Solution [104]

Published in: Journal of Materials Engineering and Performance (2018)

Link to original: https://doi.org/10.1007/s11665-018-3536-y

If you want to cite the papers, please refer to the original.

For technical reasons we could not place the page where the original quote is coming from.

Abstract-Summary

"The electrochemical study of $AlCuCrFeMnW_x$ (x = 0, 0.05, 0.1, 0.5, 1 mol) high-entropy alloys (HEA) is performed in 3.5 wt.% NaCl solution at room temperature (25 °C)."

"The potentiodynamic polarization curve of the $AlCuCrFeMnW_x$ (x = 0, 0.05, 0.1, 0.5, 1 mol) alloys obtained in 3.5 wt.% NaCl solution shows that the corrosion resistance of the tungsten (W)-free HEA is lower than the alloys containing W. It is observed that W forms solid WO_3 layer in the outer region of the film in a 3.5 wt.% NaCl."

Introduction

"These shortcomings and requirements of specific applications motivated the investigator to undertake the research to synthesize high-entropy alloys (HEAs) having optimum properties such as higher strength to weight ratio, excellent mechanical property such as hardness and superior corrosion resistance similar to stainless steel."

"The effect of the addition of Mo to the $Co_{1.5}CrFeNi_{1.5}Ti_{0.5}Mo_x$ alloys on the corrosion behavior in a 1 M NaCl solution was investigated (Ref. [89])."

"Devesh and others have reported that AlCuCrFeMnW HEA, synthesized by spark plasma sintering at 900 °C, has high hardness and potential candidates for structural applications (Ref. [90]). The effect of W on the corrosion behavior of $AlCuCrFeMnW_x$ alloys is not yet studied."

"The purpose of this study was to investigate the corrosion behavior of $AlCuCrFeMnW_x$ ($x = 0$, 0.05, 0.1, 0.5, 1 mol) alloys in 3.5 wt.% NaCl solution."

Experimental

"A circular pellet of high-entropy alloy used for electrochemical measurements has a thickness of 3 mm and a diameter of 20 mm and exposed surface area of high-entropy alloy of is 3.14 cm^2."

"Potentiodynamic polarization measurements of bulk $AlCuCrFeMnW_x$ ($x = 0$, 0.05, 0.1, 0.5, 1 mol) alloys was carried out in a typical three-electrode cell setup with the specimen as a working electrode, a saturated calomel reference electrode (SCE), and a platinum counter electrode."

"The potential of the working electrode was measured through the Luggin probe with respect to the reference electrode which was placed as close as possible to the specimen."

"Immediately thereafter, the morphology of the corroded surface of the high-entropy alloy was investigated using a FE-SEM (Nova Nano 450SEM, FEI, North America) operated at 20 kV. Infrared (IR) spectroscopy was performed to characterize the corrosion product after polarization measurement."

Results and Discussion

"With the addition of tungsten, the corrosion potential of the alloys increased from − 0.892 to − 0.313 V, and the corrosion current density decreased from 6.97×10^{-6} to 3.32×10^{-6} A/cm^2, which demonstrated that a high content of tungsten enhanced the corrosion resistance of the alloys in the 3.5 wt.% NaCl solution."

"It appears that the alloy without W is more prone to pitting corrosion in the 3.5 wt.% NaCl solution than the alloy containing W. FTIR was performed to determine the functional groups present on the surface after polarization experiment."

"AlCuCrFeMnW alloy shows much easier passivation, and it is more corrosion resistant in 3.5 wt.% NaCl solution."

"The most appropriate simple equivalent circuit consisting of a series resistor connected in series with a resistor and a constant phase element (instead of a capacitor) in parallel is presented as electrical double layer in case of proposed HEAs in 3.5 wt.% NaCl solution."

Conclusion

"The electrochemical performance of the high-entropy alloy $AlCuCrFeMnW_x$ ($x = 0, 0.05, 0.1, 0.5, 1$ mol) in a 3.5 wt.% NaCl solution was studied."

"Polarization curves and SEM images show that the alloy containing W in 3.5 wt.% NaCl solution is not easily affected by pitting corrosion."

"With the addition of tungsten, the corrosion potential of the alloys increased from − 0.892 to − 0.313 V, and the corrosion current density decreased from 6.97×10^{-6} to 3.32×10^{-6} A/cm^2, which demonstrated that a high content of tungsten enhanced the corrosion resistance of the alloys."

References

1. J.W. Yeh, S.K. Chen, S.J. Lin, J.Y. Gan, T.S. Chin, T.T. Shun, C.H. Tsau, and S.Y. Chang, Adv. Eng. Mater. 6, 299 (2004).
2. Sieradzki, K. & Newman, R. Percolation model for passivation in stainless steels. J. Electrochem. Soc. 133, 1979–1980 (1986).
3. Gupta, R. K. et al. Simultaneous improvement in the strength and corrosion resistance of Al via high-energy ball milling and Cr alloying. Mater. Des. 84, 270–276 (2015).
4. C. Ni, Y. Shi, J. Liu, and G. Huang, Characterization of Al0.5FeCu0.7NiCoCr High-Entropy Alloy Coating on Aluminum Alloy by Laser Cladding, Opt. Laser Technol., 2018, 105, p 257–263. https://doi.org/10.1016/j.optlastec.2018.01.058
5. X. Wu Qiu, M. Jun Wu, C. Ge Liu, Y. Peng Zhang, and C. Xiang Huang, Corrosion Performance of Al2CrFeCoxCuNiTi High-Entropy Alloy Coatings in Acid Liquids, J. Alloys Compd., 2017, 708, p 353–357. https://doi.org/10.1016/j.jallcom.2017.03.054
6. F. Shu, B. Yang, S. Dong, H. Zhao, B. Xu, F. Xu, B. Liu, P. He, and J. Feng, Effects of Fe-to-Co Ratio on Microstructure and Mechanical Properties of Laser Cladded FeCoCrBNiSi High-Entropy Alloy Coatings, Appl. Surf. Sci., 2018, 450, p 538–544. https://doi.org/10.1016/j.apsusc.2018.03.128
7. C.L. Wu, S. Zhang, C.H. Zhang, H. Zhang, and S.Y. Dong, Phase Evolution and Cavitation Erosion-Corrosion Behavior of FeCoCrAlNiTixhigh Entropy Alloy Coatings on 304 Stainless Steel by Laser Surface Alloying, J. Alloys Compd., 2017, 698, p 761–770. https://doi.org/10.1016/j.jallcom.2016.12.196
8. W.L. Hsu, Y.C. Yang, C.Y. Chen, and J.W. Yeh, Thermal Sprayed High-Entropy NiCo0.6Fe0.2Cr1.5SiAlTi0.2 Coating with Improved Mechanical Properties and Oxidation Resistance, Intermetallics, 2017, 89(May), p 105–110. https://doi.org/10.1016/j.intermet.2017.05.015
9. Qiu, X. W. Microstructure and properties of AlCrFeNiCoCu high entropy alloy prepared by powder metallurgy. J. Alloy Compd. 555, 246–249 (2013).
10. B. Cantor, I.T.H. Chang, P. Knight and A.J.B. Vincent, Microstructural Development in Equiatomic Multicomponent Alloys, Mater. Sci. Eng. A, 2004, 375–377, p 213–218.
11. T.K. Liu, Z. Wu, A.D. Stoica, Q. Xie, W. Wu, Y.F. Gao, H. Bei and K. An, Twinning Mediated Work Hardening and Texture Evolution in CrCoFeMnNi High Entropy Alloys at Cryogenic Temperature, Mater. Des., 2017, 131, p 419–427.

12. B. Guillaume, L.B. Mathilde, P. Loic, P. Remy, J. Jean-Marc and G. Ivan, The Fcc Solid Solution Stability in the Co-Cr-Fe-Mn-Ni Multi-Component System, Acta Mater., 2017, 128, p 327–336.
13. F. Otto, Y. Yang, H. Bei and E.P. George, Relative Effects of Enthalpy and Entropy on the Phase Stability of Equiatomic High-Entropy Alloys, Acta Mater., 2013, 61, p 2628–2638.
14. R.D. Li, P.D. Niu, T.C. Yuan, P. Cao, C. Chen and K.C. Zhou, Selective Laser Melting of an Equiatomic CoCrFeMnNi High-Entropy Alloy: Processability, Non-Equilibrium Microstructure and Mechanical Property, J. Alloy. Compd., 2018, 746, p 125–134.
15. M. Zhu, B.Z. Zhao, Y.F. Yuan, S.Y. Guo and G.Y. Wei, Study on Corrosion Behavior and Mechanism of CoCrFeMnNi HEA Interfered by AC Current in Simulated Alkaline Soil Environment, J. Electroanal. Chem., 2021, 882, 115026.
16. W. Ye, Y. Li and F.H. Wang, Effects of Nanocrystallization on the Corrosion Behavior of 309 Stainless Steel, Electrocheim. Acta, 2006, 51, p 4426–4432.
17. S. Maximovitch, Influence of Formation Conditions on Impedance Properties of Nickel Passive Layers Formed in 1 M KOH, Electrochim. Acta, 1996, 41, p 2761–2771.
18. S. Carmelo, P. Salvatore and D.Q. Francesco, Photocurrent Spectroscopic Investigations of Passive Films on Chromium, J. Electrochem. Soc., 1990, 137, p 2411–2417.
19. J.Z. Yao, D.D. Macdonald and C.F. Dong, Passive Film on 2205 Duplex Stainless Steel Studied by Photo-Electrochemistry and ARXPS Methods, Corros. Sci., 2019, 146, p 221–232.
20. Li WJ, Guo WM, Zhang HL, Xu HH, Chen L, Zeng JS, Liu B, Ding ZB (2022) Influence of Mo on the Microstructure and Corrosion Behavior of Laser Cladding FeCoCrNi High-Entropy Alloy Coatingss. Entropy 24(4):539. https://doi.org/10.3390/e24040539
21. Yeh, J. W. & Lin, S. J. Breakthrough applications of high-entropy materials. J. Mater. Res. 33, 3129–3137 (2018).
22. Y. Zhang, High-Entropy materials- A Brief Introduction (Springer, Singapore, 2019)
23. Ching, W. Y. et al. Fundamental electronic structure and multiatomic bonding in 13 biocompatible high-entropy alloys. npj Comput. Mater. 6, 1–10 (2020).
24. C. Zhang, F. Zhang, H.Y. Diao, M.C. Gao, Z. Tang, J.D. Poplawsky, and P.K. Liaw, Understanding Phase Stability of Al-Co-Cr-Fe-Ni High Entropy Alloys, Mater. Des., 2016, 109(5), p 425–433.
25. Csaki, I. et al. Researches regarding the processing technique impact on the chemical composition, microstructure and hardness of AlCrFeNiCo high entropy alloy. Rev. Chim. 67, 1373–1377 (2016).
26. Y. Shon, S.S. Joshi, S. Katakam, R. Shanker Rajamure, and N.B. Dahotre, Mater. Lett. 142, 122 (2015).
27. Li, Q. H., Yue, T. M., Guo, Z. N. & Lin, X. Microstructure and corrosion properties of AlCoCrFeNi high entropy alloy coatings deposited on AISI 1045 steel by the electrospark process. Metall. Mater. Trans. A 44, 1767–1778 (2012).
28. Zhou, P. F., Xiao, D. H. & Yuan, T. C. Microstructure, mechanical and corrosion properties of AlCoCrFeNi high-entropy alloy prepared by spark plasma sintering. Acta Metall. Sin. (English Lett.) 33, 937–946 (2020).
29. Feng, X. et al. Effect of Zr addition on microstructure and mechanical properties of CoCrFeNiZrx high-entropy alloy thin films. Appl. Nanosci. https://doi.org/10.1007/s13204-019-01057-7 (2019).
30. Zhang X, Huang T, Yang W, Xiao R, Liu Z, Li L (2016a) Microstructure and mechanical properties of laser beam-welded AA2060 Al-Li alloy. J Mater Process Technol 237:301–308
31. Zhang Z, Axinte E, Ge W, Shang C, Wang Y (2016b) Microstructure, mechanical properties and corrosion resistance of CuZrY/Al, Ti, Hf series high-entropy alloys. Mater Des 108:106–113
32. Santodonato LJ, Zhang Y, Feygenson M, Parish CM, Gao MC, Weber RJ, Neuefeind JC, Tang Z, Liaw PK (2015) Deviation from high-entropy configurations in the atomic distributions of a multi-principal-element alloy. Nat Commun 6(1):1–13
33. Seifi M, Li D, Yong Z, Liaw PK, Lewandowski JJ (2015) Fracture toughness and fatigue crack growth behavior of as-cast high-entropy alloys. J Miner Met Mater Soc 67(10):2288–2295

34. Fayomi OSI, Akande IG, Oluwole OO, Daramola D (2018) Effect of water-soluble chitosan on the electrochemical corrosion behaviour of mild steel. Chem Data Collect 17:321–326
35. Bae JW, Kim HS (2020) Towards ferrous medium-entropy alloys with low-cost and high-performance. Scripta Mater 186:169–173
36. Singh, S., Wanderka, N., Murty, B. S., Glatzel, U. & Banhart, J. Decomposition in multi-component AlCoCrCuFeNi high-entropy alloy. Acta Materialia 59, 182–190 (2011).
37. W.H. Liu, Y. Wu, J.Y. He, Y. Zhang, C.T. Liu, Z.P. Lu, JOM 66, 1973 (2014)
38. Ma, E. & Wu, X. Tailoring heterogeneities in high-entropy alloys to promote strength–ductility synergy. Nat. Commun. 10, 5623 (2019).
39. Y. Ma, G.J. Peng, D.H. Wen, T.H. Zhang, Mater. Sci. Eng. A 621, 111 (2015).
40. B. Ren, Z.X. Liu, D.M. Li, L. Shi, B. Cai, M.X. Wang, J. Alloy. Compd. 493, 148 (2010)
41. B. Ren, Z.X. Liu, B. Cai, M.X. Wang, L. Shi, Mater. Design 33, 121 (2012)
42. J.I. Hyun, K.H. Kong, K.C. Kim, W.T. Kim, D.H. Kim, Appl. Microsc. 45, 9 (2015)
43. Z.Y. Rao, X. Wang, J. Zhu, X.H. Chen, L. Wang, J.J. Si, Y.D. Wu, X.D. Hui, Intermetallics 77, 23 (2016)
44. X. Chen, S.D. Zhai, D. Gao, J. Xu, Y. Liu, Y. Liu, F.C. Yin, J. of Mater. Eng. Perform. 28, 4502 (2019)
45. S.K. Bonagani, V. Bathula, V. Kain, Corros. Sci. 131, 340 (2018)
46. Zhang, Y. et al. Microstructures and properties of high-entropy alloys. Prog. Mater. Sci. 61, 1–93 (2014).
47. Chang, S.Y.; Chen, D.S.: 10-nm-thick quinary (AlCrTaTiZr)N film as effective diffusion barrier for Cu interconnects at 900 C. Appl. Phys. Lett. 94, 231909 (2009). https://doi.org/10.1063/1.3155196
48. Azaath, L.M.; Natarajan, U.; Veerappan, G., et al.: Experimental investigations on the mechanical properties, microstructure and corrosion effect of Cu–20Al–4Ni/SiC composites synthesized using powder metallurgy route. SILICON (2021). https://doi.org/10.1007/s12633-021-01363-2
49. L.J. Santodonato, Y. Zhang, M. Feygenson, C.M. Parish, M.C. Gao, R.J.K. Weber, J.C. Neuefeind, Z. Tang and P.K. Liaw: Nature Comm., 2015, vol. 6, pp. 5964.
50. Low, S.; Machado, R R.: IOP Conf. Series: Journal of Physics: Conf. Series 1065 (2018) 062007. https://doi.org/10.1088/1742-6596/1065/6/062007
51. ISO 6508–1:2016 Metallic Materials—Rockwell Hardness Test—Part 1: Test method (Geneva: International Organization for Standardization)
52. D.H. Xiao, P.F. Zhou, W.Q. Wu, H.Y. Diao, M.C. Gao, M. Song, Mater. Des. 116, 438 (2017)
53. Y. Qiu, S. Thomas, M.A. Gibson, H.L. Fraser, K. Pohl, N. Birbilis, Corros. Sci. 133, 386 (2018)
54. Hsu, Y. J., Chiang, W. C. & Wu, J. K. Corrosion behavior of FeCoNiCrCux high-entropy alloys in 3.5% sodium chloride solution. Mater. Chem. Phys. 92, 112–117 (2005).
55. Shi Y, Yang B, Xie X, Brechtl J, Dahmen KA, Liaw PK (2017) Corrosion of Al xCoCrFeNi high-entropy alloys: Al-content and potential scan-rate dependent pitting behavior. Corros Sci 119:33–45
56. C.M. Lin and H.L. Tsai, Intermetallics 19, 288 (2011).
57. Pourbaix M (1974) Atlas of electrochemical equilibria in aqueous solutions. National Association of Corrosion, Houston
58. Pourbaix M (1973) Lectures on electrochemical corrosion. Plenum Press, New York
59. McCafferty E (2010) Introduction to corrosion science. Springer, Berlin
60. T.-T. Shun, L.-Y. Chang, and M.-H. Shiu, Microstructure and Mechanical Properties of Multiprincipal Component CoCrFeNiMox alloys, Mater. Charact., 2012, 70, p 63–67
61. W.-R. Wang, W.-L. Wang, S.-C. Wang, Y.-C. Tsai, C.-H. Lai, and J.-W. Yeh, Effects of Al Addition on the Microstructure and Mechanical Property of AlxCoCrFeNi High-Entropy Alloys, Intermetallics, 2012, 26, p 44–51
62. H.-P. Chou, Y.-S. Chang, S.-K. Chen, and J.-W. Yeh, Microstructure, Thermophysical and Electrical Properties in AlxCoCrFeNi ($0 \leq x \leq 2$) High-Entropy Alloys, Mater. Sci. Eng., B, 2009, 163(3), p 184–189

63. Singh AK, Subramaniam A (2014) On the formation of disordered solid solutions in multi-component alloys. J Alloy Compd 587:113–119. https://doi.org/10.1016/j.jallcom.2013.10.133
64. Wu, Z., Bei, H., Otto, F., Pharr, G. M. & George, E. P. Recovery, recrystallization, grain growth and phase stability of a family of FCC-structured multi-component equiatomic solid solution alloys. Intermetallics 46, 131–140 (2014).
65. K.H. Huang and J.W. Yeh, A study on multicomponent alloy systems containing equal-mole elements[D]. (National Tsing Hua University, Hsinchu, 1996).
66. J.W. Yeh, Ann. Chim. Sci. Mater. 31, 633 (2006).
67. Miracle, D. B. & Senkov, O. N. A critical review of high entropy alloys and related concepts. Acta Mater. 122, 448–511 (2017).
68. S. Guo, C. Ng, J. Lu, and C.T. Liu, J. Appl. Phys. 109, 103505 (2011).
69. Lin, C.-M., Tsai, H.-L. & Bor, H.-Y. Effect of aging treatment on microstructure and properties of high-entropy Cu0.5CoCrFeNi alloy. Intermetallics 18, 1244–1250 (2010).
70. C.L. Wu, S. Zhang, C.H. Zhang, H. Zhang, and S.Y. Dong, Surf. Coat. Technol. 315, 368 (2017).
71. Y. Guo, X. Shang, and Q. Liu, Microstructure and Properties of In-Situ TiN Reinforced Laser Cladding CoCr2FeNiTix High-Entropy Alloy Composite Coatings, Surf. Coat. Technol., 2018, 344, p 353–358. https://doi.org/10.1016/j.surfcoat.2018.03.035
72. Fu, Z. et al. A high-entropy alloy with hierarchical nanoprecipitates and ultrahigh strength. Sci. Adv. 4, eaat8712 (2018).
73. Zou, Y., Maiti, S., Steurer, W. & Spolenak, R. Size-dependent plasticity in an Nb25Mo25Ta25W25 refractory high-entropy alloy. Acta Mater. 65, 85–97 (2014).
74. C.W. Tsai, M.H. Tsai, J.W. Yeh, and C.C. Yang, J. Alloys Compd. 490, 160 (2010).
75. O.N. Senkov, J.M. Scott, S.V. Senkova, D.B. Miracle, and C.F. Woodward, J. Alloys Compd. 509, 6043 (2011).
76. A. Gali and E.P. George, Tensile Properties of High- and Medium-Entropy Alloys, Intermetallics, 2013, 39, p 74–78
77. F. Otto, A. Dlouhý, C. Somsen, H. Bei, G. Eggeler, and E.P.P. George, The Influences of Temperature and Microstructure on the Tensile Properties of a CoCrFeMnNi High-Entropy Alloy, Acta Mater., 2013, 61(5743), p 5755.
78. He, J. Y. et al. Effects of Al addition on structural evolution and tensile properties of the FeCoNiCrMn high-entropy alloy system. Acta Materialia 62, 105–113 (2014).
79. M. Wang, Y. Lu, T. Wang, C. Zhang, Z. Cao, T. Li, and P.K. Liaw, A Novel Bulk Eutectic High-Entropy Alloy with Outstanding As-Cast Specific Yield Strengths at Elevated Temperatures, Scr. Mater., 2021, 204, p 114132.
80. M.J. Yao, K.G. Pradeep, C.C. Tasan, and D. Raabe, Scripta Mater. 72, 5 (2014).
81. Y.Y. Chen, U.T. Hong, J.W. Yeh, and H.C. Shih, Selected Corrosion Behaviors of a Cu0.5NiAlCoCrFeSi Bulk Glassy Alloy in 288 °C High-Purity Water, Scr. Mater., 2006, 54, p 1997–2001.
82. Y.X. Wang, Y.J. Yang, H.J. Yang, M. Zhang, S.G. Ma, and J.W. Qiao, Microstructure and Wear Properties of Nitrided AlCoCrFeNi High-Entropy Alloy, Mater. Chem. Phys., 2017, 210, p 233.
83. Mingliang Wang, Lu. Yiping, Guojia Zhang, Hongzhi Cui, Xu. Dingfeng, Na. Wei, and Tingju Li, A Novel High-Entropy Alloy Composite Coating with Core-Shell Structures Prepared by Plasma Cladding, Vacuum, 2021, 184, p 109905.
84. Gludovatz, B. et al. A fracture-resistant high-entropy alloy for cryogenic applications. Science 345, 1153–1158 (2014).
85. L. Zhang, Y. Zhou, X. Jin, X.Y. Dua, and B.S. Li, The Microstructure and High-Temperature Properties of Novel Nano Precipitation-Hardened Face Centered Cubic High-Entropy Superalloys, Scr. Mater., 2018, 146, p 226–230.
86. X.Z. Gao, Y.P. Lu, B. Zhang, N.N. Liang, G.Z. Wu, G. Sha, J.Z. Liu, Y.H. Zhao, Acta Mater. 141, 59 (2017)

87. Li, Z., Pradeep, K. G., Deng, Y., Raabe, D. & Tasan, C. C. Metastable high-entropy dual-phase alloys overcome the strength–ductility trade-off. Nature 534, 227–230 (2016).
88. I.S. Wani, T. Bhattacharjee, S. Sheikh, Y.P. Lud, S. Chatterjee, P.P. Bhattacharjeea, S. Guo, and N. Tsujib: Mater. Res. Lett., 2016, vol. 4, pp. 174–79.
89. Chou Y L, Yeh J W and Shih H C, Corr Sci 52 (2010) 2571.
90. D. Kumar, O. Maulik, S. Kumar, Y.V.S.S. Prasad, V. Kumar, Phase and Thermal Study of Equiatomic AlCuCrFeMnW High Entropy Alloy Processed via Spark Plasma Sintering. Mat. Chem. Phys. https://doi.org/10.1016/j.matchemphys.2017.08.049 (in press)
91. Qiu, Yao; Thomas, Sebastian; Gibson, Mark A.; Fraser, Hamish L.; Birbilis, Nick Corrosion of high entropy alloys. npj Materials Degradation (2017). https://doi.org/10.1038/s41529-017-0009-y
92. Vyas, Akash; Menghani, Jyoti; Natu, Harshad Influence of WC Particle on the Metallurgical, Mechanical, and Corrosion Behavior of AlFeCuCrCoNi-WC_x High-Entropy Alloy Coatings. Journal of Materials Engineering and Performance (2021). https://doi.org/10.1007/s11665-021-05523-8
93. Zhu, M.; Zhao, B. Z.; Yuan, Y. F.; Guo, S. Y.; Pan, J. Influence of Microstructure and Passive Film on the Corrosion Behavior of CoCrFeMnNi High-Entropy Alloy Prepared by Different Cooling Methods in a Simulated Alkaline Soil Solution. Journal of Materials Engineering and Performance (2022). https://doi.org/10.1007/s11665-022-06966-3
94. Sim, R. K.; Xu, Z.; Wu, M. Y.; He, A.; Chen, D. L.; Li, D. Y. Microstructure, mechanical properties, corrosion and wear behavior of high-entropy alloy $AlCoCrFeNi_x$ (removenotallowedtagthe) and medium-entropy alloy (removenotallowedtag). Journal of Materials Science (2022). https://doi.org/10.1007/s10853-022-07256-y
95. Zhou, Chao; Wang, Shouren; Wang, Gaoqi; Zhang, Jianpeng; Liu, Wenlong Investigation on the microstructure, wear and corrosion resistance of $FeCoNiCrMo_x$ high-entropy alloy coatings deposited on 40Cr by laser cladding. Journal of Materials Science (2022). https://doi.org/10.1007/s10853-022-07684-w
96. López Ríos, M.; Socorro Perdomo, P. P.; Voiculescu, I.; Geanta, V.; Crăciun, V.; Boerasu, I.; Mirza Rosca, J. C. Effects of nickel content on the microstructure, microhardness and corrosion behavior of high-entropy AlCoCrFeNix alloys. Scientific Reports (2020). https://doi.org/10.1038/s41598-020-78108-5
97. Ayuba, Samuel Ushe; Fayomi, Ojo Sunday Isaac; Omotosho, Olugbenga Adeshola Investigation of microstructure, mechanical and corrosion resistance properties of Cu–Al–Zn–Sn high-entropy alloy for kitchen utensils and other industrial applications. Chemical Papers (2022). https://doi.org/10.1007/s11696-022-02336-4
98. Chen, Xu; Hu, Jiaxuan; Liu, Ye; Xiang, Feng Corrosion Behavior in Nitric Acid Solution and Tensile Properties of $(CuFeNiMn)_{1-x}Cr_x$ HEAs. Metals and Materials International (2020). https://doi.org/10.1007/s12540-020-00887-3
99. Veerappan, G.; Ravichandran, M.; Mohanavel, V.; Pritima, D.; Rajesh, S. Effect of Copper on Mechanical Properties and Corrosion Behavior of Powder Metallurgy Processed Ni–Co–Cr–Fe–Mn–Cu_x High Entropy Alloy. Arabian Journal for Science and Engineering (2022). https://doi.org/10.1007/s13369-022-07041-8
100. Ma, Guoliang; Zhao, Yong; Cui, Hongzhi; Song, Xiaojie; Wang, Mingliang; Lee, Kwangmin; Gao, Xiaohua; Song, Qiang; Wang, Canming Addition Al and/or Ti Induced Modifications of Microstructures, Mechanical Properties, and Corrosion Properties in CoCrFeNi High-Entropy Alloy Coatings. Acta Metallurgica Sinica (English Letters) (2021). https://doi.org/10.1007/s40195-021-01219-z
101. Zhang, Boliang; Zhang, Yi; Guo, S. M. A thermodynamic study of corrosion behaviors for CoCrFeNi-based high-entropy alloys. Journal of Materials Science (2018). https://doi.org/10.1007/s10853-018-2652-2
102. Liang, Xuewei; Su, Yunhai; Yang, Taisen; Dai, Zhiyong; Wang, Yingdi Effect of Ti Addition on the Microstructure and Corrosion Resistance of $AlCrCuFeNiNb_{0.2}$ High-Entropy Alloy. JOM (2022). https://doi.org/10.1007/s11837-022-05620-6

103. Bijnavandi, Mehrab Seifpour; Ghaderi, Armin; Dehghani, Kamran Effect of Cold Rolling on the Microstructure Evolution, Mechanical, and Corrosion Properties of $AlCoCrFeNi_{2.4}$ High-Entropy Alloy. Journal of Materials Engineering and Performance (2023). https://doi.org/10.1007/s11665-023-08101-2
104. Kumar, Devesh; Maulik, Ornov; Sharma, V. K.; Prasad, Y. V. S. S.; Kumar, Vinod Understanding the Effect of Tungsten on Corrosion Behavior of $AlCuCrFeMnW_x$ High-Entropy Alloys in 3.5 wt.% NaCl Solution. Journal of Materials Engineering and Performance (2018). https://doi.org/10.1007/s11665-018-3536-y

Chapter 9
High Entropy Alloys as Functional Materials

9.1 Introduction by Editor

Designing high entropy alloys as functional materials involves an interdisciplinary approach that combines materials science, solid-state physics, and other engineering principles. The category of functional high entropy alloys commonly consists of soft magnetic, thermoelectric, superconducting, catalytic, and hydrogen storage HEAs. This chapter briefly summarizes some of the functional properties, predicted and assessed, guiding us on the path of developing next-generation functional materials in the form of HEAs.

The origin of high entropy alloys can be traced to almost 2 decades ago to develop alloys with multi-principal elements inheriting well-known four core effects. Initially, the alloys were significantly adapted for the study from the structural application point of view as these alloys show significant strength, ductility, and other related mechanical properties. However, switching from conventional equiatomic to near equiatomic and from 3-d transition series to 4-d, 5-d, s-block, and p-block elements these alloys showed a significant potential in other properties which are of significant interest to materials physicists and chemists. As a result, the horizon of high entropy alloys made a gradual shift from materials with mechanical properties to materials with a wide range of properties including magnetic, electrical, chemical, etc. Not only this, the term "high entropy" is also adopted by core polymer and ceramic research communities to define and develop materials called high entropy polymers and high entropy ceramics which is unfortunately out of context in this book. From the metals and alloys point of view, HEAs were found accountable for enhancing the functional properties of the preexisting alloys that were contributing to this field well before the existence of HEAs. The unknown limit that was set by conventional alloys now has achieved new benchmarks and still has ample potential to evolve effectively.

Functional properties include a wide range of properties like catalytic converters, hydrogen storage mediums, batteries, superconductivity, etc. HEAs in this case have found prominent contributors in the role of substitute, modifier, or enhancer leading

S. S. Nene (ed.), *High Entropy Alloys*, https://doi.org/10.1007/978-981-99-7173-2_9

to optimized productivity or efficiency of the reaction. In the field of chemistry, HEAs are showing an effective contribution by developing excellent materials from 4 and 5d transition series with Ti for hydrogen storage. Few HEA nanoparticles are seen to be effective catalysts in delivering improved kinetics in orders of hundred. The existence of magnetic materials (soft magnets) with significant strength ductility synergy was considered to be practically impossible. HEAs proved to be a successful alternative here as well. For the superconductivity aspect, these alloys showed excellent superconductivity at comparatively high temperatures. A few of these properties along with some properties of functional importance have made high entropy alloys a breakthrough material. The limits are however not known, but HEAs still hold many possibilities for discovering new properties and the potential to change many fundamental theories with its evolution.

The development in this sector was more or less related to the core effects along with improvement in properties like transfer mechanisms, reaction mechanisms, and material kinetics altogether. This was necessary to improve the preexisting material potential by inducing various pathways, unpredicted or unidentified potential, and behavior that the material may undergo over by replacing it completely with a different methodology. Introducing HEAs here has not only contributed to advancement in properties but also made a significant change in the materials researchers' vision to look into material for targeted applications significantly contributing effectively to the scientific community.

This chapter summarizes the possible way of including this newly designed alloying approach leading to an evolution in the functional materials sector thereby incorporating the enhancement of the product efficiency and exploring the field or the potential of HEAs to another possible limit. This section will provide a glimpse of the implications of functional high entropy alloys as soft magnetic materials, thermoelectric materials, superconductors, catalytic converts, enhanced degradation materials, and hydrogen storage HEAs. Subsequently, this chapter briefly summarizes some of the functional properties, predicted and assessed, guiding us on the path for developing next-generation functional materials in the form of HEAs.

9.2 Machine Generated Summaries

Disclaimer: The summaries in this chapter were generated from Springer Nature publications using extractive AI auto-summarization: An extraction-based summarizer aims to identify the most important sentences of a text using an algorithm and uses those original sentences to create the auto-summary (unlike generative AI). As the constituted sentences are machine selected, they may not fully reflect the body of the work, so we strongly advise that the original content is read and cited. The auto generated summaries were curated by the editor to meet Springer Nature publication standards. To cite this content, please refer to the original papers.

Machine generated keywords: catalyst, hydrogen, reaction, activity, catalytic, nanoparticle, adsorption, site, functional, surface, nps, catalysis, efficiency, efficient, highly

Hydrogen Spillover-Driven Synthesis of High-Entropy Alloy Nanoparticles as a Robust Catalyst for CO_2 Hydrogenation [75]

This is a machine-generated summary of:

Mori, Kohsuke; Hashimoto, Naoki; Kamiuchi, Naoto; Yoshida, Hideto; Kobayashi, Hisayoshi; Yamashita, Hiromi: Hydrogen spillover-driven synthesis of high-entropy alloy nanoparticles as a robust catalyst for CO_2 hydrogenation [75]

Published in: Nature Communications (2021)

Link to original: https://doi.org/10.1038/s41467-021-24228-z

If you want to cite the papers, please refer to the original.

For technical reasons we could not place the page where the original quote is coming from.

Abstract-Summary

"We demonstrate that titanium dioxide (TiO_2) is a promising platform for the low-temperature synthesis of supported CoNiCuRuPd HEA nanoparticles (NPs) at 400 °C."

"This process is driven by the pronounced hydrogen spillover effect on TiO_2 in conjunction with coupled proton/electron transfer."

"The CoNiCuRuPd HEA NPs on TiO_2 produced in this work were found to be both active and extremely durable during the CO_2 hydrogenation reaction."

Introduction

"The development of new and simpler techniques for the synthesis of HEA NPs, especially those immobilized on the surfaces of conventional support materials, represents an ongoing challenge."

"Van Bokhoven and colleagues [1] reported experimental data in conjunction with theoretical calculations showing that hydrogen spillover on TiO_2 proceeds ten orders of magnitude faster than that on the non-reducible oxide Al_2O_3, and that TiO_2 provides longer migration distances from the noble metal proton sources."

"The highly specific formation of binary alloy NPs based on combinations of normally immiscible noble and base metals can be achieved with the assistance of the strong spillover effect obtained from TiO_2."

"Using this oxide allows spillover hydrogen species with high reduction potentials to be generated from noble metals (Ru or Rh) and to rapidly migrate to and reduce base metals (Ni or Cu) at low temperatures."

"CoNiCuRuPd HEA NPs on TiO_2 displayed high activity and outstanding stability during the CO_2 hydrogenation reaction."

Results

"These results suggest that the TiO_2 support ensured more rapid and homogeneous reduction at lower temperatures, allowing the formation of nuclei to provide smaller, uniform HEA NPs without segregation."

"The reduction of Ti^{4+} to Ti^{3+} together with the transfer of H atoms from Pd nuclei at the metal-support interfaces (Step 2) is accompanied by the migration of electrons from Ti^{3+} ions to neighboring Ti^{4+} ions."

"These results also indicated the initial formation of Pd_{core}–M_{shell} (M represents Co, Ni, Cu, or Ru metal) structure, which finally forms HEA NPs via atomic diffusion with increasing the reduction temperature, owing to the increase of configuration entropy."

"The reduction of other deposited metal cations by the spilled H atoms (Step 4) was further evaluated by calculating E_a for the attack of a neighboring H atom on an M^{n+}–OH species ($M = Co^{2+}$, Ni^{2+}, Cu^{2+}, Ru^{3+}, or Pd^{2+}) on the support, together with the loss of H_2O."

Discussion

"We succeeded in the facile low-temperature synthesis of supported HEA NPs, taking advantage of the hydrogen spillover that proceeds on TiO_2 via a coupled proton–electron transfer mechanism."

"Theoretical investigations also emphasized that the sluggish diffusion in these CoNiCuRuPd HEA NPs is caused by the combination of multiple metals, and that lattice distortion plays a crucial role in the superior robustness of this material."

"This study demonstrates not only an ideal heterogeneous catalyst based on HEA NPs with durability that suggests potential practical applications but also offers advanced insights into an innovative catalyst/photocatalyst architecture providing an essentially unlimited compositional space."

Methods

"The sample was reduced under a 20 mL/min flow of H_2 at 400 °C for 2 h to yield CoNiCuRuPd/TiO_2 (Pd 1.0 wt%; Co:Ni:Cu:Ru:Pd = 1:1:1:1:1 [on a molar basis])."

"Simulations for the formation mechanism of HEA NPs via H_2 spillover were performed using a $TiO_2(101)$ slab with 2×2 surface unit cell and three-layer thickness was constructed with a vacuum thickness of 20 Å, on which a square pyramidal Pd_5 cluster was loaded."

"For the cohesive energy (E_c) calculation of pure metal and HEA clusters composed of Co, Ni, Cu, Ru, and Pd, the plane wave-based program Castep was employed [2, 3]."

"Sphere-like 79 and 81 atom clusters were used for FCC and HCP metals, and the clusters were placed in a cubic cell with a side of 30 Å. For the alloy cluster preparation, Pd atoms in Pd_{79} cluster were randomly replaced by Co, Ni, Cu, and Ru atoms, and $Co_{16}Ni_{15}Cu_{16}Ru_{16}Pd_{16}$ cluster was built."

Fast Site-to-Site Electron Transfer of High-Entropy Alloy Nanocatalyst Driving Redox Electrocatalysis [76]

This is a machine-generated summary of:

Li, Hongdong; Han, Yi; Zhao, Huan; Qi, Wenjing; Zhang, Dan; Yu, Yaodong; Cai, Wenwen; Li, Shaoxiang; Lai, Jianping; Huang, Bolong; Wang, Lei: Fast site-to-site electron transfer of high-entropy alloy nanocatalyst driving redox electrocatalysis [76]

Published in: Nature Communications (2020)

Link to original: https://doi.org/10.1038/s41467-020-19277-9

If you want to cite the papers, please refer to the original.

For technical reasons we could not place the page where the original quote is coming from.

Abstract-Summary

"The $Pt_{18}Ni_{26}Fe_{15}Co_{14}Cu_{27}/C$ catalyst exhibits excellent electrocatalytic performance for hydrogen evolution reaction (HER) and methanol oxidation reaction (MOR)."

"The catalyst shows ultrasmall overpotential of 11 mV at the current density of 10 mA cm^{-2}, excellent activity (10.96 A $mg^{-1}{}_{Pt}$ at − 0.07 V vs. reversible hydrogen electrode) and stability in the alkaline medium."

"Periodic DFT calculations confirm the multi-active sites for both HER and MOR on the HEA surface as the key factor for both proton and intermediate transformation."

"The construction of HEA surfaces supplies the fast site-to-site electron transfer for both reduction and oxidation processes."

Introduction

"To improve the activity, selectivity, and stability of the catalytic reaction, it is necessary to develop high-performance advanced catalysts that can meet the needs of rapid development [4–10]."

"High-entropy alloys (HEAs) have attracted wide interest as catalytic materials in the past few years [11–14]."

"Some HEAs had used as catalysts for electrocatalytic reactions, which display superior stability and catalytic selectivity and activity compared with traditional alloys [11, 15–19]."

"The $Pt_{18}Ni_{26}Fe_{15}Co_{14}Cu_{27}$/C catalyst displays a low overpotential (11 mV at 10 mA cm^{-2}), the activity of 10.96 A $mg^{-1}{}_{Pt}$ at − 0.07 V vs. reversible hydrogen electrode (RHE) for HER (reduction reaction) in 1 M KOH solution, and through the chronoamperometric method and 10,000th CV tests, it indicates that the notable stability of the HEA catalyst."

"Theoretical calculations reveal that each element in HEA displays different contributions for the electrocatalysis process, which promotes the site-to-site electron transfer and the stabilization of the intermediates."

Results

"To further evaluate the stability of $Pt_{18}Ni_{26}Fe_{15}Co_{14}Cu_{27}$/C and Pt/C catalysts, the 1000th and 10,000th CV cycles in 1 M KOH solution were performed."

"To further explore the $Pt_{18}Ni_{26}Fe_{15}Co_{14}Cu_{27}$/C catalyst as redox bi-function electrocatalysts, we tested the MOR (oxidation reaction) activity."

"To study the stability of the $Pt_{18}Ni_{26}Fe_{15}Co_{14}Cu_{27}$/C and Pt/C catalysts in MOR, the chronoamperometry test at 0.65 V vs. RHE and 1000 CV cycles were executed."

"These test results indicate that the NPs-HEAs catalyst achieves double enhancement of catalytic activity and stability."

"The I_f/I_b (I_f: forward current density, I_b: backward current density) ratio of $Pt_{18}Ni_{26}Fe_{15}Co_{14}Cu_{27}$/C catalyst (3.26) is larger than that of Pt/C (2.31), which shows the strong CO anti-poisoning performance of $Pt_{18}Ni_{26}Fe_{15}Co_{14}Cu_{27}$/C catalyst."

"Fe shows a site-independent electronic structure within HEA, which preserves the stable adsorption of intermediates with stronger anti-poisoning capability for the MOR process."

Discussion

"It is also the effective catalysts for MOR and displayed the excellent activity (15.04 A $mg^{-1}{}_{Pt}$) and better CO anti-poisoning in alkaline solution, which is the best alkaline MOR activity among the ever reported."

"Through DFT calculations, the origin of remarkable electroactivity and durability of HEA in HER and MOR is attributed to the synergistic effect of each element for efficient electron transfer."

"The suitable electronic environment of the HEA realizes the multi-active sites for appropriate adsorption of key intermediates and efficient electron transfer during the electrocatalysis, which maximizes the utilization of surface electroactivity."

"The simple oil phase synthesis strategy proposed in this work, as well as the multi-active sites and fast site-to-site electron transfer mechanism, which is expected to lay the foundation for the preparation of other HEAs and their application in related electrocatalysis."

Methods

"Cyclic voltammograms (CVs) were performed in N_2-saturated 1 M KOH solution from 0 to 1.2 V vs. RHE at a scan rate of 20 mV s^{-1}."

"The CVs curve was collected from 0 to 1.2 V vs. RHE in 0.5 M H_2SO_4 solution with a scan rate of 20 mV s^{-1}: where n is the number of active sites, number 2 represents two electrons (produce one hydrogen molecule), Q_H represents electron transfer quantity (the hydrogen desorption peak), F is Faraday's constant (96,485.3 C mol^{-1}), and I (A) is the current measured at a specific potential during LSV measurement."

"The CVs for MOR were conducted in N_2-saturated 1 M KOH + 1 M CH_3OH solution between 0.2–1.2 V vs. RHE with a scan rate of 20 mV s^{-1}."

Highly Efficient Decomposition of Ammonia Using High-Entropy Alloy Catalysts [77]

This is a machine-generated summary of:

Xie, Pengfei; Yao, Yonggang; Huang, Zhennan; Liu, Zhenyu; Zhang, Junlei; Li, Tangyuan; Wang, Guofeng; Shahbazian-Yassar, Reza; Hu, Liangbing; Wang, Chao: Highly efficient decomposition of ammonia using high-entropy alloy catalysts [77]

Published in: Nature Communications (2019)

Link to original: https://doi.org/10.1038/s41467-019-11848-9

If you want to cite the papers, please refer to the original.

For technical reasons we could not place the page where the original quote is coming from.

Abstract-Summary

"We report on highly efficient ammonia decomposition using novel high-entropy alloy (HEA) catalysts made of earth abundant elements."

"These HEA nanoparticles demonstrate substantially enhanced catalytic activity and stability for ammonia decomposition, with improvement factors achieving > 20 versus Ru catalysts."

"Catalytic activity of HEA nanoparticles is robustly tunable by varying the Co/Mo ratio, allowing for the optimization of surface property to maximize the reactivity under different reaction conditions."

Introduction

"Bimetallic Co-Mo has been shown to be promising for ammonia decomposition, as rationalized by the computational interpolation of adsorption properties on the alloy surfaces with mixed sites [20, 21]."

"How to overcome this hurdle is of great interest for both fundamental understanding of alloy catalysts and practical improvement of the performance of earth-abundant materials in energy-conversion systems."

"We report on a new class of high-entropy alloy (HEA) catalysts made of earth-abundant elements for highly efficient decomposition of ammonia."

"These HEA nanoparticles are subjected to systematic and comprehensive catalytic studies for the NH_3 decomposition and compared to bimetallic Co-Mo and monometallic Ru catalysts."

"The obtained knowledge is further integrated with the measured reaction kinetics and simulated atomic structures of the HEA catalysts to understand the catalytic enhancement mechanisms."

Results

"The determined apparent activation energy (E_{app}) varies with the Co/Mo ratio, with the trend being consistent with that for the NH_3 conversion, namely from the lowest 64.7 kJ mol^{-1} for the most active HEA-$Co_{25}Mo_{45}$ to 123.1 kJ mol^{-1} for the least active HEA-$Co_{55}Mo_{15}$ catalyst."

"To the enhanced catalytic activities, the HEA-Co_xMo_y catalysts were further demonstrated to be highly stable under the reaction conditions for NH_3 decomposition."

"A clear trend is established for the HEA catalysts of different compositions, with the desorption temperatures (both onset and peak positions) rising at decreasing Co/

Mo ratios, suggesting the continuous tuning of nitrogen binding strength by varying the HEA composition."

"Discussion, we can attribute the high catalytic activity of NH_3 decomposition achieved with the HEA catalysts to the robust tuning of surface adsorption properties, as enabled by the continuous varying of alloy composition (Co/Mo ratio)."

Discussion

"The HEA CoMoFeNiCu nanoparticles enabled robust tuning of the Co/Mo ratio and breaking the miscibility limitation in conventional bimetallic Co-Mo catalysts."

"These novel catalysts achieved significantly enhanced catalytic activities and stabilities, with the improvement factors exceeding 20 times as compared to precious metal Ru, and even more so versus conventional Co-Mo catalysts."

"The robust tunability of alloy composition and surface adsorption property demonstrated on the HEA catalysts indicate great potential for their implementation in practical reactors with the catalytic performance optimized under various reaction conditions."

Methods

"Before measurements, the Co_xMo_y-HEA catalysts were reduced in H_2 at 600 °C for 2 h. The adventitious carbonaceous C 1 s line (284.6 eV) was used to calibrate the binding energy (BE)."

"30 mg of Co_xMo_y-HEA catalysts were loaded in a U-type sample tube, reduced in H_2 (30 mL min^{-1}) at 600 °C for 2 h, and then blown with He (30 mL min^{-1}) for 1 h to remove the hydrogen adsorbed on the catalyst surface."

"For measuring combinative N_2 desorption energy, three different ramping rates (10, 20, 30 °C min^{-1}) were used in a similar process."

"The catalyst was heated to 600 °C at a rate of 5 °C min^{-1} under H_2 (50 mL min^{-1}) for 2 h and then purged by Helium (50 mL min^{-1}) for 1 h. After this, the catalyst was cooled down to 275 °C under He atmosphere."

"The surface metal atoms of bimetallic Co-Mo and HEA-Co_xMo_y catalysts were estimated by assuming the nanoparticles as perfect FCC cubo-octahedral particles."

Superior Hydrogen Storage in High Entropy Alloys [78]

This is a machine-generated summary of:

Sahlberg, Martin; Karlsson, Dennis; Zlotea, Claudia; Jansson, Ulf: Superior hydrogen storage in high entropy alloys [78]

Published in: Scientific Reports (2016)

Link to original: https://doi.org/10.1038/srep36770

If you want to cite the papers, please refer to the original.

For technical reasons we could not place the page where the original quote is coming from.

Abstract-Summary

"This requires the design of alloys which allow for a very high H/M ratio."

"Transition metal hydrides typically have a maximum H/M ratio of 2 and higher ratios can only be obtained in alloys based on rare-earth elements."

"We demonstrate, for the first time to the best of our knowledge, that a high entropy alloy of TiVZrNbHf can absorb much higher amounts of hydrogen than its constituents and reach an H/M ratio of 2.5."

Introduction

"One possibility is to store hydrogen as a solid hydride using suitable metals or alloys."

"Metal hydrides have been widely studied as storage materials but most alloys are unable to fulfil the requirements of a competitive hydrogen storage unit that can be exploited in practical applications."

"It is well-known that metals and alloys with a BCC-type structure have a large capacity to store hydrogen (see e.g. Refs. [22–24])."

"These two factors suggest that high entropy alloys (HEA) can have excellent hydrogen storage potential due to their important lattice strain."

"The large difference in atomic radii of the elements ($\delta = 6.82\%$) suggest a significant lattice distortion that could favour a more efficient hydrogen storage capacity."

"The aim of the present work has been to investigate hydride formation in TiZrHfNbV and to evaluate its potential for use as a hydrogen storage material."

Methods

"Equimolar TiVZrNbHf samples were synthesized by arc-melting stoichiometric amounts of Ti (99.995% pure, Chempur), V (99.95% pure, MRC), Zr (99.8% pure, Chempur)."

"The chemical composition and microstructure was studied using a Zeiss 1550 scanning electron microscope (SEM) equipped with a secondary electron (SE) detector, a backscatter electron (BSE) detector and an energy dispersive X-ray spectrometer (EDS). (TiVZrNbHf)H_x was synthesized by subjecting powdered TiVZrNbHf (ball milled and sieved to < 50 μm particle size) to 20 bar H_2 at 400 °C for 48 h in an alumina crucible placed in a purpose built high pressure furnace."

"The mass of the samples were measured carefully before and after hydrogen absorption to calculate the storage capacity."

"The pressure-composition-isotherm (PCT) measurements were performed on the TiVZrNbHf in powder form using the SETARAM PCTPRO volumetric instrument."

"The structure of the sample as-synthesized was refined from the X-ray diffraction data using the Rietveld method [25] implemented in the FullProf [26] program."

Results

"The hydrogen storage properties of TiVZrNbHf were studied in two independent experiments: gravimetric measurements using a high pressure furnace (difference between the hydrogenated and initial sample) and volumetric pressure-composition-isotherm at 299 °C."

"In the gravimetric measurements made using the high pressure furnace, the TiVZrNbHf sample absorbed hydrogen easily above 200 °C."

"Gravimetric measurements made before and after hydrogen absorption gave a ratio for H/M of 2.42."

"The XRD measurements of the samples after desorption show that the hydrogen absorption/desorption is reversible and that the BCC-type structure of the sample as-synthesized was regained."

"The peaks in the XRD pattern after desorption are doublets indicating a phase separation into two different BCC phases with slightly different unit cell parameters during hydrogen desorption."

"These binary hydrides have a maximum H/M ratio of 2, while we observe an H/M ratio of 2.5 in the fully hydrogenated TiVZrNbHf alloy."

Discussion

"The observed H/M ratio of 2.5 in the hydrogenated TiVZrNbHf alloy requires that both tetrahedral sites and about 50% of the octahedral sites are filled with hydrogen."

"Alloys with both transition metals and rare earth (RE) metals (lanthanides and yttrium) are known to form hydrides with H/M > 2 but our results indicate that using an HEA could be a feasible and useful strategy to avoid inclusion of RE elements in hydrogen storage materials."

"This requires that hydrogen can be placed in both octahedral and tetrahedral sites in the HEA in contrast to other transition metal hydrides."

"The formation of a distorted FCC or BCT structure in the fully hydrogenated alloy is also similar to the structure formed in rare-earth compounds."

"It is suggested that the unprecedented hydrogen storage capacity is an effect of the strain in the distorted HEA lattice, which favours hydrogen occupying both tetrahedral and octahedral sites."

Additional Information

"How to cite this article: Sahlberg, M. and others Superior hydrogen storage in high entropy alloys."

"Publisher's note: Springer Nature remains neutral with regard to jurisdictional claims in published maps and institutional affiliations."

Unraveling the Electronegativity-Dominated Intermediate Adsorption on High-Entropy Alloy Electrocatalysts [79]

This is a machine-generated summary of:

Hao, Jiace; Zhuang, Zechao; Cao, Kecheng; Gao, Guohua; Wang, Chan; Lai, Feili; Lu, Shuanglong; Ma, Piming; Dong, Weifu; Liu, Tianxi; Du, Mingliang; Zhu, Han: Unraveling the electronegativity-dominated intermediate adsorption on high-entropy alloy electrocatalysts [79]

Published in: Nature Communications (2022)

Link to original: https://doi.org/10.1038/s41467-022-30379-4

If you want to cite the papers, please refer to the original.

For technical reasons we could not place the page where the original quote is coming from.

Abstract-Summary

"We report a conceptual and experimental approach to overcome the limitations of single-element catalysts by designing a FeCoNiXRu (X: Cu, Cr, and Mn) High-entropy alloys system with various active sites that have different adsorption capacities for multiple intermediates."

"The electronegativity differences between mixed elements in HEA induce significant charge redistribution and create highly active Co and Ru sites with optimized energy barriers for simultaneously stabilizing OH^* and H^* intermediates, which greatly enhances the efficiency of water dissociation in alkaline conditions."

"This work provides an in-depth understanding of the interactions between specific active sites and intermediates, which opens up a fascinating direction for breaking scaling relation issues for multistep reactions."

Introduction

"Scaling relation issues applicable to many multistep reactions require the stabilization of multiple intermediates during the reaction, but the limited number of active sites with monometallic catalysts cannot achieve simultaneous stabilization of all intermediates on the available active sites with optimal binding energies [27, 28]."

"The relationships between multiple active sites in HEAs and reaction intermediates are still obscure with the lack of guidelines for designing reasonable active sites."

"We propose a conceptual and experimental approach to overcome the limitations of single-element catalysts by designing a FeCoNiXRu (X: Cu, Cr, and Mn) HEA system with two kinds of active sites that have different adsorption capacities for multiple intermediates."

"Constructing controllable multifunctional active sites on HEA surfaces allows for different interactions with related intermediates, which is a powerful option for solving this scaling relation challenge."

Results

"No separated XRD peaks from Fe, Co, Ni, Mn, Ru, or metal oxides were observed, suggesting the formation of a single-phase HEA."

"The above results indicated that the Fe, Co, Ni, and Ru sites in HEAs play different roles in the HER, and we further present a series of FeCoNiXRu (X = Mn, Cr, Cu) HEAs by using fifth elements (X) with different electronegativities to understand the relationship between electronegativity and HER activity."

"In the FeCoNiMnRu HEA, the electronegativity differences in Fe (1.83), Co (1.88), Ni (1.91), Mn (1.55), and Ru (2.20) induce significant charge redistribution and create the most active Co and Ru sites with optimized energy barriers for simultaneously stabilizing OH^* and H^* intermediates, greatly promoting the HER efficiency in alkaline solution."

"In a FeCoNiMnRu NP, the Co site was the preferred active site with the lowest energy barriers (0.34 eV) for water dissociation when compared with Fe, Ni, and Ru sites."

Discussion

"FeCoNiMnRu HEA NPs were designed and synthesized in electrospun CNFs by combining the electrospinning technique and graphitization process."

"In the FeCoNiMnRu HEA, the Co site facilitated H_2O dissociation, and the Ru sites simultaneously accelerated the combination of H^* to H_2."

"The ability to stabilize multiple intermediates on various active sites in HEAs was verified experimentally and theoretically."

"Adjustments of the HER activities of HEA catalysts were shown experimentally and theoretically by tailoring the electronegativities of the composition."

Methods

"X-ray diffraction (XRD) patterns were taken via a Smartlab 9kw advance powder X-ray Cu K_α radiation diffractometer ($\lambda = 1.5406$ Å) in the 2θ range of 20–80° at the scanning rate of 0.5 or 10° min^{-1}, with the Cu K_α source operating at 40 kV and 40 mA. X-ray photoelectron (XPS) spectra were acquired by Thermo Scientific K-Alpha with the Al (mono) K_α source (1486.6 eV) operating at 12 kV and 6 mA. The binding energies were calibrated with C 1 s (284.8 eV) as the standard."

"For the comparison of mass activities and eliminating the difference in electrode structure, Pt/C, RuO_2, IrO_2, and FeCoNiMnRu/CNFs powder were dropped onto GCE, and the calculation process is shown as follow: $J_{mass} = J_{geo} \times 0.07069\ cm^2/m$, $m = 3\ mg \times \omega \times (5\ \mu L/1000\ \mu L)$, where m and ω are the mass loading and mass fraction of noble metal (Ru, Pt, or Ir)."

Subnanometer High-Entropy Alloy Nanowires Enable Remarkable Hydrogen Oxidation Catalysis [80]

This is a machine-generated summary of:

Zhan, Changhong; Xu, Yong; Bu, Lingzheng; Zhu, Huaze; Feng, Yonggang; Yang, Tang; Zhang, Ying; Yang, Zhiqing; Huang, Bolong; Shao, Qi; Huang, Xiaoqing: Subnanometer high-entropy alloy nanowires enable remarkable hydrogen oxidation catalysis [80]

Published in: Nature Communications (2021)

Link to original: https://doi.org/10.1038/s41467-021-26425-2

If you want to cite the papers, please refer to the original.

For technical reasons we could not place the page where the original quote is coming from.

Abstract-Summary

"We synthesize unique PtRuNiCoFeMo HEA subnanometer nanowires (SNWs) for alkaline hydrogen oxidation reaction (HOR)."

"The mass and specific activities of HEA SNWs/C reach $6.75\ A\ mg_{Pt+Ru}{}^{-1}$ and $8.96\ mA\ cm^{-2}$, respectively, which are 2.8/2.6, 4.1/2.4, and 19.8/18.7 times higher than those of HEA NPs/C, commercial PtRu/C and Pt/C, respectively."

"Density functional theory calculations reveal that the strong interactions between different metal sites in HEA SNWs can greatly regulate the binding strength of proton and hydroxyl, and therefore enhances the HOR activity."

Introduction

"The ideal catalysts for HOR should simultaneously fulfill the features of high activity, high durability as well as low cost."

"As promising catalysts in heterogeneous catalysis, HEAs generally consist of five or more metals, which display the structure-dependent synergistic effects and enhanced catalytic performance in terms of activity and durability [29, 30]."

"It is highly desirable to develop facile strategies for fabricating PtRu-based HEAs."

"We fabricated PtRuNiCoFeMo HEA subnanometer nanowires (SNWs) as highly active and durable catalyst for HOR."

"The mass activity of HEA SNWs for HOR is 4.1 and 19.8 times higher than that of commercial PtRu/C and Pt/C, respectively."

"Density functional theory (DFT) calculations confirm that the strong interactions between different metals in HEA SNWs can regulate the electronic structures of different metals and thus enhance the HOR activity."

Results

"To reveal the mechanism for the enhanced HOR performance over HEA SNWs/C, the surface properties of catalyst was studied by X-ray photoelectron spectroscopy (XPS) measurement."

"Considering that only Ru^0 3p XPS spectrum negatively shifts by 0.36 eV, we might conclude that the positive shifts of Ni^0 2p, Co^0 2p, Fe^0 2p, and Mo^{4+} 3d XPS spectra were attributed to the electron transfer from Ni, Co, Fe and Mo to Ru in HEA SNWs."

"With respect to the electronic structures, the strong coupling between the bonding and anti-bonding orbitals on the HEA surfaces delivers a high HOR activity with efficient site-to-site electron transfer between different metal sites."

"It is noted that the overall Pt-5d orbital is further away from the Fermi level compared to the pure Pt metal (black color), indicating that HEA SNWs has an appropriate level of Pt-5d band for H_2 adsorption and therefore exhibits enhanced HOR activity."

Discussion

"We have demonstrated a successful synthesis of PtRuNiCoFeMo HEA SNWs with abundant surface atomic steps and facet boundaries for efficient HOR."

"Compared with the state-of-the-art PtRu/C and Pt/C catalysts, HEA SNWs/C exhibits advantages, including high activity, enhanced stability, and promising resistance to CO poisoning, being a promising catalyst for alkaline HOR."

"The mass activity and specific activity of HEA SNWs reach 6.75 A $mg_{Pt+Ru}{}^{-1}$ and 8.96 mA cm^{-2}, which are 2.8/2.6, 4.1/2.4, and 19.8/18.7 times higher than those of HEA NP/C, commercial PtRu/C and Pt/C, respectively."

Methods

"In a typical synthesis of PtRuNiCoFeMo HEA SNWs, 9.8 mg Pt(acac)$_2$, 5.0 mg Ru(acac)$_3$, 3.8 mg Ni(acac)$_2$, 7.0 mg Co(acac)$_3$, 7.0 mg Fe(acac)$_3$, 10 mg Mo(CO)$_6$, 60 mg glucose, and 39.3 mg STAB were dissolved in 5 mL OAm, followed by ultrasonication for 2 h. The mixture was then heated in an oil bath at 210 °C for 5 h. The cooled product was collected by centrifugation and washed three times with a cyclohexane/ethanol (v/v = 1:9) mixture."

"HOR measurements were performed in 0.1 M KOH solutions purged with high-purity H_2."

"The element ratios are consistent with the ICP-OES results of experiments, which show the component of 28.5/6.8/18.6/15.2/10.7/20.2 for Pt/Ru/Ni/Fe/Co/Mo. In this work, the slab model is constructed based on the random arrangements coding methods."

Development of a Novel High-Entropy Alloy with Eminent Efficiency of Degrading Azo Dye Solutions [81]

This is a machine-generated summary of:

Lv, Z. Y.; Liu, X. J.; Jia, B.; Wang, H.; Wu, Y.; Lu, Z. P.: Development of a novel high-entropy alloy with eminent efficiency of degrading azo dye solutions [81]

Published in: Scientific Reports (2016)

Link to original: https://doi.org/10.1038/srep34213

If you want to cite the papers, please refer to the original.

For technical reasons we could not place the page where the original quote is coming from.

Abstract-Summary

"We report that high entropy alloys, which have been emerging as a new class of metallic materials in the last decade, have excellent performance in degradation of azo dyes."

"The newly developed AlCoCrTiZn high-entropy alloy synthesized by mechanical alloying exhibits a prominent efficiency in degradation of the azo dye (Direct Blue 6: DB6), as high as that of the best metallic glass reported so far."

"The excellent capability of our high-entropy alloys in degrading azo dye is attributed to their unique atomic structure with severe lattice distortion, chemical composition effect, residual stress and high specific surface area."

"Our findings have important implications in developing novel high-entropy alloys for functional applications as catalyst materials."

Introduction

"Zero-valent metals have been extensively exploited in industrial applications because of their low cost, high degradation capability and simple operations, but their efficiency decays dramatically as a result of their rapid corrosion behavior in water, such as zero-valent-Mg [31]."

"It was reported that metallic glasses showed excellent capability in degrading azo dyes due to their thermodynamically metastable feature of atomic structures [31–34]."

"High-entropy alloys (HEAs) are emerging as a new class of metallic materials."

"In view of the good chemical and catalytic properties of glassy alloys [35, 36], it is expected that the metastable HEAs could also possess excellent properties in degrading contaminants."

"The investigation demonstrates that the HEAs powders have higher reaction efficiency in degrading azo dyes than commercial zero-valent ion powders."

"The newly developed AlCoCrTiZn HEA powder exhibits an excellent performance, which is comparable to the best performance reported in metallic glasses so far."

Results

"To test the degradation capacity of the resultant HEAs, we put 0.1 g HEA powders into 10 ml azo dye solution with a concentration of 0.2 g/L Direct Blue 6 (DB6, $C_{32}H_{20}N_6S_4O_{14}Na_4$) each time."

"It was reported that the BM G-MgZn is about 20 times faster than the BM Fe-based glassy powder (BM G-Fe, $t_0 = 16$ min) and 1000 times faster than the gas-atomization Fe-based glassy powder (GA G-Fe, $t_0 = 810$ min) in degradation of DB6 [31]."

"It should be pointed out that the degradation efficiency of the BM G-Fe is about 200 times faster than that of the widely used commercial Fe powder [32]."

"These results indicate that the newly developed S1 HEA powder can reduce the energy barrier for the degradation reaction effectively and thus make the occurrence of reaction easier as compared with other materials such as the glassy Fe-based powders."

Discussion

"Effects of surface chemical segregation on the degradation performance can be negligible in the case of as-milled HEA powders."

"The results indicate that the ion concentrations of all the five constituents in the treated solutions are below their detection limits of the machine, i.e., 0.067, 0.016, 0.017, 0.005 and 0.012 μg/ml for Al, Co, Cr, Ti and Zn, respectively, clearly demonstrating that the degradation of azo dyes by using HEA powders as agents does not induce secondary pollution."

"We demonstrated that the HEA powders synthesized by mechanical alloying, particularly the newly developed AlCoCrTiZn HEA, exhibited high efficiency in degradation of azo dye, comparable with the best metallic glasses."

"The combination of unique atomic structure with severe lattice distortion, chemical composition effect, residual stress and high specific surface area is responsible for the observed excellent capability of HEAs in degrading azo dye."

Methods

"Al, Ti, Cr, Mn, Fe, Co, Ni, Zn elemental powders with high purity (> 99.5%) and particle size of less than 45 μm were used as raw materials to synthesize equal-atomic AlCoCrTiZn, AlCoCrFeNi and CoCrFeMnNi HEAs by mechanical alloying."

"The milling was carried out in high energy planetary ball miller at 300 rpm with a ball to powder ratio of 15:1."

"After 60 h of milling the powders were completely taken out for further characterization and testing."

"For all the degradation tests, 0.1 g HEA powders were dissolved into 10 ml of azo dyes solution for reaction."

"The TEM samples were prepared by dropping powders dispersed in absolute ethanol upon copper grids with carbon film as voids support."

"The morphology and size distribution were studied by SEM (Zeiss Supra 55), and the compositions of the milled powders were examined by EDX equipped on the SEM."

Ab Initio to Activity: Machine Learning-Assisted Optimization of High-Entropy Alloy Catalytic Activity [82]

This is a machine-generated summary of:

Clausen, Christian M.; Nielsen, Martin L. S.; Pedersen, Jack K.; Rossmeisl, Jan: Ab Initio to Activity: Machine Learning-Assisted Optimization of High-Entropy Alloy Catalytic Activity [82]

Published in: High Entropy Alloys & Materials (2022)

Link to original: https://doi.org/10.1007/s44210-022-00006-4

If you want to cite the papers, please refer to the original.

For technical reasons we could not place the page where the original quote is coming from.

Abstract-Summary

"High-entropy alloys are slowly making their debut as a platform for catalyst discovery, but conventional methods, theoretical as well as experimental, may fall short of screening the vast composition space inhabited by this class of materials."

"New theoretical approaches are needed to gauge the catalytic activity of high-entropy alloys and optimize the alloy composition within a feasible time frame as a prerequisite for further experimental studies."

"We establish a workflow for simulations of catalysis on high-entropy alloy surfaces."

"This result shows that an unbiased in silico pre-screening and discovery of catalyst candidates is viable and will help scale the otherwise insurmountable challenge of searching for high-entropy alloy catalysts."

"It is our hope that our computational framework, which is freely available on GitHub, will aid other research groups to efficiently identify promising high-entropy alloy catalysts."

Introduction

"The canonical process of catalytic theory has been to utilize quantum–mechanical calculations to model the relative energies of specific surface structures and adsorption sites."

"The energies have been used to establish theoretical frameworks to guide our understanding of catalytic activity and surface-adsorbate interaction e.g. the Brønsted–Evans–Polanyi principle [37–39] or scaling relations [40, 41]."

"We instead consider the probability distribution of possible sites and their differing surface properties, such as the adsorption energy of reaction intermediates, which can then be coupled to an expected catalytic activity via established theory [42]."

"We propose that a precise adsorption energy regression algorithm, together with effects of co-adsorption of multiple species, will make it possible to rapidly gauge the catalytic activity of an HEA surface given a specific composition."

"From sampling the composition space with density functional theory (DFT) calculations to a predictive measure of the net catalytic activity and lastly optimization of the alloy composition."

Sampling the HEA Surface

"By geometry optimizing each slab this way, any strain correction of energies is avoided when regressing the adsorption energies of sites on surfaces with other elemental compositions than what each slab was sampled from."

"These archetype motifs are known as ensembles [43, 44] and can be viewed as discrete adsorption energies broadened to Gaussian-like distributions of adsorption energies by the electronic effect of the local atomic environment surrounding the binding site."

"Having predicted the net adsorption energy distributions of adsorbates on the surface we link those distributions to an expected catalytic activity, which is kinetically modeled assuming the associative oxygen reduction with the following steps and Gibbs' free reaction energies: where ΔG_{*ads} is the adsorption energy of the corresponding adsorbate, e is the elementary charge and U the applied potential vs. the reversible hydrogen electrode (RHE)."

Conclusion

"The quinary Ag–Ir–Pd–Pt–Ru alloy and the quaternary sub alloys were represented as $3 \times 3 \times 5$ atom-sized surface slabs and a vacuum of 10.0 Å was added above and below the slab."

"The atoms in the two bottom layers were held fixed and the structures were optimized until the maximum force on any atom was below at least 0.1 eV/Å. The wave functions were expanded in plane waves with an energy cut-off set to 400 eV, and the Brillouin zone was sampled with a Monkhorst–Pack grid [45] of $4 \times 4 \times 1$ k-points."

"Exceptions to this are the Dirichlet-sampled dataset where each slab had its own set of probabilities drawn from a Dirichlet distribution with uniform density in the 5-dimension simplex space ($\alpha = 1$) and the atom identities were subsequently picked from that set of probabilities."

"By allocating the predicted surface atoms adsorption energy to their corresponding array indices, we obtained a gross adsorption grid."

Superconductivity in a Uranium Containing High Entropy Alloy [83]

This is a machine-generated summary of:

Nelson, W. L.; Chemey, A. T.; Hertz, M.; Choi, E.; Graf, D. E.; Latturner, S.; Albrecht-Schmitt, T. E.; Wei, K.; Baumbach, R. E.: Superconductivity in a uranium containing high entropy alloy [83]

Published in: Scientific Reports (2020)

Link to original: https://doi.org/10.1038/s41598-020-61666-z

If you want to cite the papers, please refer to the original.

For technical reasons we could not place the page where the original quote is coming from.

Abstract-Summary

"These systems exhibit remarkable functionality, often along several distinct axes: e.g., the examples $[TaNb]_{1-x}(TiZrHf)_x$ are high strength and damage resistant refractory metals that also exhibit superconductivity with large upper critical fields."

"Similar to the Zr-analogue, this material crystallizes in a body-centered cubic lattice with the lattice constant a = 3.41(1) Å and exhibits phonon mediated superconductivity with a transition temperatures $T_c \approx 3.2$ K and upper critical fields $H_{c2} \approx 6.4$ T. These results expand this class of materials to include actinide elements, shows that superconductivity is robust in this sub-group, and opens the path towards leveraging HEAs as functional waste forms for a variety of radioisotopes."

Introduction

"High entropy alloys (HEA) are multicomponent mixtures of randomly combined elements where, instead of producing disordered mixtures of lower order crystalline compounds (binaries, ternaries, etc.), solid solutions with simple structures (e.g., body center cubic, face center cubic, or hexagonal close packed) are formed [46–50]."

"Bulk electrical transport, magnetization, heat capacity, thermopower, and thermal conductivity measurements reveal other similarities to the Zr-analogues, where we find that the f-electrons are delocalized and that the chemical disorder results in a large and nearly temperature independent electrical resistivity and a small thermal conductivity. $[TaNb]_{0.31}(TiUHf)_{0.69}$ thus emerges as a representative of what likely is a much larger family of actinide materials that would be of importance for both technological (e.g., development of durable waste forms) and basic science (reservoir for superconductivity) reasons."

Results

"In the inset, the transition has a width of approximately $\Delta = 0.4$ K, and together with the large ρ_0 this suggests that the chemical disorder impacts both the normal state behavior and superconductivity."

"We suggest that the significant disorder in the system broadens the superconducting transition, with the bulk transition occurring near $T_c \approx 2.4$ K. From the resistive transitions, an upper limit for $H_{c2}(0)$ is determined using the expression $H_{c2}(T) = H_{c2}(0)(1 - (T/T_c)^n)$, where $H_{c2}(0) = 6.4$ T and n = 1.27."

"This value of $H_{c2}(0)$ is similar to Pauli-paramagnetic limit for weak electron–phonon coupling, $H_{c2}(0) = 1.84T_c \approx 5.7$ K. As suggested earlier for the Zr-based analogue, this may suggest that strong spin–orbit coupling plays a role in determining the properties of the superconducting state [51]."

Discussion and Conclusion

"T_c and H_{c2} for $[TaNb]_{0.31}(TiUHf)_{0.69}$ are suppressed from the values extrapolated from the Zr analogue; i.e., the presence of the uranium ions has the effect of weakening the superconducting state."

"We also point out that earlier work in the d-electron analogues shows that there are several other important parameters that can impact the superconductivity, including the electron count and the degree of disorder."

"Further work will be needed to address these questions; e.g., including X-Ray absorption spectroscopy to quantify the uranium valence and scanning tunneling electron microscopy to image the disorder. $[TaNb]_{0.31}(TiUHf)_{0.69}$ thus emerges as an example of a small but growing family of f-electron materials with both practical and fundamental implications."

Methods

"DC electrical resistivity ρ measurements were performed in a four wire configuration for flattened specimens using a Quantum Design Physical Properties Measurement Systems with He3 option down to temperatures $T > 0.4$ K and magnetic fields $H < 9$ T. Specific heat C measurements were also performed in the same system using a conventional thermal relaxation technique."

"The pressure was determined by the shift in ruby fluorescence peaks as measured below $T = 10$ K. The thermal conductivity was measured using the conventional one-heater-two-thermometer method."

"A step-wise increasing heater power (dP) to the chip resistor generated step-wise thermal gradient across the sample."

"The temperature gradient (dT) and the corresponding thermoelectric voltage (dV) were simultaneously measured when dT was stabilized at different heater power."

Influence of Mechanical Alloying on Structural, Thermal, and Magnetic Properties of $Fe_{50}Ni_{10}Co_{10}Ti_{10}B_{20}$ High Entropy Soft Magnetic Alloy [84]

This is a machine-generated summary of:

Panigrahi, Mrutyunjay; Avar, Baris: Influence of mechanical alloying on structural, thermal, and magnetic properties of $Fe_{50}Ni_{10}Co_{10}Ti_{10}B_{20}$ high entropy soft magnetic alloy [84]

Published in: Journal of Materials Science: Materials in Electronics (2021)

Link to original: https://doi.org/10.1007/s10854-021-06612-z

If you want to cite the papers, please refer to the original.

For technical reasons we could not place the page where the original quote is coming from.

Abstract-Summary

"A multicomponent with functional $Fe_{50}Ni_{10}Co_{10}Ti_{10}B_{20}$ (at.%) high entropy soft magnetic alloy powders were produced from the elemental powders by mechanical alloying (MA)."

"With prolonging milling time, the amorphous phase appeared after 20 h of MA."

"In the final stage of MA (50 h), the saturation magnetization (Ms) and the coercivity (Hc) were 89.7 emu/g and 32.5 Oe, respectively, proposing the alloy as a very good high entropy soft magnet in nature."

Introduction

"Depending on the multiple elements, HEAs are a novel class of alloys that provide improved performance by adjusting the composition and managing the process, where the phase composition is transformed from a single-phase solid solution to several complex phase compositions."

"In this research, the Fe-based HESM alloys were prepared using the MA process with high-energy mills because of its simple nature, reasonably low-cost, and controlled process parameters [52, 53]."

"The MA is an efficient powder production method to create an extensive range of stable and metastable materials such as intermetallics, supersaturated solid solution, nanocrystalline, quasicrystalline, and amorphous alloys."

"The present work was taken up with the intention of developing a Fe-based HESM amorphous alloys targeting to reduce the application of the valuable alloying elements with high thermal stability."

"The design of Fe-based HESM alloys with excellent properties is of tremendous significance for the growth of the application of high-temperature environment."

Strategy of the alloy design

"Pure iron (Fe) was decided to use as a primary constituent to decrease the total cost of the HESM alloys."

"Due to the above-observed enhancement, it motivated the authors to prepare various Fe-based HESM alloys."

"The Cobalt (Co) addition in Fe-based HESM alloys is familiar to form intermetallic forming elements."

"The corrosion resistance of the HESM alloys can increase with the addition of Ti content [54]."

"For metalloids such as Boron (B), an inexpensive material can be added to the Fe-based HESM alloys to increase the wear resistance and hardness by the formation of borides [55, 56]."

"Increase in the addition of B in the Fe-based HESM alloys indicate the decrease of the corrosion resistance property [57]."

"Taking into above consideration, the $Fe_{50}Ni_{10}Co_{10}Ti_{10}B_{20}$ (at.%) category of Fe-based HESM alloys was therefore designed."

Experimental procedure

"Every 20 min of MA process was followed by a suspension of 20 min to prevent excessive heating."

"The MA process was interrupted at predetermined intervals and a small amount of HESM alloy powder was taken for further characterization."

"The morphological change of the powder particles was analyzed by a scanning electron microscopy and energy-dispersive X-ray spectroscopy (SEM/EDX) using a FEI-Quanta FEG 450 microscope operated at 15 kV in secondary electron mode."

"The thermal behavior of the milled powders was determined by differential thermal analysis (DTA) using a SII Exstar TG/DTA 7200 in flowing N_2 atmosphere, temperature in the range of 300–950 °C at a heating rate of 20 °C/min."

Results and discussion

"After further increase of the milling duration up to 50 h, all crystalline peaks disappear and only a broad hump exists which estimates the formation of a high percentage of amorphous phase-like structure of the $Fe_{50}Ni_{10}Co_{10}Ti_{10}B_{20}$ HESM alloy powders [58]."

"As the milling time of the mixed powders increases up to 20 h, an additional broad peak was traced at lower temperature around 542 °C in addition to the distinct peak traced at around 642 °C, which suggest the transformation of the amorphous phase from the α-Fe solid solution phase during continuous heating."

"Further increasing the milling time of the $Fe_{50}Ni_{10}Co_{10}Ti_{10}B_{20}$ HESM alloy powder at 50 h exhibits the amorphous phase [59]."

"The Hc was determined as 13.3, 14.5, and 72.2 Oe, and while heat treated, the 50-h milled $Fe_{50}Ni_{10}Co_{10}Ti_{10}B_{20}$ HESM alloy powder at 300, 500, and 700 °C respectively."

Conclusions

"The highlight of the research is that the crystalline phases of the $Fe_{50}Ni_{10}Co_{10}Ti_{10}B_{20}$ HESM alloy were successively decreased and amorphism nature was enhanced with respect to the milling duration."

"The XRD analysis of 50-h milled $Fe_{50}Ni_{10}Co_{10}Ti_{10}B_{20}$ HESM alloy powder revealed the complete disappearance of the crystalline phases."

"The DTA curves showed the formation of crystallinity of the amorphous phase of the 50-h milled $Fe_{50}Ni_{10}Co_{10}Ti_{10}B_{20}$ HESM alloy powder."

"The increased Ms value (89.7 emu/g) in the 50-h milled $Fe_{50}Ni_{10}Co_{10}Ti_{10}B_{20}$ HESM alloy powder exhibits the amorphous phase."

"The heat treatment of the 50-h milled $Fe_{50}Ni_{10}Co_{10}Ti_{10}B_{20}$ HESM amorphous alloy powder at 300 °C increased the behavior of the saturation magnetization with high Ms value (101.9 emu/g) and reasonably less (38.3 emu/g) at 700 °C, which indicates full crystallization of the $Fe_{50}Ni_{10}Co_{10}Ti_{10}B_{20}$ HESM alloy."

High Critical Current Density and High-Tolerance Superconductivity in High-Entropy Alloy Thin Films [85]

This is a machine-generated summary of:

Jung, Soon-Gil; Han, Yoonseok; Kim, Jin Hee; Hidayati, Rahmatul; Rhyee, Jong-Soo; Lee, Jung Min; Kang, Won Nam; Choi, Woo Seok; Jeon, Hye-Ran; Suk, Jaekwon; Park, Tuson: High critical current density and high-tolerance superconductivity in high-entropy alloy thin films [85]

Published in: Nature Communications (2022)
Link to original: https://doi.org/10.1038/s41467-022-30912-5

If you want to cite the papers, please refer to the original.
For technical reasons we could not place the page where the original quote is coming from.

Abstract-Summary

"The feasibility of practical applications of HEA superconductors has not yet been demonstrated because the critical current density (J_c) for HEA superconductors has not yet been adequately characterized."

"We report the fabrication of high-quality superconducting (SC) thin films of Ta–Nb–Hf–Zr–Ti HEAs via a pulsed laser deposition."

"The superconductivity of the HEA films is more than 1000 times more resistant to displacement damage than that of other promising superconductors with technological applications, such as MgB_2, Nb_3Sn, Fe-based superconductors, and high-T_c cuprate superconductors."

"These results demonstrate that HEA superconductors have considerable potential for use under extreme conditions, such as in aerospace applications, nuclear fusion reactors, and high-field SC magnets."

Introduction

"The superconductivity of HEAs was discovered in 2014 in Ta–Nb–Hf–Zr–Ti multi-principal element alloy, attracting considerable interest in their pairing mechanism; generally, a high level of disorder in crystalline superconductors limits the formation of Cooper pairs partially owing to the decrease in the density of states or the increase in effective Coulomb repulsion between paired electrons [60, 61]."

"The unique properties of HEAs, such as their high hardness, high strength, and excellent irradiation tolerance, are advantageous for use under extreme conditions such as those in aerospace applications, irradiation environments, and superconducting (SC) rotating machines [50, 62–65]."

"We report high-J_c Ta–Nb–Hf–Zr–Ti HEA SC thin films fabricated by pulsed laser deposition (PLD) and the extraordinarily robust superconductivity against irradiation-induced disorder verified using 200 keV Kr-ion irradiation."

"The superconductivity of Ta–Nb–Hf–Zr–Ti HEA SC thin films was found to be approximately 1000 times more resistant to irradiation damage than that of other promising SC materials, including high-T_c cuprate superconductors."

Results and discussion

"The range of the dpa values for the HEA SC thin films differed significantly from those for the other superconductors: the superconductivity of the HEA SC thin films was over 1000 times more resistant to displacement damage than that of the other superconductors."

"Our irradiation study of Ta–Nb–Hf–Zr–Ti HEA SC thin films manifests the excellent robustness of the SC phase against irradiation damage, which is expected to open the door to many potential applications of HEA superconductors."

"The superconductivity of the HEA SC thin films was over 1000 times more robust to displacement damage than that of other representative SC materials, such as Nb_3Sn, MgB_2, Fe-based superconductors, and high-T_c cuprates."

"With the wide range of growth temperatures for high-quality thin films, these results demonstrate the great potential of HEA superconductors for practical applications under various extreme environments, such as aerospace applications, high-field SC magnets, and irradiation environments."

Methods

"High-quality Ta–Nb–Hf–Zr–Ti HEA SC thin films were fabricated on c-cut Al_2O_3 substrates (10 mm $\times$ 10 mm) using a PLD technique."

"By controlling the growth time, we obtained Ta–Nb–Hf–Zr–Ti HEA SC thin films with thicknesses in the range of 115–700 nm."

"Kr ions with a beam energy of 200 keV were used to irradiate Ta–Nb–Hf–Zr–Ti HEA SC thin films with a thickness of 115 nm at the Korea Multi-Purpose Accelerator Complex (KOMAC) in Gyeongju."

"The crystal structure of the fabricated Ta–Nb–Hf–Zr–Ti HEA SC thin films was investigated using an X-ray diffractometer (Rigaku miniflex-600 diffractometer, Cu-$K_{\alpha 1}$ radiation, $\lambda = 1.541$ Å)."

"The SC transition temperature (T_c) of the fabricated Ta–Nb–Hf–Zr–Ti HEA SC thin films was evaluated from the temperature dependence of the electrical resistivity (ρ) and magnetization (M) using a physical property measurement system (PPMS 9 T, Quantum Design) and a magnetic property measurement system (MPMS 5 T, Quantum Design), respectively."

High Entropy Liquid Electrolytes for Lithium Batteries [86]

This is a machine-generated summary of:

Wang, Qidi; Zhao, Chenglong; Wang, Jianlin; Yao, Zhenpeng; Wang, Shuwei; Kumar, Sai Govind Hari; Ganapathy, Swapna; Eustace, Stephen; Bai, Xuedong; Li,

Baohua; Wagemaker, Marnix: High entropy liquid electrolytes for lithium batteries [86]

Published in: Nature Communications (2023)

Link to original: https://doi.org/10.1038/s41467-023-36075-1

If you want to cite the papers, please refer to the original.

For technical reasons we could not place the page where the original quote is coming from.

Abstract-Summary

"We show this strategy in liquid electrolytes for rechargeable lithium batteries, demonstrating the substantial impact of raising the entropy of electrolytes by introducing multiple salts."

"Unlike all liquid electrolytes so far reported, the participation of several anionic groups in this electrolyte induces a larger diversity in solvation structures, unexpectedly decreasing solvation strengths between lithium ions and solvents/anions, facilitating lithium-ion diffusivity and the formation of stable interphase passivation layers."

"These findings, rationalized by the fundamental relationship between entropy-dominated solvation structures and ion transport, bring forward high-entropy electrolytes as a composition-rich and unexplored space for lithium batteries and beyond."

Introduction

"This motivates investigation as to how the main electrolyte functional properties are impacted, including redox stability, ion conductivity, charge transfer and solid electrolyte interphase (SEI) formation."

"For liquids, excess entropy scaling has been proposed and empirically demonstrated to be the relationship between the entropy of the system and its kinetic behavior, suggesting that increasing the entropy can result in improved diffusivity [66, 67], which has never been explored for electrolytes."

"Results show that more anionic groups can participate in the solvation structures (similar to highly concentrated electrolytes) in the low salt concentration electrolyte, however, the interactions of lithium ions with both solvent and anionic groups are much weaker in this HE electrolyte, a direct result from the higher disorder, which is never observed in any liquid electrolytes."

"This suggests that raising the entropy of mixing through introducing multiple salts in solvent provides a general strategy to tailor the functional properties for the development of electrolytes."

Results

"A downfield shift for the 0.6 M HE-DME electrolyte is observed, at − 0.68 ppm, demonstrating a relatively lower shielding of the lithium-ions, which may promote lithium-ion diffusivity based on a weaker solvation interaction [68]."

"The 0.6 M HE-DME shows an oxidation stable potential up to ~ 4.51 V, higher than the single salt (4.36 V for 0.6 M LiFSI-DME), where the subsequent capacity increase is suggested to be due to the formation of a cathode electrolyte interphase (CEI) on the surface of the cathode."

"The presence of these species could be responsible for a more facile and homogeneous lithium-ion supply, supporting dense lithium metal growth [69], and less decomposition of solvent species in 0.6 M LiFSI-DME electrolyte."

"This compositional analysis indicates that the CEI that formed in 0.6 M HE-DME electrolyte consists of components that support a high stability, in combination with a high conductivity."

Discussion

"The MD simulations demonstrate that the introduction of several salts into the HE electrolyte leads to a much larger diversity in solvation structures, expressing the higher entropy, and to a higher lithium-ion mobility, as compared to single salt electrolytes with the same salt concentration."

"In the HE electrolyte, the inorganic rich SEI/CEI and improved lithium-ion kinetics are attributed to the increasing entropy of mixing, resulting in denser lithium metal growth, despite the low concentration of HE electrolyte."

"The results show that more salt anions participate in the solvation structures of the low-concentration HE electrolyte, resulting in a thinner and inorganic-rich SEI compared to the single salt electrolyte."

"The observed weaker solvation strengths between lithium ions and the solvents/anions, as compared to the commonly single-salt low- and high-concentration electrolytes, is held responsible for the enhanced charge transfer and the improved electrolyte conductivity."

Methods

"Li||$LiFePO_4$ cells were cycled in the galvanostatic mode, whereas a voltage range of 2.5–5.0 V was used to gauge the oxidation stabilities of the different electrolytes for which $LiFePO_4$ cathode doesn't show the extra redox reaction from Fe^{2+} to Fe^{3+} above ~ 3.8 V. For electrochemical rate capabilities of Li||$Li_4Ti_5O_{12}$ and Li||NMC811 cells, the areal capacity of 2 mAh cm^{-2} for $Li_4Ti_5O_{12}$ (12 mm diameter) and capacity of 2 mAh cm^{-2} for NMC811 were used with lithium metal foils having a thickness of 250 μm as counter electrode."

"For cryo-TEM sample preparation and transfer, cells were disassembled immediately in an argon-filled glovebox after cycling and then both lithium metal anodes and NCM811 cathodes were rinsed with pure DME three times to remove lithium salts, followed by drying in the glove box for one hour at room temperature to remove the residual solvent."

Atomic Study of Hydrogen Behaviors at ∑3(111) Grain Boundary in Equiatomic CoCrNi and CoCrNiFe Alloys [87]

This is a machine-generated summary of:

Guan, Hua-Qing; Jing, Yan-Mei; Huang, Shao-Song: Atomic study of hydrogen behaviors at ∑3(111) grain boundary in equiatomic CoCrNi and CoCrNiFe alloys [87]

Published in: Tungsten (2022)

Link to original: https://doi.org/10.1007/s42864-022-00152-7

If you want to cite the papers, please refer to the original.

For technical reasons we could not place the page where the original quote is coming from.

Abstract-Summary

"High-entropy alloys are potential advanced structural materials for applications in nuclear energy due to their proved high irradiation performance."

"The solution energy, binding energy and diffusion barrier of H atoms at ∑3 GBs (GBs is short for grain boundaries) are presented."

"Based on the DFT data, we found that Cr limits the H atom to a specific potential position, thus suppresses H segregation."

"The lattice distortion effect-induced trapping for H provides higher H diffusion barriers at ∑3 GBs than that in pure Ni."

Introduction

"Zhou and others investigated H behaviors in CoCrNiFeMn HEAs by DFT and fatigue tests, and indicated that the stress fields of dislocation and fracture energies are increased to resist HE [70]."

"Investigating the H behavior at GBs and the interaction between H and solute atoms is beneficial to further understanding the reasons for the HEAs resistance to HE."

"Few experimental studies considered H atoms interactions at GBs in SP-CSAs, which is partly due to the difficulty for excluding effects of other structural imperfections, such as point defects, dislocations, or impurities, which can all influence the H behavior measurements [71]."

"In this study, the H behaviors at $\sum 3$ GBs in CoCrNi and CoCrNiFe alloys are systematically investigated using first-principles calculations to shed some lights on the GB-related HE mechanisms in HEA."

Methods

"The convergence criterions were performed until the force on each atom became less than 0.2 eV nm^{-1}, and the energy convergence of each atom less than 1×10^{-4} eV. The GB supercell consists of an orthorhombic cell with 144 atoms and contains two GBs separated by ten atomic (111) layers."

"The climbing-image nudged elastic band (CI-NEB) method was applied to find minimum energy path, and the diffusion barrier (DB) of H atoms at the $\sum 3(111)$ GBs in different diffusion paths was investigated [72, 73]."

"The solution energy of an interstitial H atom at the GBs is defined as The E_{GB+H}, and E_{GB} are the total energies of the GBs with a H atom, and without H atom."

"A negative solution energy indicates that the process of solution H atom at the GBs is exothermic."

Results and discussion

"The average solution energy of an interstitial H atom at Oct sites is lower than those at Tet sites in the alloys, suggesting the Oct sites are favorable positions for H atoms at the $\sum 3(111)$ GBs."

"It is interesting that the octahedral interstitial sites of Σ3 GB in CoCrNiFe can spontaneously dissolve H with lower solution energy than that in CoCrNi, indicating the GBs in CoCrNiFe can trap more H atoms."

"At the $\sum 3(111)$ GBs in CoCrNi and CoCrNiFe, two H atoms could not stay together at the same Oct or Tet site, with one of them would move to its nearest neighbor interstitial site."

"The sluggish effect originated from the fluctuation in the lattice potential energy [74], and reduced the diffusion of H. An additional important factor is that Cr and Fe atoms lose electrons to H atoms, which stabilized the H atom at the interstitial sites."

Conclusion

"First-principles calculations are carried out to investigate the H behavior, including solution, binding, and diffusion at $\sum 3$ (111) GBs in CoCrNi and CoCrNiFe alloys."

"The results show that although in CoCrNiFe alloy, H can spontaneously dissolve in the Oct sites at the $\sum 3$ (111) GBs, Cr limits the H atom to a specific potential position, thus suppresses H segregation."

"The interaction between Cr, Fe and H is much stronger than that between H and other atoms."

"The dipole moment interaction between H and Fe atoms weakens the binding of H atoms."

References

1. Karim, W. et al. Catalyst support effects on hydrogen spillover. Nature 541, 68–71 (2017).
2. Payne, M. C., Teter, M. P., Allan, D. C., Arias, T. A. & Joannopoulos, J. D. Iterative minimization techniques for ab initio total-energy calculations: molecular dynamics and conjugate gradients. Rev. Mod. Phys. 64, 1045–1097 (1992).
3. Milman, V. et al. Electronic structure, properties, and phase stability of inorganic crystals: a pseudopotential plane-wave study. Inter. J. Quantum Chem. 77, 895–910 (2000).
4. Li, L., Wang, P., Shao, Q. & Huang, X. Metallic nanostructures with low dimensionality for electrochemical water splitting. Chem. Soc. Rev. 49, 3072–3106 (2020).
5. Lai, J., Nsabimana, A., Luque, R. & Xu, G. 3D porous carbonaceous electrodes for electrocatalytic applications. Joule 2, 76–93 (2018).
6. Xie, C. et al. Insight into the design of defect electrocatalysts: from electronic structure to adsorption energy. Mater. Today 31, 47–68 (2019).
7. Jin, H. et al. Emerging two-dimensional nanomaterials for electrocatalysis. Chem. Rev. 118, 6337–6408 (2018).
8. Chen, Y. et al. Phase engineering of nanomaterials. Nat. Rev. Chem. 4, 243–256 (2020).
9. Tian, X., Lu, X. F., Xia, B. Y. & Lou, X. W. Advanced electrocatalysts for the oxygen reduction reaction in energy conversion technologies. Joule 4, 45–68 (2020).
10. Yuan, Y. et al. Zirconium nitride catalysts surpass platinum for oxygen reduction. Nat. Mater. 19, 282–286 (2020).
11. Batchelor, T. A. A. et al. High-entropy alloys as a discovery platform for electrocatalysis. Joule 3, 834–845 (2019).
12. George, E. P., Raabe, D. & Ritchie, R. O. High-entropy alloys. Nat. Rev. Mater. 4, 515–534 (2019).
13. Chen, P. C. et al. Polyelemental nanoparticle libraries. Science 352, 1565–1569 (2016).
14. Koo, W. T., Millstone, J. E., Weiss, P. S. & Kim, I. D. The design and science of polyelemental nanoparticles. ACS Nano 14, 6407–6413 (2020).
15. Zhang, G. et al. High entropy alloy as a highly active and stable electrocatalyst for hydrogen evolution reaction. Electrochim. Acta 279, 19–23 (2018).
16. Glasscott, M. W. et al. Electrosynthesis of high-entropy metallic glass nanoparticles for designer, multi-functional electrocatalysis. Nat. Commun. 10, 2650 (2019).
17. Jin, Z. et al. Nanoporous Al–Ni–Co–Ir–Mo high-entropy alloy for record-high water splitting activity in acidic environments. Small 15, 1904180 (2019).
18. Qiu, H. J. et al. Nanoporous high-entropy alloys for highly stable and efficient catalysts. J. Mater. Chem. A 7, 6499–6506 (2019).
19. Xie, P. et al. Highly efficient decomposition of ammonia using high-entropy alloy catalysts. Nat. Commun. 10, 4011 (2019).
20. Jacobsen, C. J. H. Novel class of ammonia synthesis catalysts. Chem. Commun. 12, 1057–1058 (2000).
21. Jacobsen, C. J. H., et al. Catalyst design by interpolation in the periodic table: Bimetallic ammonia synthesis catalysts. J. Am. Chem. Soc. 123, 8404–8405 (2001).
22. Sakintuna, B., Lamari-Darkrim, F. & Hirscher, M. Metal hydride materials for solid hydrogen storage: A review. Int. J. Hydrogen Energ. 32(9), p. 1121–1140, https://doi.org/10.1016/j.ijhydene.2006.11.022 (2007).
23. Akiba, E. & Iba, H. Hydrogen absorption by Laves phase related BCC solid solution. Intermetallics. 6(6), p. 461–470. (1998).
24. Akiba, E. & Nakamura, Y. Hydrogenation properties and crystal structures of Ti–Mn-V BCC solid solution alloys. Met. Mater. Int. 7(2), p. 165–168, https://doi.org/10.1007/BF03026955 (2001).
25. Rietveld, H. M. A Profile Refinement Method for Nuclear and Magnetic Structures. J. Appl. Cryst. 2, https://doi.org/10.1107/S0021889869006558 (1969).
26. Rodríguez-Carvajal, J. Recent advances in magnetic structure determination by neutron powder diffraction. Physica B. 192, https://doi.org/10.1016/0921-4526(93)90108-I (1993).

27. Wen, L. et al. Cr-dopant induced breaking of scaling relations in CoFe layered double hydroxides for improvement of oxygen evolution reaction. Small 15, 1902373 (2019).
28. Li, Y. & Sun, Q. Recent advances in breaking scaling relations for effective electrochemical conversion of CO2. Adv. Energy Mater. 6, 1600463 (2016).
29. Shi, P. et al. Enhanced strength-ductility synergy in ultrafine-grained eutectic high-entropy alloys by inheriting microstructural lamellae. Nat. Commun. 10, 489 (2019).
30. Alfred, L. Maximize mixing in highly polyelemental solid solution alloy nanoparticles. Matter 4, 2100–2101 (2021).
31. Wang, J. Q. et al. Rapid Degradation of Azo Dye by Fe-Based Metallic Glass Powder. Advanced Functional Materials 22, 2567–2570, https://doi.org/10.1002/adfm.201103015 (2012).
32. Wang, J. Q. et al. Excellent capability in degrading azo dyes by MgZn-based metallic glass powders. Scientific reports 2, 418, https://doi.org/10.1038/srep00418 (2012).
33. Zhao, Y. F., Si, J. J., Song, J. G., Yang, Q. & Hui, X. D. Synthesis of Mg–Zn–Ca metallic glasses by gas-atomization and their excellent capability in degrading azo dyes. Materials Science and Engineering: B 181, 46–55, https://doi.org/10.1016/j.mseb.2013.11.019 (2014).
34. Tang, Y. et al. Insight into the high reactivity of commercial Fe–Si–B amorphous zero-valent iron in degrading azo dye solutions. RSC Adv. 5, 34032–34039, https://doi.org/10.1039/c5ra02870a (2015).
35. Katona, T. & Molnar, A. Amorphous Alloy Catalysis-VII. Activation and Surface Characterization of an Amorphous Cu-Ti Alloy Catalyst Precursor in the Dehydrogenation of 2-Propanol and Comparison with Cu-Zr1. Journal of Catalysis 153, 333–343 (1995).
36. Inoue A and Takeuchi A, Acta Mater 59 (2011) 2243.
37. N. Brønsted, Acid and basic catalysis. Chem. Rev. 5, 231 (1928)
38. M.G. Evans, N.P. Polanyi, Inertia and driving force of chemical reactions. Trans. Faraday Soc. 34, 11–24 (1938)
39. J.K. Nørskov, T. Bligaard, A. Logadottir, S. Bahn, L.B. Hansen, M. Bollinger, H. Bengaard, B. Hammer, Z. Sljivancanin, M. Mavrikakis, Y. Xu, S. Dahl, C.J.H. Jacobsen, Universality in heterogeneous catalysis. J. Catal. 209(2), 275–278 (2002)
40. F. Abild-Pedersen, J. Greeley, F. Studt, J. Rossmeisl, T.R. Munter, P.G. Moses, E. Skulason, T. Bligaard, J.K. Nørskov, Scaling properties of adsorption energies for hydrogen-containing molecules on transition-metal surfaces. Phys. Rev. Lett. 99(1), 016105 (2007)
41. J. Rossmeisl, A. Logadottir, J.K. Nørskov, Electrolysis of water on (oxidized) metal surfaces. Chem. Phys. 319(1–3), 178–184 (2005)
42. J. Rossmeisl, G.S. Karlberg, T. Jaramillo, J.K. Nørskov, Steady state oxygen reduction and cyclic voltammetry. Faraday Discuss. 140, 337–346 (2008)
43. T. Löffler, A. Ludwig, J. Rossmeisl, W. Schuhmann, What makes high-entropy alloys exceptional electrocatalysis. Angew. Chem. Int. Ed. 60(52), 26894–26903 (2021)
44. H. Li, K. Shin, G. Henkelman, Effects of ensembles, ligand, and strain on adsorbate binding to alloy surfaces. J. Chem. Phys. 149(17), 174705 (2018)
45. Monkhorst, H. J. & Pack, J. D. Special points for brillouin-zone integrations. Phys. Rev. B 13, 5188 (1976).
46. George, E. P., Raabe, D. & Ritchie, R. O. High Entropy Alloys. Nature Reviews 4, 515 (2019).
47. Chen, J. et al. A review on fundamental of high entropy alloys with promising high–temperature properties. J. Alloys Compd. 760, 15–30 (2018).
48. Miracle, D. B. & Senkov, O. N. A critical review of high entropy alloys and related concepts. Acta Mater. 122, 448–511 (2017).
49. M.H. Tsai and J.W. Yeh: Mater. Res. Lett., 2014, vol. 2, pp. 107–23.
50. Sun, L. & Cava, R. C. High entropy alloy superconductors: Status, opportunities, and challenges. Physical Review Materials 3, 090301 (2019).
51. von Rohr, F., Winiarski, M. J., Tao, J., Klimczuk, T. & Cava, R. J. Effect of electron count and chemical complexity in the Ta-Nb-Hf-Zr-Ti high-entropy alloy superconductor. Proceedings of the National Academy of Sciences USA 113, E7144 (2016).
52. B. Avar, T. Simsek, S. Ozcanc, A.K. Chattopadhyay, B. Kalkan, Structural stability of mechanically alloyed amorphous (FeCoNi)70Ti10B20 under high-temperature and high-pressure. J. Alloys Compds. 860(1–9), 158528 (2020)

53. W. Li, P. Liu, P.K. Liaw, Microstructures and properties of high-entropy alloy films and coatings: a review. Mater. Res. Lett. 6(4), 199–229 (2018)
54. X. Wang, Q. Liu, Y. Huang, L. Xie, Q. Xu, T. Zhao, Effect of Ti content on the microstructure and corrosion resistance of CoCrFeNiTix high entropy alloys prepared by laser cladding. Materials 13(4–13), 2209 (2020)
55. S. Ma, J. Zhang, Wear resistant high boron cast alloy—a review. Rev. Adv. Mater. Sci. 44, 54–62 (2016)
56. L. Zong, N. Guo, R. Li, H. Yu, Effect of B content on microstructure and wear resistance of Fe-3Ti-4C hardfacing alloys produced by plasma-transferred arc welding. Coatings 9(2–11), 265 (2019)
57. Lee, C. P., Chen, Y. Y., Hsu, C. Y., Yeh, J. W. & Shih, H. C. The effect of boron on the corrosion resistance of the high entropy alloys Al0.5CoCrCuFeNiBx. J. Electrochem. Soc. 154, C424–C430 (2007).
58. L. Sang, Y. Xu, Amorphous behavior of ZrxFeNiSi0.4B0.6 high entropy alloys synthesized by mechanical alloying. J. Non-Cryst. Solids 530(1–7), 119854 (2020)
59. Y.J. Liu, I.T.H. Chang, Effects of oxide phases on thermomagnetisation of mechanically alloyed amorphous Fe–(Co, Ni)–Zr–B powder. Mater. Sci. Eng. A 375–377, 1092–1096 (2004)
60. Anderson, P. W., Muttalib, K. A. & Ramakrishnan, T. V. Theory of the "universal" degradation of Tc in high-temperature superconductors. Phys. Rev. B 28, 117–120 (1983).
61. Kirkpatrick, T. R. & Belitz, D. Suppression of superconductivity by disorder. Phys. Rev. Lett. 68, 3232–3235 (1992).
62. Wang, X., Guo, W. & Fu, Y. High-entropy alloys: Emerging materials for advanced functional applications. J. Mater. Chem. A 9, 663–701 (2021).
63. Kitagawa, J., Hamamoto, S. & Ishizu, N. Cutting edge of high-entropy alloy superconductors from the perspective of materials research. Metals 10, 1078 (2020).
64. Shi, T. et al. Current development of body-centered cubic high-entropy alloys for nuclear applications. Tungsten 3, 197–217 (2021).
65. George, E. P., Curtin, W. A. & Tasan, C. C. High entropy alloys: a focused review of mechanical properties and deformation mechanisms. Acta Materialia 188, 435–474 (2020).
66. Rosenfeld, Y. Relation between the transport coefficients and the internal entropy of simple systems. Phys. Rev. A 15, 2545–2549 (1977).
67. Dyre, J. C. Perspective: excess-entropy scaling. J. Chem. Phys. 149, 210901 (2018).
68. Qiao, B. et al. Supramolecular regulation of anions enhances conductivity and transference number of lithium in liquid electrolytes. J. Am. Chem. Soc. 140, 10932–10936 (2018).
69. Zhang, Z. et al. Capturing the swelling of solid-electrolyte interphase in lithium metal batteries. Science 375, 66–70 (2022).
70. Zhou X, Curtin WA. First principles study of the effect of hydrogen in austenitic stainless steels and high entropy alloys. Acta Mater. 2020;200:932.
71. Stefano DD, Mrovec M, Elsässer C. First-principles investigation of hydrogen trapping and diffusion at grain boundaries in nickel. Acta Mater. 2015;98:306.
72. Sheppard D, Xiao P, Chemelewski W, Johnson DD, Henkelman G. A generalized solid-state nudged elastic band method. J Chem Phys. 2012;136(7): 074103.
73. Hooshmand MS, Niu C, Trinkle DR, Ghazisaeidi M. First-principles prediction of oxygen diffusivity near the (101–2) twin boundary in titanium. Acta Mater. 2018;156:11.
74. K.Y. Tsai, M.H. Tsai, and J.W. Yeh, Acta Mater. 61, 4887 (2013).
75. Mori, Kohsuke; Hashimoto, Naoki; Kamiuchi, Naoto; Yoshida, Hideto; Kobayashi, Hisayoshi; Yamashita, Hiromi Hydrogen spillover-driven synthesis of high-entropy alloy nanoparticles as a robust catalyst for CO_2 hydrogenation. Nature Communications (2021). https://doi.org/10.1038/s41467-021-24228-z
76. Li, Hongdong; Han, Yi; Zhao, Huan; Qi, Wenjing; Zhang, Dan; Yu, Yaodong; Cai, Wenwen; Li, Shaoxiang; Lai, Jianping; Huang, Bolong; Wang, Lei Fast site-to-site electron transfer of high-entropy alloy nanocatalyst driving redox electrocatalysis. Nature Communications (2020). https://doi.org/10.1038/s41467-020-19277-9

77. Xie, Pengfei; Yao, Yonggang; Huang, Zhennan; Liu, Zhenyu; Zhang, Junlei; Li, Tangyuan; Wang, Guofeng; Shahbazian-Yassar, Reza; Hu, Liangbing; Wang, Chao Highly efficient decomposition of ammonia using high-entropy alloy catalysts. Nature Communications (2019). https://doi.org/10.1038/s41467-019-11848-9
78. Sahlberg, Martin; Karlsson, Dennis; Zlotea, Claudia; Jansson, Ulf Superior hydrogen storage in high entropy alloys. Scientific Reports (2016). https://doi.org/10.1038/srep36770
79. Hao, Jiace; Zhuang, Zechao; Cao, Kecheng; Gao, Guohua; Wang, Chan; Lai, Feili; Lu, Shuanglong; Ma, Piming; Dong, Weifu; Liu, Tianxi; Du, Mingliang; Zhu, Han Unraveling the electronegativity-dominated intermediate adsorption on high-entropy alloy electrocatalysts. Nature Communications (2022). https://doi.org/10.1038/s41467-022-30379-4
80. Zhan, Changhong; Xu, Yong; Bu, Lingzheng; Zhu, Huaze; Feng, Yonggang; Yang, Tang; Zhang, Ying; Yang, Zhiqing; Huang, Bolong; Shao, Qi; Huang, Xiaoqing Subnanometer high-entropy alloy nanowires enable remarkable hydrogen oxidation catalysis. Nature Communications (2021). https://doi.org/10.1038/s41467-021-26425-2
81. Lv, Z. Y.; Liu, X. J.; Jia, B.; Wang, H.; Wu, Y.; Lu, Z. P. Development of a novel high-entropy alloy with eminent efficiency of degrading azo dye solutions. Scientific Reports (2016). https://doi.org/10.1038/srep34213
82. Clausen, Christian M.; Nielsen, Martin L. S.; Pedersen, Jack K.; Rossmeisl, Jan Ab Initio to Activity: Machine Learning-Assisted Optimization of High-Entropy Alloy Catalytic Activity. High Entropy Alloys & Materials (2022). https://doi.org/10.1007/s44210-022-00006-4
83. Nelson, W. L.; Chemey, A. T.; Hertz, M.; Choi, E.; Graf, D. E.; Latturner, S.; Albrecht-Schmitt, T. E.; Wei, K.; Baumbach, R. E. Superconductivity in a uranium containing high entropy alloy. Scientific Reports (2020). https://doi.org/10.1038/s41598-020-61666-z
84. Panigrahi, Mrutyunjay; Avar, Baris Influence of mechanical alloying on structural, thermal, and magnetic properties of $Fe_{50}Ni_{10}Co_{10}Ti_{10}B_{20}$ high entropy soft magnetic alloy. Journal of Materials Science: Materials in Electronics (2021). https://doi.org/10.1007/s10854-021-06612-z
85. Jung, Soon-Gil; Han, Yoonseok; Kim, Jin Hee; Hidayati, Rahmatul; Rhyee, Jong-Soo; Lee, Jung Min; Kang, Won Nam; Choi, Woo Seok; Jeon, Hye-Ran; Suk, Jaekwon; Park, Tuson High critical current density and high-tolerance superconductivity in high-entropy alloy thin films. Nature Communications (2022). https://doi.org/10.1038/s41467-022-30912-5
86. Wang, Qidi; Zhao, Chenglong; Wang, Jianlin; Yao, Zhenpeng; Wang, Shuwei; Kumar, Sai Govind Hari; Ganapathy, Swapna; Eustace, Stephen; Bai, Xuedong; Li, Baohua; Wagemaker, Marnix High entropy liquid electrolytes for lithium batteries. Nature Communications (2023). https://doi.org/10.1038/s41467-023-36075-1
87. Guan, Hua-Qing; Jing, Yan-Mei; Huang, Shao-Song Atomic study of hydrogen behaviors at $\sum 3(111)$ grain boundary in equiatomic CoCrNi and CoCrNiFe alloys. Tungsten (2022). https://doi.org/10.1007/s42864-022-00152-7

SPRINGER NATURE

GPSR Compliance

The European Union's (EU) General Product Safety Regulation (GPSR) is a set of rules that requires consumer products to be safe and our obligations to ensure this.

If you have any concerns about our products, you can contact us on ProductSafety@springernature.com

In case Publisher is established outside the EU, the EU authorized representative is:

Springer Nature Customer Service Center GmbH
Europaplatz 3
69115 Heidelberg, Germany

The manufacturer's authorised representative in the EU is Springer Nature Customer Service Centre GmbH, Europaplatz 3, 69115 Heidelberg, Germany. If you have any concerns regarding our products, please contact ProductSafety@springernature.com

Printed and bound by CPI Group (UK) Ltd, Croydon, CR0 4YY
01/10/2026
02225906-0009